The Management of Early Alzheimer's Disease

The Management of Early Alzheimer's Disease

Biological and Technological Advances

Edited by

Dennis Chan

Institute of Cognitive Neuroscience, University College London, London, United Kingdom

Academic Press is an imprint of Elsevier
125 London Wall, London EC2Y 5AS, United Kingdom
525 B Street, Suite 1650, San Diego, CA 92101, United States
50 Hampshire Street, 5th Floor, Cambridge, MA 02139, United States

Copyright © 2025 Elsevier Inc. All rights are reserved, including those for text and data mining, AI training, and similar technologies.

For accessibility purposes, images in electronic versions of this book are accompanied by alt text descriptions provided by Elsevier. For more information, see https://www.elsevier.com/about/accessibility.

Publisher's note: Elsevier takes a neutral position with respect to territorial disputes or jurisdictional claims in its published content, including in maps and institutional affiliations.

No part of this publication may be reproduced or transmitted in any form or by any means, electronic or mechanical, including photocopying, recording, or any information storage and retrieval system, without permission in writing from the publisher. Details on how to seek permission, further information about the Publisher's permissions policies and our arrangements with organizations such as the Copyright Clearance Center and the Copyright Licensing Agency, can be found at our website: www.elsevier.com/permissions.

This book and the individual contributions contained in it are protected under copyright by the Publisher (other than as may be noted herein).

Notices

Knowledge and best practice in this field are constantly changing. As new research and experience broaden our understanding, changes in research methods, professional practices, or medical treatment may become necessary.

Practitioners and researchers must always rely on their own experience and knowledge in evaluating and using any information, methods, compounds, or experiments described herein. In using such information or methods they should be mindful of their own safety and the safety of others, including parties for whom they have a professional responsibility.

To the fullest extent of the law, neither the Publisher nor the authors, contributors, or editors, assume any liability for any injury and/or damage to persons or property as a matter of products liability, negligence or otherwise, or from any use or operation of any methods, products, instructions, or ideas contained in the material herein.

ISBN: 978-0-12-822240-9

> For information on all Academic Press publications visit our website at https://www.elsevier.com/books-and-journals

Publisher: Megan Ball
Acquisitions Editor: Joslyn Chaiprasert-Paguio
Editorial Project Manager: Sara Pianavilla
Production Project Manager: Sharmila Kirouchenadassou
Cover Designer: Christian Bilbow

Typeset by TNQ Technologies

Contents

Contributors   xi

1. **Introduction**   1
   Dennis Chan
   References   7

2. **A century of Alzheimer's disease: Merging neuropathology with in vivo biomarkers**   9
   Theresa M. Harrison and William J. Jagust

   Overview   9
   The modern beginning of "Alzheimer's disease"   10
   The era of clinical-pathological correlation   11
   The rise of genetics and the molecular biology of AD   15
   Challenges to the status quo   16
   The advent of biomarkers   17
   The research framework and a biological definition of AD   20
   Challenges to the research framework   22
   With a view to the future   24
   References   25

3. **Early Alzheimer's disease: Clinical overview**   35
   James Selwood and Elizabeth Coulthard

   Alzheimer's disease pathology   35
   Early Alzheimer's disease pathology   35
   Amyloid-beta   35
   Tau   36
   Co-pathology   37
   The syndrome of mild cognitive impairment   38
   Subjective cognitive decline   39
   Mild behavioral impairment   39
   Sub-classification mild cognitive impairment   39
   Conversion to dementia   40
   Management of mild cognitive impairment   41
   Clinical assessment   41

	Treatment	43
	References	46

4. The need for early diagnosis—Clinical, societal and health economic drivers   51
Ingrid van Maurik and Wiesje M. van der Flier

Epidemiology of dementia	51
The aging population — prevalence and incidence of dementia and dementia subtypes	51
Global time trends and inequalities in dementia prevalence and incidence	52
Risk factors	54
Dementia costs	57
Access to diagnosis	60
Future perspectives	62
Conclusion	63
References	63

5. An overview of current diagnostic strategies   69
Stefano F. Cappa and Chiara Cerami

Clinical assessment	70
Neuropsychological testing	73
Screening tools	73
Cognitive domains testing	74
Functional assessment instruments	78
Diagnostic tools to support clinical diagnosis	78
Cerebrospinal fluid	79
Magnetic Resonance Imaging	80
Positron emission tomography imaging	82
Conclusions	83
References	85

Chaper 6
Novel approaches to diagnosis   97

6a. Opportunities arising from the tech revolution   99
Giedrė Čepukaitytė and Dennis Chan

Introduction	99
Digital testing	100
App-based active testing	101
Passive sensing	102
Big data	105
Machine learning	108
Personalized diagnostics	110

	Risks associated with usage of digital tools	112
	Conclusions	113
	References	113

6b. Advances in imaging   123
Emrah Düzel and Jose Bernal

Neurodegeneration and MRI	123
Reserve and resilience	124
Brain reserve	125
Cognitive reserve	125
Structural traits	126
Synaptic dysfunction	127
Cerebrovascular imaging	128
Time of flight (TOF) imaging	128
Perfusion and permeability imaging	129
Glymphatic imaging	131
Perivascular spaces	131
Inferior frontal sulcal hyperintensities	133
White matter hyperintensities	135
Summary	137
References	137

6c. Advances in fluid-based biomarkers   149
Henrik Zetterberg

Introduction	149
Fluid biomarkers for Aβ pathology	150
Biomarkers for tau pathology	151
Fluid biomarkers for neurodegeneration	153
From research tools to clinical implementation	153
Limitations of fluid-based biomarkers	154
Biomarkers for AD—how can they be used in the most effective manner now that we have approved disease-modifying therapies?	154
Acknowledgments	155
References	155

6d. Advances in cognitive testing   161
David Berron

Novel cognitive tests sensitive to early Alzheimer's disease	161
Progression of AD pathology and the functional architecture of episodic memory and spatial navigation	162
Object memory	162
Feature binding and associative memory	163

Contents

Spatial memory and navigation	164
Comparison to current clinical standards	165
Digital cognitive assessments in Alzheimer's disease	165
In-clinic supervised digital cognitive assessments using tablets and personal computers	165
Remote unsupervised assessments using mobile devices in the home environment	167
Need for validation of novel digital assessments	169
Other approaches to infer cognitive status	170
References	172

Chapter 7
Challenges for the new approach   185

7a. Ethical challenges associated with the early detection of Alzheimer's disease   187
Richard Milne, Alessia Costa and Carol Brayne

Introduction	187
Early what?	188
The right to know and the consequences of early detection for the individual	189
The value of knowing (and knowing what)	189
The potential harms of predictive information	190
The social consequences of early detection	192
Stigma and discrimination	192
Justice and equity considerations	194
The future of early detection	195
Early detection and genetic risk	195
Expanding early detection and screening	197
Conclusion	198
References	199

7b. Privacy   207
Abhirup Ghosh, Jagmohan Chauhan and Cecilia Mascolo

Introduction	207
Privacy issues	208
Types of adversaries	209
Privacy attacks and mitigation techniques	210
Conclusion and future directions	213

Chapter 8
An eye to the future 215

8a. Beyond diagnosis: how to use such approaches for intervention 215

8a(i) Clinical trials 217
Sarah Gregory, Stina Saunders and Craig W. Ritchie

Introduction 217
Early disease identification and the predictive value of biomarkers 218
A shift from late-stage clinical presentation to early disease detection 218
The Alzheimer's disease continuum 219
The neuropathology of Alzheimer's disease 219
Methods for investigating Alzheimer's disease pathology 220
Genetics 221
Conclusion 221
Current drugs development and past successes/failures 222
Symptomatic treatments: Past, present and future 223
Disease modifying treatments: Past, present and future 227
Future directions e.g., novel outcome measures 230
Measuring treatment success 230
Clinical meaningfulness in Alzheimer's disease outcomes 231
Electronic person specific outcome measure (ePSOM) development program 231
Computerized assessment methods 232
Conclusion 233
Clinical trials designs, considerations, major initiatives 233
Randomized control trials 233
Database discoveries 234
Platform trials 234
Stratified medicine 236
Conclusion 236
References 237

8a(ii) Non-pharmacological interventions 249
Jill G. Rasmussen, Shireen Sindi and Miia Kivipelto

Overview of risk and protective factors 251
Identification of "at risk" individuals 251
Non-pharmacological management strategies 252
Lifestyle interventions 252
Exercise 254

	Multi-domain interventions	254
	Psychological symptoms	254
	Sleep	256
	Hearing impairment	257
	Cognitive training/stimulation	257
	Role of health and care professionals	258
	Conclusions	258
	Funding sources	260
	References	261
8b.	**Next generation technologies for diagnosis**	**265**
	Arlene J. Astell, Tamlyn Watermeyer and James Semple	
	Introduction	265
	Advancing current approaches	265
	Imaging	266
	Optical markers	268
	Behavioral markers	269
	Virtual and augmented reality markers	271
	Digital biomarkers	273
	Artificial Intelligence	274
	Direct brain technologies	277
	What do we want?	280
	References	280
Index		289

Contributors

Arlene J. Astell, Psychology Department, Northumbria University, Newcastle upon Tyne, United Kingdom; Department of Occupational Sciences & Occupational Therapy & Department of Psychiatry, University of Toronto, Toronto, ON, Canada; KITE Research, University Health Network, Toronto, ON, Canada

Jose Bernal, Institute of Cognitive Neurology and Dementia Research, Otto-von-Guericke University Magdeburg, Magdeburg, Germany; German Center for Neurodegenerative Diseases, Magdeburg, Germany; Centre for Clinical Brain Sciences, University of Edinburgh, Edinburgh, United Kingdom

David Berron, Clinical Cognitive Neuroscience Group, German Center for Neurodegenerative Diseases (DZNE), Magdeburg, Germany; Clinical Memory Research Unit, Department of Clinical Sciences Malmö, Lund University, Lund, Sweden

Carol Brayne, Cambridge Public Health, University of Cambridge, Cambridge, United Kingdom

Stefano F. Cappa, University Institute for Advanced Studies (IUSS), Pavia, Italy; IRCCS Mondino Foundation, Pavia, Italy

Giedrė Čepukaitytė, Institute of Cognitive Neuroscience, University College London, London, United Kingdom

Chiara Cerami, University Institute for Advanced Studies (IUSS), Pavia, Italy; IRCCS Mondino Foundation, Pavia, Italy

Dennis Chan, Institute of Cognitive Neuroscience, University College London, London, United Kingdom

Jagmohan Chauhan, Electronics and Computer Science, University of Southampton, Southampton, United Kingdom

Alessia Costa, Engagement and Society, Wellcome Connecting Science, Hinxton, United Kingdom; Kavli Centre for Ethics, Science, and the Public, University of Cambridge, Cambridge, United Kingdom

Elizabeth Coulthard, University of Bristol, North Bristol NHS Trust, Bristol, United Kingdom

Emrah Düzel, Institute of Cognitive Neurology and Dementia Research, Otto-von-Guericke University Magdeburg, Magdeburg, Germany; German Center for Neurodegenerative Diseases, Magdeburg, Germany; Institute of Cognitive Neuroscience, University College London, London, United Kingdom

Abhirup Ghosh, School of Computer Science, University of Birmingham, Birmingham, United Kingdom

Sarah Gregory, Centre for Clinical Brain Sciences, University of Edinburgh, Edinburgh, United Kingdom; Scottish Brain Sciences, Edinburgh, United Kingdom

Theresa M. Harrison, Department of Neuroscience, University of California, Berkeley, CA, United States

William J. Jagust, Department of Neuroscience, University of California, Berkeley, CA, United States; Lawrence Berkeley National Laboratory, Berkeley, CA, United States

Miia Kivipelto, Division of Clinical Geriatrics, Center for Alzheimer Research, Department of Neurobiology, Care Sciences and Society (NVS), Karolinska Institutet, Stockholm, Sweden; Aging Epidemiology (AGE) Research Unit, School of Public Health, Imperial College London, London, United Kingdom; Institute of Public Health and Clinical Nutrition, University of Eastern Finland, Kuopio, Finland; Theme Inflammation and Aging, Karolinska University Hospital, Stockholm, Sweden

Cecilia Mascolo, Department of Computer Science and Technology, University of Cambridge, Cambridge, United Kingdom

Richard Milne, Engagement and Society, Wellcome Connecting Science, Hinxton, United Kingdom; Kavli Centre for Ethics, Science, and the Public, University of Cambridge, Cambridge, United Kingdom

Jill G. Rasmussen, Psi-napse, Surrey, United Kingdom

Craig W. Ritchie, Centre for Clinical Brain Sciences, University of Edinburgh, Edinburgh, United Kingdom; Scottish Brain Sciences, Edinburgh, United Kingdom; Mackenzie Institute, University of St Andrews, St Andrews, United Kingdom

Stina Saunders, Centre for Clinical Brain Sciences, University of Edinburgh, Edinburgh, United Kingdom; Linus Health Europe, Edinburgh, United Kingdom

James Selwood, University of Bristol, Devon Partnership Trust, Exeter, United Kingdom

James Semple, Formerly GSK Clinical Pharmacology Unit, Addenbrookes Hospital, Cambridge, United Kingdom

Shireen Sindi, Division of Clinical Geriatrics, Center for Alzheimer Research, Department of Neurobiology, Care Sciences and Society (NVS), Karolinska Institutet, Stockholm, Sweden; Aging Epidemiology (AGE) Research Unit, School of Public Health, Imperial College London, London, United Kingdom

Wiesje M. van der Flier, Alzheimer Center Amsterdam, Neurology, Vrije Universiteit Amsterdam, Amsterdam UMC, Amsterdam, The Netherlands; Epidemiology and Data Science, Vrije Universiteit Amsterdam, Amsterdam UMC, Amsterdam, The Netherlands; Amsterdam Neuroscience, Neurodegeneration, Amsterdam, The Netherlands

Ingrid van Maurik, Alzheimer Center Amsterdam, Neurology, Vrije Universiteit Amsterdam, Amsterdam UMC, Amsterdam, The Netherlands; Epidemiology and Data Science, Vrije Universiteit Amsterdam, Amsterdam UMC, Amsterdam, The Netherlands; Northwest Academy, Northwest Clinics Alkmaar, Alkmaar, The Netherlands

Tamlyn Watermeyer, Psychology Department, Northumbria University, Newcastle upon Tyne, United Kingdom; Edinburgh Dementia Prevention, Centre for Clinical Brain Sciences, College of Medicine & Veterinary Sciences, University of Edinburgh, Edinburgh, United Kingdom

Henrik Zetterberg, Department of Psychiatry and Neurochemistry, Institute of Neuroscience & Physiology, Sahlgrenska Academy at the University of Gothenburg, Mölndal, Sweden; Clinical Neurochemistry Laboratory, Sahlgrenska University Hospital, Mölndal, Sweden; Department of Neurodegenerative Disease, UCL Institute of Neurology, London, United Kingdom; UK Dementia Research Institute at UCL, London, United Kingdom; Hong Kong Center for Neurodegenerative Diseases, Clear Water Bay, Hong Kong, China; Wisconsin Alzheimer's Disease Research Center, University of Wisconsin School of Medicine and Public Health, University of Wisconsin—Madison, Madison, WI, United States

Chapter 1

Introduction

Dennis Chan
Institute of Cognitive Neuroscience, University College London, London, United Kingdom

Year 2024 CE may in time be considered a watershed year in the history of Alzheimer's disease (AD) management, as a result of several interlinked developments. The first is the clinical rollout, following initial approvals by the US Food and Drug Administration with subsequent approvals in other countries such as China and Japan, of anti-amyloid immunotherapies for the treatment of early AD. These recommendations for the clinical use of the monoclonal antibodies lecanemab and donanemab followed on from trials reporting positive outcomes in terms of treatment-related reduction in rates of cognitive and functional decline. While at present there remain major reservations about their benefits in terms of treatment effect size and the side effect profile, with many experts voicing concerns about the benefit: risk ratio and cost-effectiveness, these trial results nevertheless have served to shift the current widespread perspective of clinicians, patients and public of AD from a disorder with an immovably poor prognosis to one which may in the future be amenable to treatments with the potential to influence disease progression (even if this may not the case with the current generation of antibodies) and deliver better long term clinical outcomes.

The second development is the emergence of blood-based testing of β-amyloid and tau biomarkers of AD as a method for early detection of AD that has the potential to be applied at scale without the need for major infrastructural support, thus overcoming the limitations of current cerebrospinal fluid- or PET scan-based biomarker testing in terms of invasiveness, high cost and limited availability even in high income countries. The availability of blood tests with high sensitivity and specificity for AD will have major consequences for current diagnostic pathways, primarily based in specialist services at present, given the opportunity for primary care and community-based diagnosis. But the impact goes beyond refashioning of clinical services. Blood tests could be undertaken by individuals outside of clinical services via commercial providers, and the implications of this in terms of ethical issues and clinical management are only now beginning to be addressed. A third issue, longer in its evolution than the above two but of huge importance in terms of global healthcare planning for aging populations, is the increasing emphasis placed on primary prevention of dementia. The 2024 Lancet Commission report [1] on dementia prevention, intervention, and care concluded that modification of 14 risk factors at different stages of life, from educational attainment in childhood through to social isolation in late adulthood, could prevent or delay dementia onset by up to 40%. Paralleling this, large scale projects such as the FINGER study [2] investigate the effects of non-pharmacological interventions such as dietary modification and management of vascular risk factors in preventing the onset of cognitive impairment. The identification of potentially modifiable lifestyle factors, backed up by robust evidence of lifestyle intervention effect, will have major implications for future public health programmes aimed at maintaining brain health into later life. Given the current problem of global inequalities in dementia diagnosis and treatment that will be amplified by high cost biological therapies, it is of crucial significance that such primary prevention initiatives are applicable in low and middle income countries as well as in high income countries.

All of this work is unfolding on a backdrop of a radical reframing of AD as a diagnostic entity. Historically defined as a clinical disorder characterized by the presence of dementia, enshrined in the 1984 NINDS-ADRDA diagnostic guidelines [3], and reflected in the perception of AD by the general population (and a not inconsiderable proportion of more traditionally minded medical practitioners with slower refresh rates, exemplified by clinic letters containing comments such as "the mild cognitive impairment has progressed to AD"), there has been a gradual shift over several years in favor of AD being defined as a biologically determined disease centered on the identification of β-amyloid and tau pathology. Within this biological framework, the occurrence of cognitive impairment becomes a secondary, downstream phenomenon whose presence is no longer considered necessary to diagnose AD but can be used to establish the severity of disease.

This shifting perspective was initially articulated in two sets of updated criteria, published within months of each other. In 2010 an International Working Group for New Research Criteria for the diagnosis of AD, building on a framework first articulated by the group in 2007, published criteria which highlighted the move away from the presence of dementia as the central defining feature of AD [4]. As well as a criterion for markers providing in vivo evidence for an underlying AD pathobiological process ("biomarkers"), the 2010 IWG criteria described a "prodromal" stage of AD in which individuals had a profile of mild cognitive impairment consistent with AD pathology, specifically relating to defective episodic memory, but not at the level severe enough to result in dementia. By comparison, the 2011 US National Institute of Aging- Alzheimer's Association guidelines described three clinical stages of AD: a preclinical stage with evidence of AD molecular pathology but in the absence of symptoms [5], mild cognitive impairment [6], and finally AD dementia [7]. As with the 2010 IWG criteria, the 2011 guidelines crucially differed from earlier guidelines in the inclusion within the diagnostic process biomarker evidence of AD pathology, notably those centered on identification of β-amyloid or tau pathology. These were followed by the 2018 publication by the US National Institute of Aging/Alzheimer's Association A/T/N criteria [8], in which the emphasis shifted strongly toward the pathobiological underpinnings of AD and away from the clinical characterization. These guidelines, published with the stated intention of aiding research studies - which need more stringent guidelines than those currently applied in clinical practice to minimize risk of incorrect diagnoses confounding study results—define AD in terms of biomarker evidence of β-amyloid (A) and tau (T) pathology alongside evidence of neurodegeneration (N), such as atrophy on MRI scanning. In striking contrast to prior diagnostic guidelines, cognitive impairment was no longer listed as a criterion of primary importance.

The latest iterations of these guidelines were published in 2024, solidifying opinions into two contrasting camps. Jack et al. [9], writing on behalf of a US Alzheimer's Association Workgroup, consider AD as a biological process in which PET- and/or fluid-based identification of core amyloid/tau biomarkers is sufficient to establish the diagnosis of AD. In For this formulation, the presence of cognitive impairment is not necessary for diagnosis but will aid clinical decision making, for example to determine whether the biomarker-determined AD pathology is the cause of the clinical presentation. Published a few months later in response to the Jack et al. [9] guidelines, the International Working Group led by Dubois et al. [10] propose instead that AD is defined in terms of a clinical-biological construct in which amyloid/tau biomarker presence in isolation is considered a risk factor for AD and insufficient for diagnosis, and—in crucial contrast to the Jack et al. [9] guidelines—the presence of objective cognitive impairment is a diagnostic requirement.

It need not take a latter-day Delphic oracle to foresee vigorous debates within the AD community about these contrasting diagnostic perspectives, but it is critically important that these be held and future issues to be discussed will range from the semantic (when is AD a "disease", when is it an "illness" and how do these differ?) to the mechanistic (do β-amyloid and phosphorylated tau pathology cause AD and are they necessary and sufficient for causation?) and operational (if "objective" cognitive impairment is a criterion for diagnosis, what is the objective criterion, how is this to be measured given the plethora of tools for testing cognition and how is the cut-point for diagnosis established?).

All of these developments are welcome at both individual and international level. AD is the commonest cause of dementia and in line with rising life expectancy the prevalence of dementia is gradually rising, to a level that risks overwhelming health economies. The associated costs are sobering, even for an audience inoculated to the concept of multi-billion sums as a result of global financial crizes and pandemics. US sources have calculated that by 2050 the cost of their national dementia care will be $1 trillion per annum.

This cost of care cannot be supported by any economy, and the need to deflect this has prompted in part the introduction of dementia initiatives in the last 2 decades by various heads of state, including those of France, UK and US.

One consequence of these advances is the increasing importance placed on early diagnosis of AD. The consensus opinion, supported by evidence from AD trials, is that treatments will be most effective when applied earlier in the disease process. This then places under scrutiny existing diagnostic methods and their fitness for this particular purpose. Clinic-based cognitive testing represents the primary tool for AD diagnosis but has several major limitations in the context of early disease. The pen and paper tests used in clinic for the last 50 years, such as the Mini Mental State Examination (MMSE) [11], Montreal Cognitive Assessment [12] or the longer Addenbrooke's Cognitive Examination [13] used as standard in UK memory clinics, have value primarily in assisting the diagnosis of established dementia or tracking clinical progression in already-diagnosed patients. However, they lack diagnostic specificity and, critically, have low sensitivity for early AD as well as low predictive ability when applied in early stages of disease, illustrated by a systematic review [14] which found that the ability of the MMSE to predict progression from mild cognitive impairment to dementia was only around 50%. In addition, performance on these tests is affected by cultural, linguistic and educational factors, restricting their application to ethnically and demographically diverse communities and aggravating issues of healthcare inequality. Detailed neuropsychological testing, probing a range of cognitive domains affected in dementia disorders, overcomes many of these limitations but still has cultural biases, testing is long and the requirement for trained neuropsychologists to apply the tests means that these are not used routinely in clinical practice and are restricted only to small numbers of patients seen in specialist services.

The need for fresh diagnostic approaches to overcome the limitations of legacy tests is therefore both necessary and timely. Happily, major advances in the fields of neuroscience and technology now provide the opportunity to usher in a new era of diagnostic practice with the capability to detect the earliest stages of disease using methods that can in principle be applied at scale to diverse populations worldwide. While understanding of brain-behavior relationships remains incomplete, sufficient progress has been made in cognitive neuroscience to inform the design of novel behavioral tests with the ability to probe the functions of brain regions vulnerable to the earliest stages of disease. The advent of wearable technologies permits monitoring of behaviors known to be affected in early disease, such as gait and sleep, but which cannot be measured using the pen-and-paper tests currently applied in clinic. As well as the potential added value for early diagnosis, especially in terms of discriminating between different dementia-causing diseases, tools capable of capturing such "real life" behaviors would have the added advantage over legacy cognitive tests in terms of ecological validity and suitability for application to diverse communities given that these behaviors are common to all human populations and analysis of such behavioral data is largely free of cultural and linguistic biases. Finally, application of machine learning algorithms to the high dimensional datasets arising from these digital devices may extract additional disease features that would be invisible to traditional data analytic approaches, potentially yielding extra diagnostic signal.

However, as with any introduction of potentially transformative and disruptive innovations, especially in clinical practice, careful consideration needs to be given to potential risks and to risk mitigation strategies. On a global level, the move toward defining AD as a biological or clinical-biological disorder, and the central role of biomarker testing for early diagnosis, has the potential to aggravate national and international inequalities in diagnostic provision given the limited availability of biomarker-based tests outside of specialized centers that are for now almost exclusively based in high income countries. The anticipated arrival of blood biomarkers will offset this problem to some extent, but even with lower cost blood testing there is a need for access to labs with appropriate equipment for biomarker assays. This has major implications for the vast majority of memory clinics worldwide in which AD diagnosis is made on clinical grounds and any pivot toward biomarker-based diagnosis will require considerable operational change and additional investment in the necessary resources. Given the probability that funding for such resources may not be available, especially in low and middle income countries, there is the risk of creating a two-tier system with biomarker-based diagnosis available only to a relative minority of individuals in high income countries fortunate enough to have access to clinical services with the appropriate diagnostic facilities.

Similarly, while there are many potential benefits arising from the use of wearable tech for diagnostic purposes, it is imperative that any plans for their future clinical deployment take into account their associated challenges. Any tech-based approach needs to address the risk of obsolescence, not only arising from the continuous process of hardware and software upgrades but also the risk that the technology providers themselves may not be able to support the clinical usage, for example in the event of changes in business strategy or lack of capacity. These risks can be mitigated by ensuring that any digital diagnostic method is device-agnostic, in other words that the measures of interest can be collected irrespective of device upgrades or even the individual devices themselves. This would also have the benefit of reducing dependence on external agencies such as commercial tech providers and the associated risks for healthcare that may arise from changes in provider viability, changes in corporate strategy and issues around intellectual property rights.

With the above as backdrop, the aim of this book is to deliver an updated overview on approaches to the diagnosis and management of early AD. In keeping with a narrative flow from the past to the future, earlier chapters will outline the historical changes in AD diagnosis and the clinical context within which individuals with AD-related cognitive impairment are evaluated, while later chapters look ahead to prospective treatment options and the potential impact of next generation technologies. In between, individual chapters will cover a variety of the key issues pertinent to early diagnosis and treatment, and the implications of the transition from legacy analog to new digital methods. Chapters on medical assessment and investigation are designed to provide non-clinicians with insights into the clinical management of patients with MCI, while more technical chapters on technologies and data science are designed to provide clinicians and those less well versed in computer science with an overview of the complexities and challenges of digital diagnostics. In between, sections addressing societal, economic and ethical issues provide the crucial broader perspective within which changes in clinical and technological evaluations will need to be embedded. Finally, leaders in the fields of AD prevention, treatment and technology look ahead to near future options with the potential to deliver transformative advances in AD management that elude current generation approaches.

Any book of this kind is obsolete to a greater or lesser extent by the time of public release, given the emergence of new information during the (inevitably longer than planned) time taken for writing and publication, and this is particularly the case with books covering technological developments. However, the timing of this book's launch coincides with that of transformative changes in the landscape of early AD diagnosis and management and the subjects covered - such as clinical presentation, diagnosis and treatment—given their generality will remain relevant into the future even though specifics may change in line with advances in each field. In providing this breadth of coverage it is hoped that this book may benefit clinicians, researchers and the interested non-expert public in terms of delivering insights into diagnostic and technological advances and their near future impact on the management of early AD.

The prevailing perception of Alzheimer's disease as a relentless untreatable dementia blighting the lives of older people will alter dramatically as knowledge accumulates about the underlying mechanisms and associated manifestations of AD pathophysiology years before the onset of cognitive impairment, accelerating the introduction of future therapeutic interventions with ever-greater efficacy. These changes in the perception of AD, especially in its earlier stages, both in the professional community and in the wider population, will greatly benefit the clinicians and scientists working on the disease whose shared ultimate objective is the avoidance of dementia. The resulting dividend for individuals, their families and healthcare systems worldwide will be considerable.

References

[1] Livingston G, Huntley J, Liu K, et al. Dementia prevention, intervention, and care: 2024 report of the Lancet Commission. Lancet 2024;404:572−628.

[2] Ngandu T, Lehtisalo J, Solomon A, et al. A 2 year multidomain intervention of diet, exercise, cognitive training, and vascular risk monitoring versus control to prevent cognitive decline in at-risk elderly people (FINGER): a randomised controlled trial. Lancet 2015;385:2255−63.

[3] McKhann G, Drachman D, Folstein M, et al. Clinical diagnosis of Alzheimer's disease: report of the NINCDS-ADRDA work group under the auspices of department of health and human services Task Force on Alzheimer's disease. Neurology 1984;34:939−44.

[4] Dubois B, Feldman H, Jacova C, et al. Revising the definition of Alzheimer's disease: a new lexicon. Lancet Neurol 2010;9:1118−27.

[5] Sperling R, Aisen P, Beckett L, et al. Toward defining the preclinical stages of Alzheimer's disease: recommendations from the National Institute on Aging-Alzheimer's Association workgroups on diagnostic guidelines for Alzheimer's disease. Alzheimer's Dementia 2011;7:280−92.

[6] Albert M, DeKosky S, Dickson D, et al. The diagnosis of mild cognitive impairment due to Alzheimer's disease: recommendations from the National Institute on Aging-Alzheimer's Association workgroups on diagnostic guidelines for Alzheimer's disease. Alzheimer's Dementia 2011;7:270−9.

[7] McKhann G, Knopman D, Chertkow H, et al. The diagnosis of dementia due to Alzheimer's disease: recommendations from the National Institute on Aging-Alzheimer's Association workgroups on diagnostic guidelines for Alzheimer's disease. Alzheimer's Dementia 2011;7:263−9.

[8] Jack C, Bennett D, Blennow K, et al. NIA-AA Research Framework: toward a biological definition of Alzheimer's disease. Alzheimer's Dementia 2018;14:535−62.

[9] Jack C, Andrews J, Beach T, et al. Revised criteria for diagnosis and staging of Alzheimer disease: Alzheimer's Association Workgroup. Alzheimer's Dementia 2024;20:5143−69.

[10] Dubois B, Villain N, Schneider L, et al. Alzheimer disease as a clinical-biological construct— an international working group recommendation. JAMA Neurol 2024. https://doi.org/10.1001/jamaneurol.2024.3770.

[11] Folstein M, Folstein S, McHugh P. "Mini-mental state". A practical method for grading the cognitive state of patients for the clinician. J Psychiatr Res 1975;12:189−98.

[12] Nasreddine Z, Phillips N, Bédirian V, et al. The Montreal Cognitive Assessment, MoCA: a brief screening tool for mild cognitive impairment. J Am Geriatr Soc 2005;53:695−9.

[13] Hsieh S, Schubert S, Hoon C, et al. Validation of the Addenbrooke's Cognitive Examination III in frontotemporal dementia and Alzheimer's disease. Dement Geriatr Cogn Disord 2013;36:242−50.

[14] Arevalo-Rodriguez I, Smailagic N, Roque I, et al. Mini-Mental State Examination (MMSE) for the detection of Alzheimer's disease and other dementias in people with mild cognitive impairment (MCI). Cochrane Database Syst Rev 2015;2015(3):CD010783.

Chapter 2

A century of Alzheimer's disease: Merging neuropathology with in vivo biomarkers

Theresa M. Harrison[a] and William J. Jagust[a,b]
[a]Department of Neuroscience, University of California, Berkeley, CA, United States; [b]Lawrence Berkeley National Laboratory, Berkeley, CA, United States

Overview

Our understanding of Alzheimer's disease (AD) has undergone substantial evolution from the time of its discovery to the present day. Reviewing these changes over a period of more than a century poses obvious limitations on content. For this reason, this chapter will review how the definition and conceptualization of the disease has changed over this time (Fig. 2.1), with a predominant theme of how the description and understanding of the neuropathology of AD and its relationship to symptoms has shaped our thinking. Thus, we begin with early efforts at description, shift to more detailed studies of clinical-pathological relationships, then more mechanistic studies of molecular biology, and finally conclude with efforts to measure AD pathology during life. In so doing, we ignore extensive research that has included epidemiology, molecular neuroscience, sociological and economic factors, behavioral neuroscience and many other important areas of research that may have a major bearing on causation or, ultimately, therapeutics. Our goal is not to produce a definitive history of the science of AD. Rather, we wish to explain the major factors involved in how the current thinking about AD as a biological disorder has come about in order to place the current era into perspective.

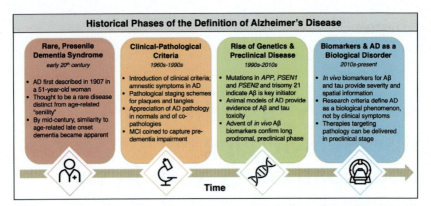

FIG. 2.1 Historical phases of the definition of Alzheimer's disease. A schematic illustrating different eras over the last 115 years during which the conceptualization and definition of Alzheimer's disease underwent significant changes. *Aβ,* beta-amyloid; *AD,* Alzheimer's disease; *MCI,* mild cognitive impairment.

The modern beginning of "Alzheimer's disease"

At a scientific conference in 1906, Dr. Alois Alzheimer described the disease that now bears his name, followed by a brief but colorful published report in 1907 [1]. The patient, Auguste Deter, was unusual at that time and her clinical course would still be somewhat unusual today. She presented at age 51 with neurobehavioral and cognitive symptoms that were already in a fairly advanced stage, culminating in death only 5 years later. Her neuropathological evaluation, in addition to cerebral atrophy, demonstrated fibrils which we now recognize as pathological accumulations of tau neurofibrillary tangles, and widely dispersed "miliary foci" composed of a "peculiar substance in the cerebral cortex" which was, in retrospect, amyloid-β (Aβ), the major constituent of the neuritic plaque. The subsequent years of descriptive neuropathology produced both confusion and argument in defining the nosology of dementing illnesses, but by the middle of the 20th century, Alzheimer's disease (AD) was relatively well established as an uncommon "presenile dementia" (technically defined as occurring before age 65). Older individuals with dementia were often described in numerous ways, including senile dementia, senile cerebral arteriosclerosis, organic brain syndrome, and other descriptive but uninformative terminology.

By the late 1960s, careful clinical-pathological observation began to clarify the similarities and differences between different forms of dementia and aging. Two seminal papers by Tomlinson, Blessed, and Roth [2,3] described the neuropathology of aging and dementia, making observations that have been repeatedly confirmed to this day. The brains of older people without dementia often showed evidence of both plaques and tangles but, compared to people with dementia, plaques were generally less frequent and tangles were generally confined to the medial temporal lobe. In contrast, in individuals with dementia tangles were seen throughout the association cortices. Cerebral infarction also frequently accompanied plaques and tangles in dementia. These reports served to establish the crucial fact that the neuropathology of the presenile form of AD was identical to the neuropathology of the most common form of senile dementia, leading to the observation by Robert Katzman in 1976 that AD was in fact not a rare form of midlife dementia, but a common form of late life dementia and a major public health problem [4]. The differentiation of AD was further established when Vladimir Hachinski introduced the term "multi-infarct dementia" to point out that dementia associated with cerebrovascular disease was an entirely different pathology from plaque and tangle dementia [5]. Thus, by the mid 1970s, AD was established as a pathologically unique and common form of late life dementia.

The era of clinical-pathological correlation

At the same time, beginning in the second half of the 20th century, a number of events collectively altered the course of Alzheimer's disease research [6]. In 1974, the National Institute on Aging (NIA) was created in the United States, increasingly drawing attention to the shifting age demographics of the US (and other economically developed societies), to the many problems of older Americans, and especially to concerns about the growing prevalence of AD that paralleled these demographic shifts. The first NIA Alzheimer's Disease Research Centers (ADRCs), funded in 1984, were focused on diagnosis, research, and care of patients with AD. ADRCs established an infrastructure for following patients as research subjects through the course of their illness and for obtaining autopsies for neuropathological diagnosis. This research infrastructure was particularly well suited to clinical-pathological correlation. In fact, at the start of this enterprise there was not widespread agreement on how AD should be diagnosed, nor as to whether diagnostic criteria were reliable and valid.

In 1984, coincident with the founding of the ADRCs, the first modern criteria for the clinical diagnosis of AD were proposed and widely accepted. The "McKhann Criteria" were notable for an emphasis on clinical and cognitive signs and symptoms and minimal laboratory testing which was performed to exclude illnesses that were unlikely to be AD [7]. A hierarchy of definite, probable, and possible AD was established to manage the uncertainty inherent in a diagnostic framework without any confirmatory biomarkers, with definite AD reserved for pathologically verified AD. This scheme thus cemented neuropathology as the only valid way of diagnosing the disease with certainty. These criteria were followed up by inclusion in the Diagnostic and Statistical Manual of Mental Disorders, third edition (revised) of a category for dementia of the Alzheimer type, with similar criteria [8]. A major effort over the next 20 years was marked by innumerable studies designed to improve upon clinical diagnosis by generating more refined neuropsychological and clinical measures and better laboratory tests that might detect aspects of AD pathology. Major efforts were also devoted to establishing pathological criteria for AD. Braak and Braak devised a grading scheme for neurofibrillary pathology that is still widely applied with a transentorhinal stage (I/II), a limbic stage (III/IV) and an isocortical stage (V/VI) [9]. These stages were subsequently combined with measures of plaque pathology into a probabilistic framework of low, intermediate, and high likelihood AD that formed the basis for the pathological diagnosis of AD until more recent criteria [10].

These efforts were effective at improving diagnosis, but problems remained. On the one hand, better and more careful characterization of clinical presentation led to the ability to predict variations in neuropathology at postmortem. A major example is the recognition of extrapyramidal and other distinctive features of a group of AD patients whose brains at postmortem demonstrated Lewy bodies, aggregates of the α-synuclein protein that had previously been associated with Parkinson's disease. This ultimately lead to the development of a separate set of clinical criteria for dementia with Lewy bodies (DLB) [11]. On the other hand, the diagnostic criteria for AD developed in the 1980s suffered from inadequate reliability and validity as well as insensitivity to very early disease stages. A review in 2001 noted reliability of AD diagnosis that was only moderate, with generalized kappa scores from 0.5–0.7, and validity compared to neuropathology that was also moderate with sensitivity averaging 81% and specificity of 70% [12]. Careful prospective clinical-pathological series revealed the complexity of establishing antemortem dementia diagnoses not only because of problems of sensitivity and specificity, but also because neuropathology of dementia is frequently multifactorial. For example, in one series, 90% of people diagnosed with probable AD had high or intermediate likelihood AD pathology; however 40% also had cerebral infarcts, 17% had Lewy bodies, and almost half had mixed pathologies [13].

Clinical-pathological studies performed during this time were also important in establishing the ubiquity of AD pathology and the ways in which the pathology was, and was not, related to cognitive symptoms and dementia. From the earlier work of Tomlinson et al. [2,3], it was anticipated that neuropathology of AD would be present in cognitively normal individuals. As clinico-pathological series of normal individuals followed to autopsy began to appear, the high prevalence of plaques and tangles in cognitively normal people drew considerable attention [14,15]. These findings ultimately generated interest in concepts like preclinical AD, or the long "incubation period" for the disease, as well as cognitive reserve, accounting for the ability of some individuals to maintain normal cognition in the presence of neuropathology.

The drive to improve diagnostic accuracy was a strong motivating factor in attempts to bring laboratory investigations to dementia diagnosis. However, during the 1980s and 1990s there were relatively few in vivo investigative modalities available, none of which could directly reveal the plaque and tangle pathology that constituted definitive evidence of the disease. Structural brain imaging, at first with X-ray computed tomography (CT) and subsequently with magnetic resonance imaging (MRI), was widely applied in dementia research as methods for identifying brain volume loss, as was functional imaging using positron emission tomography (PET) to measure cerebral glucose metabolism and blood flow, and single photon emission computed tomography (SPECT) to measure cerebral blood flow. The initial use of structural imaging was for the exclusion of pathologies, such as brain tumor and stroke, that might shift a diagnosis away from AD. In attempts to find a positive signal for AD, extensive early data initially with CT and then with MRI showed that on a group level, brain atrophy was greater in patients with dementia than in normal controls [16]. However, it soon became apparent that normal aging and many different dementing illnesses were also associated with atrophy such that whole brain atrophy was neither sensitive nor specific for a diagnosis of AD. Subsequently, a number of laboratories used both CT and MRI to show reduced hippocampal volume in patients with AD compared to controls [17—19]. Similarly, a characteristic pattern of atrophy in temporal and parietal cortex has also been proposed as an "AD signature" and may have predictive value for cognitive decline in older people [20]. While numerous clinical studies went on to show sensitivity and specificity of hippocampal atrophy for the detection of AD pathology in vivo, subsequent autopsy studies linked these findings more definitively [21,22]. Over time, however, it became apparent that this finding was not specific for AD, being associated with hippocampal sclerosis, frontotemporal lobar degeneration, and another ubiquitous pathological aggregated protein TAR DNA binding protein 43 (TDP-43) [23,24]. In addition, MRI biomarkers associated with AD perform poorly as prognostic indicators in early stages of the disease [25]. Thus, because of its inability to predict decline or to define a sensitive and specific biomarker of AD neuropathology, the primary use of MRI remained excluding non-AD structural brain lesions. Most recently, its utility in staging neurodegeneration severity is part of the current biomarker research framework, described in detail in a subsequent section of this chapter [26].

At almost the same time, functional brain imaging of blood flow and glucose metabolism with SPECT and PET followed a similar trajectory, initially defining a pattern of temporoparietal hypometabolism or hypoperfusion as characteristic of AD [27—31]. This pattern was widely viewed as reasonably sensitive and specific for the detection of Alzheimer's disease, and a series of imaging-pathologic studies showed reasonable agreement between this pattern and definite AD [32—35]. Nevertheless, this metabolic or perfusion pattern has also been shown to be non-specific since it can be associated with non-AD pathology. Because of the lack of specificity, the role of functional brain imaging has also been largely limited to the measurement of neurodegeneration in more recent diagnostic schemes [36].

In the latter part of the 20th century, as diagnostic acumen improved, clinicians became aware that patients frequently presented with more subtle cognitive findings, particularly isolated memory impairment, that did not fulfill criteria for AD [7]. A number of different rating scales and diagnostic terms were applied to such individuals, but by 1999 the term Mild Cognitive Impairment (MCI) had become popular in describing people with cognitive deficits that were more severe than expected for normal aging but which did not meet criteria for dementia [37]. These MCI criteria fulfilled a major need and were important in expanding the diagnosis of AD to encompass earlier, pre-dementia, stages of disease. Over time, the clinical criteria for MCI were broadened to include a wider array of cognitive deterioration than just memory impairment, and numerous research groups came up with different ways of defining the level and type of cognitive impairment necessary to meet MCI criteria. Among the most important studies were those with longitudinal observation to define the incidence rate of dementia in MCI cohorts. Rates of progression to dementia varied widely, dependent on such factors as diagnostic criteria and recruitment source [38], but could be 10%−15% per year in memory clinics [39]. Imaging findings of AD-typical regional brain atrophy and regional functional alterations increased the likelihood of decline to dementia in many studies [40,41]. MCI has remained a very useful diagnostic category in dementia research, particularly as new therapeutic approaches are aimed at earlier disease stages.

Finally, clinical-pathological correlations were crucially important in leading to the first approved therapies for AD patients. By the mid-1970s a number of investigators reported alterations in cortical measures of cholinergic function in the brains of AD patients. Davies and Maloney, examining post-mortem tissue from three AD patients, found substantial reductions in the synthetic enzyme choline acetyltransferase and proposed a specific cholinergic system failure as a feature of the disease [42]; similar findings were reported almost simultaneously by other groups [43,44]. Pathological examinations also revealed degeneration of the basal forebrain, the source of cortical cholinergic innervation, linking the pathology of AD to the neurochemical lesion [45]. This evidence implied a pathophysiological process that scientists at the time thought might be similar to Parkinson's disease — degeneration of a nucleus leading to loss of a neurotransmitter that, if replaced, might ameliorate symptoms. In 1992, the cholinesterase inhibitor tacrine was reported to show reduction in the rate of decline in AD patients [46]. While this drug was ultimately abandoned because of its side effects profile, the approach was replicated with several cholinesterase inhibitors [47−49] that were well tolerated and, despite generally small and suboptimal effects, were the mainstay of drug therapy for AD until very recently.

The rise of genetics and the molecular biology of AD

Well before groundbreaking discoveries began to demystify the molecular biology and genetics of AD, familial factors were known to play a major role in the disease [50]. In addition to relatively uncommon clusters of apparent autosomal dominant inheritance, an association with Down's syndrome also pointed to a genetic etiology in some cases [51]. These associations, however, did not have a major impact until 1987, when a locus for early onset familial AD was found on chromosome 21, where the gene that encodes the amyloid precursor protein (APP) resides [52]. The first missense mutation in a family was described in this gene in 1991 [53], leading to the subsequent description of multiple mutations that have been etiologically linked to AD. Shortly after the discovery of *APP* mutations, several laboratories reported other causal mutations on chromosome 14 in the gene coding for the presenilin 1 protein, *PSEN1* [54,55]. This protein was subsequently identified as a component of the γ-secretase enzyme involved in the processing of APP to Aβ, which had already been identified as the major constituent of the plaque. At least 200 mutations in *PSEN1* have now been described in different families in association with AD. A third major genetic locus for mutations leading to AD was described in 1995 in the gene coding for the presenilin 2 protein, *PSEN2* [56,57]. Each of these mutations is unusual in being causally linked to AD, associated with early age of onset, and relatively rare. However, discovery of a risk-related polymorphism, the ε4 allele of the apolipoprotein E (*APOE*) gene [58], defined a genetic factor that was related to a high proportion of late onset AD cases. While these four genes have been extremely important to advancing our understanding of AD, the genetics of AD remains complex and incompletely understood. It is likely that late onset AD is related to many genes with small effects, some of which have been described [59]. However, the majority of the heritability of AD is still unexplained [60]. As a final interesting coda to this story, analysis of tissue from Auguste Deter's brain was, relatively recently, discovered to harbor a mutation in *PSEN1*, tying the molecular explanation of AD back to its descriptive origins [61].

These genetic discoveries had a profound effect on the understanding of the pathophysiology of AD, leading to a widely held and still dominant model described as "the amyloid hypothesis" of AD shortly after the start of the 21st century [62]. The argument essentially holds that genetic mutations leading to increased Aβ deposition recapitulate the full pathological and clinical spectrum of AD, in particular leading to alterations in tau and the production of neurofibrillary pathology, thus driving the argument that Aβ deposition is the inciting event in all cases of AD. An ensuing deluge of research using human transgenes in experimental animals to study how Aβ may exert its effects has provided further support for molecular mechanisms that might underlie this chain of events [63]. However, there is considerable debate about whether a mechanistic model derived from rare autosomal dominantly inherited cases and tested in incomplete animal models truly reflects the sporadic late onset forms of AD in humans.

Challenges to the status quo

The research reviewed to this point established AD as a clinical syndrome, usually beginning as amnesia and developing into dementia (marked by global impairment) associated with plaque and tangle pathology and an underlying biology that positioned Aβ deposition as the initiating event. This represented a remarkable advance from the rudimentary knowledge about the disease available in the mid 20th century. However, the shortcomings of this state of knowledge became apparent even as new insights developed. While we have alluded to some of these issues, here we describe how they have slowly led to another controversial but widely used reconceptualization of AD as a biological disorder rather than a clinical-pathological syndrome.

The main shortcoming of the clinical-pathological syndromic nature of AD is the poor diagnostic specificity of the clinical syndrome for the pathological findings. As noted, sensitivity and specificity of clinical diagnosis for plaque and tangle pathology is modest [12]. This problem became more apparent as less common behavioral variants of AD became recognized, particularly forms presenting with language, visuospatial, or executive dysfunction rather than amnesia [64–66]. These forms of AD, while behaviorally unusual, and presenting at generally early ages, were nevertheless neuropathologically identical to the more common late-onset AD and could present diagnostic challenges. As clinical-pathological characterization advanced, AD became better differentiated from the frontotemporal lobar degenerations (FTLD). However, the syndromic differentiation of AD and FTLD was also not necessarily straightforward, particularly outside of specialized centers. The importance of multiple co-pathologies, particularly evident in pathological series, pointed to the frequent co-occurrence of AD pathology with cerebrovascular disease, α-synuclein, and TDP-43 pathologies. Finally, diagnostic uncertainty was even more problematic in people presenting with mild clinical syndromes, either MCI or even more subtle amnestic syndromes that could reflect very early AD but were difficult to diagnose. The presence of substantial plaque pathology in cognitively normal older people and the existence of non-AD etiologies of primary memory impairment compounded the problem. The result was that the definition of AD as a clinical syndrome was not sufficiently accurate to predict plaque and tangle pathology.

Further compounding the problem of diagnostic uncertainty was the increasing interest in biologically-driven therapies — such as those designed to lower brain Aβ - along with an understanding that very early initiation of therapies would be necessary to achieve positive outcomes. A number of early Aβ lowering therapies failed, notably a phase 3 trial of the anti-Aβ immunotherapy, bapineuzumab, which proved instructive to the scientific community. PET scanning with an Aβ-targeted ligand showed clear drug-related reduction of brain Aβ load without clinical benefit; however because measurement of brain Aβ was not an entry criterion of the study, 6.5% of ApoE4 carriers and 36.1% of non-carriers did not have high levels of brain Aβ at the start of the study [67]. Thus, clinical diagnosis was inadequate to recruit participants to a study that employed a therapeutic agent targeted to a presumptive biological mechanism. Subsequently, investigators proposed targeting therapy to individuals with biomarker evidence of pathology but with very mild symptoms, or even to those without symptoms [68]. A syndromic description of AD, centered on cognitive symptomatology, would be unfit for assessing drug efficacy in these trials.

The advent of biomarkers

As the limitations of clinically defined AD became apparent, attention returned to the importance of neuropathology as the defining feature of the disease, and a new emphasis on diagnostic biomarker discovery took hold. AD-specific biomarkers were needed to help identify and track AD in patients and, theoretically, in cognitively normal older adults living with AD in its preclinical phase. As AD is defined neuropathologically by Aβ and tau pathology, in vivo measures of these proteins represented key additions to the earlier but neuropathologically non-specific MR and PET surrogate measures of neurodegeneration.

The earliest approaches to measuring Aβ in the brain used samples of cerebrospinal fluid (CSF) to measure Aβ42, a 42-residue long Aβ peptide that decreases in the CSF as Aβ plaques accumulate. The first studies reporting decreases in Aβ42 in AD patients compared to controls were published in the mid-to-late 1990s [69,70]. Around the same time, studies examining total tau and phosphorylated tau levels in CSF showed increases in AD patients compared to controls [71,72]. These early CSF biomarkers were developed with ELISA technology in specific laboratories and suffered from interlaboratory standardization problems. In addition, while large scale studies of CSF Aβ42 and tau showed differences between AD patients and controls, substantial overlap between groups limited sensitivity and specificity [73]. CSF AD biomarker standardization problems have been diminished with the development of fully automated methods of measuring Aβ and tau in the CSF [74,75], and early issues relating to diagnostic specificity in retrospect likely reflected the frequent presence of Aβ pathology in unimpaired people. CSF has thus become a widely used method of detection of Aβ and tau pathology, although the invasiveness of a spinal tap and the inability of CSF measures to provide information on the location of AD pathology within the brain remain limitations of CSF AD biomarkers.

A major breakthrough in the AD biomarker effort came with the invention of the PET tracer Pittsburgh compound B (PiB) in 2004 [76]. PiB was developed by modifying thioflavin-T, a histological dye used for Aβ plaques in neuropathological studies, to create an Aβ-specific binding molecule that could be tagged with a radioisotope and cross the blood brain barrier. PiB-PET imaging allowed not only in vivo quantification of Aβ pathology but also corresponding spatial information about where Aβ pathology had accumulated in the brain. The success of the ^{11}C PiB tracer helped pave the way for the development of a group of ^{18}F tracers for Aβ including florbetaben, florbetapir, flutemetamol, and flutafuranol which, like PiB, showed good correspondence with postmortem measurements of Aβ in PET imaging-to-autopsy studies [77–80]. The longer half-life of ^{18}F tracers allowed them to be shipped from manufacturing sites and used at a broad range of research centers across the world without need for a local cyclotron, which was a requirement for production of the short half-life ^{11}C PiB tracer. Increases in availability and accumulating evidence supporting Aβ-PET tracers as accurate and reliable has led to the widespread use of Aβ-PET in clinical trials of anti-Aβ therapies, in observational longitudinal research and in clinical settings following approval by regulatory agencies (e.g., the US Food and Drug Administration and the European Medicines Agency).

In 2013 tau-PET arrived and was poised to further revolutionize the field with radiotracers that bound to paired helical filament tau in neurofibrillary tangles [81]. Neuropathological data suggested that tau, unlike Aβ, was highly predictive of clinical symptoms, and that the spatial topography of tau deposition would be key to interpreting these associations in terms of established functional neuroanatomy [82]. Indeed studies have shown that flortaucipir (FTP; previously AV1451, T807), the first tau-PET tracer to be widely adopted and the first approved for clinical use, is associated with clinical disease severity and even memory performance in unimpaired older adults [83–86]. Critically, the spatial pattern of FTP binding predicts future atrophy patterns, is inversely related to hypometabolism as measured by FDG-PET and is related to AD clinical syndromic presentation [84,87,88]. Perhaps related to the high throughput screens used to identify candidates, tau-PET tracers suffer from problematic off target binding [89–93]. Subsequent research in this area has been extensive, with the availability of multiple additional tau-PET tracers [94]. Despite this active development of tracers, all have some degree of varying off-target binding and there remains interest in optimizing tracers for imaging tau. Nevertheless, existing tau-PET tracers have all proven to be very useful in charting the course of AD.

Although tracers for both Aβ and tau are available and many have regulatory approval, they have only just begun to receive clinical application in view of the fact that biologically specific disease-modifying therapies are just becoming available. The recent successful clinical trials of amyloid-lowering therapies is a testament to the model of AD based on the importance of Aβ and the use of biomarkers in therapeutic development [95,96].

To date, CSF and PET biomarkers of Aβ and tau are the best established and both have been integrated into diagnostic and research criteria for AD. Expertise in CSF or PET analysis tends to reside in distinct research groups, so collaborative work to explore relationships between biomarkers across the two techniques has been essential. CSF biomarkers have been shown to be similarly accurate and reliable compared to PET biomarkers [97,98] and automated methods for CSF biomarker quantification may further improve the concordance [75].

Plasma biomarkers for AD have been an area of intense interest throughout the history of AD biomarker research because of the relative ease of obtaining plasma samples, compared with the invasive and high-cost CSF and PET methods. Initial studies of various measures of both plasma Aβ and tau have shown diagnostic accuracy for AD and high correlation with PET measures [99–101]. As methods for assaying blood-based biomarkers improve, future work will focus further on the diagnostic and prognostic utility of plasma Aβ and tau biomarkers, on their relationship with CSF and PET biomarkers, and on their utility for selecting participants for clinical trials of biologically oriented therapies (see Chapter 7 for further discussion of blood-based biomarkers).

The rapid progress in AD biomarker research has driven changes in diagnostic and research criteria. The 1984 "McKhann Criteria" for the diagnosis of AD relied primarily on a clinical presentation and included laboratory testing only to rule out other potential causes of the symptoms [7]. A meaningful update to the McKhann criteria did not come until over 20 years later with the publication of research criteria by an international working group (IWG) of experts [102]. The original IWG criteria incorporated biomarkers as supporting information but still relied on determination of the clinical syndrome for AD diagnosis. The IWG criteria were updated in 2014 with IWG2 criteria, in which the diagnostic utility of biomarkers was formalized by the specification that the presence of Aβ and tau biomarkers were necessary, alongside clinical symptoms, for a definite AD diagnosis. In parallel to the IWG criteria, a group convened by the National Institute on Aging (NIA) and the Alzheimer's Association (AA) published three papers outlining updated guidelines for three different clinical stages of AD: preclinical AD, MCI due to AD, and AD dementia [103–105]. These guidelines also incorporated biomarkers, but clinical presentation was still an essential component of AD diagnosis which was still qualified with a 'possible' or 'probable' designation. Importantly, the guidelines for preclinical AD, in contrast to those for MCI and AD, were not meant to guide clinical care or diagnosis but rather formed a basis for common description of individuals who were cognitively normal with positive AD biomarkers. While this was an important step in moving AD toward a biological definition rather than a clinical one, these criteria have been supplanted by the more recent framework, which is discussed in the next section.

The movement toward a more biological definition of AD as opposed to a clinical/syndromic one has been facilitated by improved understanding and a testable model defining the pathophysiological basis of the disease. Decades of clinical description, genetic studies, neuropathological research and biomarker development have informed a model of AD pathogenesis that, broadly, begins with accumulation of Aβ in multiple cortical regions. This Aβ interacts, in ways that are poorly understood, with the medial temporal lobe (MTL) tau pathology, in the form of neurofibrillary tangles, that is ubiquitous in older people. This Aβ-tau interaction prompts the spread of tau out of the MTL throughout neocortex, which appears to be a defining feature of AD since this tau spread reflects a point at which cognitive decline intensifies [106]. Tau, compared to Aβ, is relatively more associated with downstream neurodegeneration, which typically occurs in a pattern resembling that of the preceding tau pathology. Neurodegeneration or atrophy of cortical and subcortical areas then results in corresponding cognitive decline and, eventually, the Alzheimer's clinical syndrome. This model illustrates the continuum nature of AD, with early pathological events uncoupled from clinical symptoms. The model provides several important features that have guided the transformation in the definition of AD during life: (1) It is based on a series of biochemical events and not a clinical syndrome (2) It strongly features Aβ and tau pathology (3) The main features of the disease — Aβ, pathological tau, and neurodegeneration — are measurable with in vivo biomarkers.

The research framework and a biological definition of AD

The advent of AD biomarkers resulted in a shift from conceptualizing AD as a clinical-pathological entity, to be confirmed at autopsy, to a pathological process that could be identified in vivo, leading to the 2018 publication of the NIA-AA research framework [26]. This framework was the result of the collaborative effort that began in 2011 with the development of the diagnostic criteria by the NIA-AA work group and the subsequent need to update criteria to reflect the rapidly advancing knowledge about the biological basis of AD. In addition, the 2011 criteria suffered from problems relating to the syndromic identification of AD (which could be unreliable and invalid as discussed in previous sections) and did not define a clear path for the inclusion of asymptomatic people in clinical trials. One key objective of the framework was to improve detection of AD earlier in the disease process, in mildly symptomatic or asymptomatic individuals, facilitating interventional research aimed at reducing disease progression and thus delaying or preventing the onset of dementia. To this end, the framework classified AD pathology in terms of biomarkers of Aβ (A), pathological tau (T), and neurodegeneration

(N) [107]. Individuals can be classified as positive or negative for each ATN biomarker, leading to eight distinct categories.

In any set of guidelines semantics are critical and there are five terms defined in the research framework that are essential to its understanding. First, the term 'Alzheimer's clinical syndrome' is used to describe the classic amnestic syndromal presentation of AD and applies to both mildly and severely impaired patients. Next, 'Alzheimer's disease' is defined as the presence of positive biomarkers for Aβ and tau, and 'Alzheimer's pathologic change' is defined by positive biomarkers for Aβ in the absence of tau pathology. The concept of the 'Alzheimer's continuum' was formally presented in the research framework to encompass both biomarker-defined AD and Alzheimer's pathologic change, in order to represent the full spectrum of AD from early, preclinical Aβ accumulation to symptomatic AD patients with positive Aβ and tau biomarkers. Finally, 'non-AD pathologic change' is defined by the presence of tau pathology or notable neurodegeneration in individuals who are negative for Aβ.

A key difference between the 2011 NIA-AA diagnostic recommendations and the research framework is that the framework can be applied to any phase of the AD pathological process, without specific variations depending on disease stage or clinical features. The framework defines AD in biological terms; in other words the presence of Aβ alone is sufficient to establish someone as on the Alzheimer's continuum, and the presence of Aβ and tau pathology establishes someone as having AD regardless of the type, or even presence, of clinical symptoms. This approach allows for the enrollment of individuals with AD into clinical trials who are mildly symptomatic or asymptomatic. The research framework does permit the definition of stages of disease based on self-reported or clinically measured symptoms, but these clinical measures are independent from the biomarker categories. Terminology for combining biomarker evidence with clinical assessment is suggested, such as the term "AD with dementia" for biomarker-defined AD in a severely clinically impaired person. Overall, it is anticipated that the clinical staging scheme, combined with biomarker measurements, will facilitate enrollment of appropriate cohorts into intervention trials and provide a common language for describing participants in research studies.

The research framework is focused on pathology which, for most of the history of AD research, could only be reliably measured at autopsy and was the gold standard for diagnosis. The framework relies on the fact that we have the ability to measure Aβ and tau pathology in vivo with reasonable accuracy. It is not, however, formally based on the "amyloid hypothesis"; Aβ positivity defines AD pathologic change not because the authors of the framework believe the presence of Aβ to be sufficient or critical in the *progression* of AD, but rather that Aβ is critical in the *diagnosis* of AD and often accumulates earlier than neocortical tau pathology. Further, tau pathology is present in many non-AD neurodegenerative diseases while Aβ is more specific to AD.

Rooting the diagnosis of AD in pathology is not new: what is new is the idea that this pathology can be measured in vivo as opposed to only at autopsy. The research framework's departure from a symptom-based definition of AD parallels other disorders such as diabetes mellitus which can be diagnosed and treated early before the emergence of clinical symptoms.

The research framework is so-called because it is not intended for use in a clinical setting. The authors acknowledge that the framework needs to evaluated, validated and modified by the greater AD research community before it could be adopted in clinical practice and such efforts are underway. This is in line historically with other sets of research guidelines, like the IWG research criteria from 2007 to 2010, which were integrated and tested by the AD research community before any integration into clinical practice. The potential future adoption of the research framework in clinical practice will come with some challenges which are discussed in the next section.

Challenges to the research framework

There are many challenges to universal implementation of the research framework. First, PET and CSF biomarkers, the most well-established to date, are costly, invasive and dependent on access to infrastructure such as nuclear imaging facilities and staff trained in lumbar puncture. This presents obstacles in research settings where cost and infrastructural support are limiting factors. The ongoing development of plasma biomarkers will in time remove many of these obstacles. Second, sensitivity and specificity of AD detection is limited by the sensitivity and specificity of currently available biomarkers. It follows that there is an incentive to increase the sensitivity and specificity of AD biomarkers with continued research and development. Third, thresholds that distinguish abnormal from normal biomarker measurements are themselves active areas of research. Related to the previous two points, a lack of uniform positivity thresholds across laboratories and across different tracers or assays for the same marker with varying sensitivity and specificity means harmonization of biomarker positivity in large, combined datasets is challenging. The now widely-used centiloid approach, an analytical approach to transforming Aβ tracer measurements to a common unit and scale, was developed to help combine data across Aβ PET tracers, including across ^{11}C and ^{18}F labeled tracers [108].

Some of these challenges, including biomarker availability, may further perpetuate existing inequities in research in the US and Europe as well as around the world. If biomarker measurements, which are currently expensive, are required to study AD then only laboratories with extensive resources could pursue state of the art research. Further, research efforts around the world may suffer if studies that use a clinical AD diagnosis without biomarkers are penalized. The authors of the research framework explicitly state they expect studies without biomarkers to continue and to have an important role, especially when these studies are large or population-based. Further, some communities have generally been excluded from access to AD biomarkers, and some cultural groups may be reluctant to undergo testing with radioactive PET ligands or lumbar puncture. These issues will compound the known problem that human AD studies have suffered from exclusion and a lack of diversity in study cohorts. There are, therefore, valid concerns about generalizability of published and ongoing AD biomarker studies, and it is imperative that adoption of the research framework does not result in worsening this problem by reducing opportunities for research in underrepresented groups. While these concerns are important for both equity and generalizability of results, there is cause for optimism in view of the explosion of knowledge about plasma biomarkers. As noted above, measurement of Aβ and tau in plasma is now a reality. While these measurements must be better understood and standardized, they offer the potential for widespread adoption in under-resourced environments and for communities that have been excluded from AD research.

Another key criticism of the research framework is the focus on Aβ and tau. From one perspective, this is a strength because the framework aims to identify the continuum of AD, which is a specific neuropathological disease defined by abnormal Aβ and tau accumulation. In real life, however, AD often occurs in the context of comorbidities and "pure" AD is relatively rare [13]. At present, there is a lack of biomarkers for common AD comorbidities such as non-AD proteinopathies (e.g., TDP-43, α-synuclein) and cerebrovascular disease. Without validated approaches to measure these non-AD disease processes, any observed association between AD as defined by the framework and clinical presentation may be mediated or potentiated by non-AD pathologic changes in the brain. Furthermore, the presence of Aβ and tau does not necessarily mean that these pathologies are the cause of the dementia syndrome; other pathologies may be more relevant to the clinical presentation. Studies have shown that age is negatively correlated with pathological burden in AD patients such that older patients have less plaque and tangle pathology relative to their disease severity [109,110]. One reason for this may be that multifactorial dementia is more likely with advancing age, when the risk of comorbidities increases alongside the increased risk for AD pathology. Thus, as biomarkers for common AD comorbidities become available it will be important to incorporate them into future AD research criteria.

Another set of challenges centers on the future adoption of the research framework in clinical practice. The cost of biomarkers and their inaccessibility poses a major challenge at present. But even if plasma biomarkers become widely available, there are other concerns. At present, biomarkers can give an answer about diagnosis without opening the door to therapy. This will change if effective treatments appear, and there is certainly cause for optimism on this front [95,96]. There are also substantial concerns for asymptomatic or very mildly symptomatic individuals who would not likely benefit from a biomarker-defined AD diagnosis and could suffer negative effects, such as an adverse psychological reaction or insurance discrimination. Clinicians may be reluctant to tell a patient that they have AD when symptoms are absent, especially if accurate prognostic information is unavailable. Clinicians may also hesitate to share AD biomarker findings when symptoms are present if they suspect that the AD pathology is not the main driver of the clinical presentation (i.e., a coexisting, unmeasured process is more important). These and other concerns have prompted a reworking of the IWG criteria, which require clinical evidence of a neurobehavioral syndrome consistent with AD along with biomarker evidence in order to diagnose AD [111]. Another concern about clinical adoption of the research framework centers on prevalence of AD and whether universal adoption of the framework would starkly impact our understanding of AD incidence and prevalence and dramatically change current estimates. This could have many downstream effects on medical insurance, research funding and public awareness. However, if AD prevalence increases because of significant improvement in AD detection, these downstream effects are unavoidable. The framework does not create more cases of AD, it just potentially includes more instances of AD by suggesting that AD is a biological entity that can be identified even in the absence of symptoms.

With a view to the future

AD is increasingly considered a disease that emerges in the brain slowly, giving rise to both preclinical or asymptomatic and clinical or symptomatic phases of a single disease spectrum. AD biomarker research, especially over the last 20 years, has supported this recasting of AD as a biological entity rather than a clinical syndrome. To date, the biological view of AD has been incompletely applied to clinical practice, but studies in clinical cohorts have shown that biomarker data has a significant effect on clinical management [112]. In addition, biomarkers may have potent prognostic ability that creates both opportunities for early intervention and risks of disclosure. Two recent studies, for example, note that unimpaired people with positive biomarkers for both Aβ and tau are at very high risk of developing AD over several years [113,114]. This empirical data combined with clinician hesitation to give biomarker results in the absence of effective treatment options makes disclosure of biomarker status to asymptomatic individuals difficult [115]. Current practice is evolving and increasing data on the importance of biomarker results is affecting the practice of both clinicians and researchers, leading toward more widespread return of biomarker results to individuals, especially those in clinical trials.

Enrolling individuals with biomarker-defined AD into clinical trials will be essential to avoid repeating past mistakes, like treating Aβ negative patients with experimental anti-Aβ therapies. Implementation of the research framework could lead to large, trial-ready cohorts with biomarker data which will speed up the formal clinical trial evaluation process for new drugs. This will be even more feasible using plasma biomarkers, which are now a reality but not yet widely utilized. Surely in the coming years this will change given the scalability of plasma assays, and it will have a massive impact on clinical trials and observational research. Plasma could become the preferred fluid biomarker over CSF assays, while PET biomarkers will complement plasma biomarkers by adding information about spatial patterns of deposition in the brain. Indeed, the field is rapidly evolving with a new set of research criteria published as this volume goes to press [116]. These criteria further develop the biomarker-based definition of the disease by focusing on Core 1 biomarkers of Aβ sufficient to establish the diagnosis, and by incorporating plasma biomarkers. If these biomarker advancements are paired with the development of effective therapies that attack the fundamental biology of AD, a true revolution in the diagnosis and management of the disease will have occurred. It does not seem overly optimistic to consider this outcome.

Since at least the 1970s, the answer to the question, "what is Alzheimer's disease?" has always ultimately distilled down to Aβ and tau. For decades the inaccessibility of the ground truth neuropathology during life led to a wealth of research and discovery focused on clinical features, molecular biology and genetics. This work has informed a huge proportion of what we currently know about how AD manifests itself, leading to cognitive decline, and much of this work remains crucial if we are to develop effective treatments. It is possible that the future will reveal a causal role for some factor that we are currently unaware of, and completely alter our thinking about the etiology and definition of AD. But none of this has changed the fact that today AD is defined by the pathological accumulation of Aβ and tau in the brain. While neuropathology studies will continue to play an important role, biomarkers for these hallmark proteins have ushered in a new era of AD research in which definitive diagnosis can occur during life, ending the era of neuropathological mysteries revealed only at autopsy.

References

[1] Alzheimer A. A characteristic disease of the cerebral cortex. In: Schultze E, Snell O, editors. Allgemeine Zeitschrift fur Psychiatrie un Psychisch-Gerichtliche Medizin. Berlin: Georg Relmer; 1907. p. 146–8.

[2] Tomlinson BE, Blessed G, Roth M. Observations on the brains of non-demented old people. J Neurol Sci 1968;7:331–56.

[3] Tomlinson BE, Blessed G, Roth M. Observations on the brains of demented old people. J Neurol Sci 1970;11:205–42.

[4] Katzman R. Editorial: the prevalence and malignancy of Alzheimer disease. A major killer. Arch Neurol 1976;33:217–8.

[5] Hachinski VC, Lassen NA, Marshall J. Multi-infarct dementia. A cause of mental deterioration in the elderly. Lancet 1974;2:207–10.

[6] Fox PJ, Kelly SE, Tobin SL. Defining dementia: social and historical background of Alzheimer disease. Genet Test 1999;3:13−9.

[7] McKhann G, Drachman D, Folstein M, Katzman R, Price D, Stadlan EM. Clinical diagnosis of Alzheimer's disease: report of the NINCDS-ADRDA work group* under the auspices of department of health and human services task force on Alzheimer's disease. Neurology 1984;34. http://www.neurology.org/content/34/7/939. [Accessed 15 July 2014].

[8] DSM-III-R. Diagnostic and statistical manual of mental disorders. 3rd ed. Washington, D.C.: American Psychiatric Association; 1987.

[9] Braak H, Braak E. Neuropathological stageing of Alzheimer-related changes. Acta Neuropathol 1991;82:239−59.

[10] Hyman BT, Trojanowski JQ. Consensus recommendations for the postmortem diagnosis of alzheimer disease from the national Institute on aging and the Reagan Institute working group on diagnostic criteria for the neuropathological assessment of Alzheimer disease. J Neuropathol Exp Neurol 1997;56:1095−7.

[11] Gibb WR, Esiri MM, Lees AJ. Clinical and pathological features of diffuse cortical Lewy body disease (Lewy body dementia). Brain 1987;110(Pt 5):1131−53.

[12] Knopman DS, DeKosky ST, Cummings JL, Chui H, Corey-Bloom J, Relkin N, Small GW, Miller B, Stevens JC. Practice parameter: diagnosis of dementia (an evidence-based review). Report of the quality standards subcommittee of the American academy of neurology. Neurology 2001;56:1143−53.

[13] Schneider JA, Arvanitakis Z, Leurgans SE, Bennett DA. The neuropathology of probable Alzheimer disease and mild cognitive impairment. Ann Neurol 2009;66:200−8.

[14] Davis DG, Schmitt FA, Wekstein DR, Markesbery WR. Alzheimer neuropathologic alterations in aged cognitively normal subjects. J Neuropathol Exp Neurol 1999;58:376−88.

[15] Bennett DA, Schneider JA, Arvanitakis Z, Kelly JF, Aggarwal NT, Shah RC, Wilson RS. Neuropathology of older persons without cognitive impairment from two community-based studies. Neurology 2006;66:1837−44.

[16] Lee BCP, Mintun M, Buckner RL, Morris JC. Imaging of Alzheimer's disease. J Neuroimaging 2003;13:199−214. https://doi.org/10.1111/j.1552-6569.2003.tb00179.x. [Accessed 11 December 2020].

[17] Seab JP, Jagust WJ, Wong ST, Roos MS, Reed BR, Budinger TF. Quantitative NMR measurements of hippocampal atrophy in Alzheimer's disease. Magn Reson Med 1988;8:200−8.

[18] de Leon MJ, George AE, Stylopoulos LA, Smith G, Miller DC. Early marker for Alzheimer's disease: the atrophic hippocampus. Lancet 1989;2:672−3.

[19] Jack Jr CR, Petersen RC, O'Brien PC, Tangalos EG. MR-based hippocampal volumetry in the diagnosis of Alzheimer's disease. Neurology 1992;42:183−8.

[20] Bakkour A, Morris JC, Dickerson BC. The cortical signature of prodromal AD: regional thinning predicts mild AD dementia. Neurology 2009;72:1048−55.

[21] Jobst KA, Smith AD, Szatmari M, Molyneux A, Esiri ME, King E, Smith A, Jaskowski A, McDonald B, Wald N. Detection in life of confirmed Alzheimer's disease using a simple measurement of medial temporal lobe atrophy by computed tomography. Lancet 1992;340:1179−83.

[22] Jack Jr CR, Dickson DW, Parisi JE, Xu YC, Cha RH, O'Brien PC, Edland SD, Smith GE, Boeve BF, Tangalos EG, Kokmen E, Petersen RC. Antemortem MRI findings correlate with hippocampal neuropathology in typical aging and dementia. Neurology 2002;58:750−7.

[23] Jagust WJ, Zheng L, Harvey DJ, Mack WJ, Vinters HV, Weiner MW, Ellis WG, Zarow C, Mungas D, Reed BR, Kramer JH, Schuff N, DeCarli C, Chui HC. Neuropathological basis of magnetic resonance images in aging and dementia. Ann Neurol 2008;63:72−80.

[24] Josephs KA, Dickson DW, Tosakulwong N, Weigand SD, Murray ME, Petrucelli L, Liesinger AM, Senjem ML, Spychalla AJ, Knopman DS, Parisi JE, Petersen RC, Jack Jr CR, Whitwell JL. Rates of hippocampal atrophy and presence of post-mortem TDP-43 in patients with Alzheimer's disease: a longitudinal retrospective study. Lancet Neurol 2017;16:917−24.

[25] Lombardi G, Crescioli G, Cavedo E, Lucenteforte E, Casazza G, Bellatorre AG, Lista C, Costantino G, Frisoni G, Virgili G, Filippini G. Structural magnetic resonance imaging for the early diagnosis of dementia due to Alzheimer's disease in people with mild cognitive impairment. Cochrane Database Syst Rev 2020. https://www.cochranelibrary.com/cdsr/doi/10.1002/14651858.CD009628.pub2/full. [Accessed 3 February 2021].

[26] Jack CR, et al. NIA-AA Research Framework: toward a biological definition of Alzheimer's disease. Alzheimer's Dementia 2018;14:535−62. [Accessed 7 January 2019].

[27] Ferris SH, de Leon MJ, Wolf AP, Farkas T, Christman DR, Reisberg B, Fowler JS, Macgregor R, Goldman A, George AE, Rampal S. Positron emission tomography in the study of aging and senile dementia. Neurobiol Aging 1980;1:127−31.

[28] Frackowiak RS, Pozzilli C, Legg NJ, Du Boulay GH, Marshall J, Lenzi GL, Jones T. Regional cerebral oxygen supply and utilization in dementia. A clinical and physiological study with oxygen-15 and positron tomography. Brain 1981;104:753−78.

[29] Benson DF, Kuhl DE, Hawkins RA, Phelps ME, Cummings JL, Tsai SY. The fluorodeoxyglucose 18F scan in Alzheimer's disease and multi-infarct dementia. Arch Neurol 1983;40:711−4.

[30] Foster NL, Chase TN, Fedio P, Patronas NJ, Brooks RA, Di Chiro G. Alzheimer's disease: focal cortical changes shown by positron emission tomography. Neurology 1983;33:961−5.

[31] Friedland RP, Budinger TF, Ganz E, Yano Y, Mathis CA, Koss B, Ober BA, Huesman RH, Derenzo SE. Regional cerebral metabolic alterations in dementia of the Alzheimer type: positron emission tomography with [18F]fluorodeoxyglucose. J Comput Assist Tomogr 1983;7:590−8.

[32] Hoffman JM, Welsh-Bohmer KA, Hanson M, Crain B, Hulette C, Earl N, Coleman RE. FDG PET imaging in patients with pathologically verified dementia. J Nucl Med 2000;41:1920−8.

[33] Jagust W, Thisted R, Devous Sr MD, Van Heertum R, Mayberg H, Jobst K, Smith AD, Borys N. SPECT perfusion imaging in the diagnosis of Alzheimer's disease: a clinical-pathologic study. Neurology 2001;56:950−6.

[34] Jagust W, Reed B, Mungas D, Ellis W, Decarli C. What does fluorodeoxyglucose PET imaging add to a clinical diagnosis of dementia? Neurology 2007;69:871−7.

[35] Silverman DH, et al. Positron emission tomography in evaluation of dementia: regional brain metabolism and long-term outcome. JAMA 2001;286:2120−7.

[36] Lesman-Segev OH, et al. Diagnostic accuracy of amyloid versus [18] F-fluorodeoxyglucose positron emission tomography in <scp>Autopsy-Confirmed</scp> dementia. Ann Neurol 2020:25968. https://onlinelibrary.wiley.com/doi/10.1002/ana.25968. [Accessed 11 December 2020].

[37] Petersen RC, Smith GE, Waring SC, Ivnik RJ, Tangalos EG, Kokmen E. Mild cognitive impairment: clinical characterization and outcome. Arch Neurol 1999;56:303−8.

[38] Farias ST, Mungas D, Reed BR, Harvey D, DeCarli C. Progression of mild cognitive impairment to dementia in clinic- vs community-based cohorts. Arch Neurol 2009;66:1151−7.
[39] Petersen RC, Roberts RO, Knopman DS, Boeve BF, Geda YE, Ivnik RJ, Smith GE, Jack Jr CR. Mild cognitive impairment: ten years later. Arch Neurol 2009;66:1447−55.
[40] Chetelat G, Desgranges B, de la Sayette V, Viader F, Eustache F, Baron JC. Mild cognitive impairment: can FDG-PET predict who is to rapidly convert to Alzheimer's disease? Neurology 2003;60:1374−7.
[41] Heister D, Brewer JB, Magda S, Blennow K, McEvoy LK, Alzheimer's Disease Neuroimaging I. Predicting MCI outcome with clinically available MRI and CSF biomarkers. Neurology 2011;77:1619−28.
[42] Davies P, Maloney AJ. Selective loss of central cholinergic neurons in Alzheimer's disease. Lancet 1976;2:1403.
[43] Bowen DM, Smith CB, White P, Davison AN. Neurotransmitter-related enzymes and indices of hypoxia in senile dementia and other abiotrophies. Brain 1976;99:459−96.
[44] Perry EK, Gibson PH, Blessed G, Perry RH, Tomlinson BE. Neurotransmitter enzyme abnormalities in senile dementia. Choline acetyltransferase and glutamic acid decarboxylase activities in necropsy brain tissue. J Neurol Sci 1977;34:247−65.
[45] Whitehouse PJ, Price DL, Struble RG, Clark AW, Coyle JT, Delon MR. Alzheimer's disease and senile dementia: loss of neurons in the basal forebrain. Science 1982;215:1237−9.
[46] Davis KL, Thal LJ, Gamzu ER, Davis CS, Woolson RF, Gracon SI, Drachman DA, Schneider LS, Whitehouse PJ, Hoover TM, et al. A double-blind, placebo-controlled multicenter study of tacrine for Alzheimer's disease. The Tacrine Collaborative Study Group. N Engl J Med 1992;327:1253−9.
[47] Rogers SL, Farlow MR, Doody RS, Mohs R, Friedhoff LT. A 24-week, double-blind, placebo-controlled trial of donepezil in patients with Alzheimer's disease. Neurology 1998;50:136−45. [Accessed 27 February 2023].
[48] Rösler M, Bayer T, Anand R, Cicin-Sain A, Gauthier S, Agid Y, Dal-Bianco P, Stähelin HB, Hartman R, Gharabawi M. Efficacy and safety of rivastigmine in patients with Alzheimer's disease: international randomised controlled trialCommentary: another piece of the Alzheimer's jigsaw. BMJ 1999;318:633−40. [Accessed 27 February 2023].
[49] Olin J, Schneider L. Galantamine for Alzheimer's disease. Cochrane Database Syst Rev 2001:CD001747. http://www.ncbi.nlm.nih.gov/pubmed/11279727. [Accessed 4 February 2021].
[50] Lowenberg K, Waggoner RW. Familial organic psychosis (Alzheimer's type). Arch Neurol Psychiatr 1934;31:737−54.
[51] Ellis WG, McCulloch JR, Corley CL. Presenile dementia in Down's syndrome. Ultrastructural identity with Alzheimer's disease. Neurology 1974;24:101−6.
[52] Tanzi RE, Gusella JF, Watkins PC, Bruns GA, St George-Hyslop P, Van Keuren ML, Patterson D, Pagan S, Kurnit DM, Neve RL. Amyloid beta protein gene: cDNA, mRNA distribution, and genetic linkage near the Alzheimer locus. Science 1987;235:880−4.
[53] Goate A, Chartier-Harlin MC, Mullan M, Brown J, Crawford F, Fidani L, Giuffra L, Haynes A, Irving N, James L. Segregation of a missense mutation in the amyloid precursor protein gene with familial Alzheimer's disease. Nature 1991;349:704−6. https://doi.org/10.1038/349704a0. [Accessed 5 September 2014].
[54] Schellenberg G, Bird T, Wijsman E, Orr H, Anderson L, Nemens E, White J, Bonnycastle L, Weber J, Alonso M, et al. Genetic linkage evidence for a familial

Alzheimer's disease locus on chromosome 14. Science 1992;258:668−71. [Accessed 22 September 2014].

[55] Van Broeckhoven C, Backhovens H, Cruts M, De Winter G, Bruyland M, Cras P, Martin JJ. Mapping of a gene predisposing to early-onset Alzheimer's disease to chromosome 14q24.3. Nat Genet 1992;2:335−9.

[56] Levy-Lahad E, Wasco W, Poorkaj P, Romano DM, Oshima J, Pettingell WH, Yu CE, Jondro PD, Schmidt SD, Wang K, et al. Candidate gene for the chromosome 1 familial Alzheimer's disease locus. Science 1995;269:973−7.

[57] Rogaev EI, Sherrington R, Rogaeva EA, Levesque G, Ikeda M, Liang Y, Chi H, Lin C, Holman K, Tsuda T. Familial Alzheimer's disease in kindreds with missense mutations in a gene on chromosome 1 related to the Alzheimer's disease type 3 gene. Nature 1995;376:775−8. https://doi.org/10.1038/376775a0. [Accessed 4 September 2014].

[58] Corder EH, Saunders AM, Strittmatter WJ, Schmechel DE, Gaskell PC, Small GW, Roses AD, Haines JL, Pericak-Vance MA. Gene dose of apolipoprotein E type 4 allele and the risk of Alzheimer's disease in late onset families. Science 1993;261:921−3. [Accessed 13 August 2014].

[59] Andrews SJ, Fulton-Howard B, Goate A. Interpretation of risk loci from genome-wide association studies of Alzheimer's disease. Lancet Neurol 2020;19:326−35. [Accessed 4 February 2021].

[60] Bertram L, Lill CM, Tanzi RE. The genetics of Alzheimer disease: back to the future. Neuron 2010;68:270−81.

[61] Muller U, Winter P, Graeber MB. A presenilin 1 mutation in the first case of Alzheimer's disease. Lancet Neurol 2013;12:129−30.

[62] Hardy J, Selkoe DJ. The amyloid hypothesis of Alzheimer's disease: progress and problems on the road to therapeutics. Science 2002;297:353−6. [Accessed 10 July 2014].

[63] Edwards FA. A unifying hypothesis for Alzheimer's disease: from plaques to neurodegeneration. Trends Neurosci 2019;42:310−22. [Accessed 11 December 2020].

[64] Mendez MF, Ghajarania M, Perryman KM. Posterior cortical atrophy: clinical characteristics and differences compared to Alzheimer's disease. Dement Geriatr Cogn Disord 2002;14:33−40.

[65] Gorno-Tempini ML, et al. Classification of primary progressive aphasia and its variants. Neurology 2011;76:1006−14.

[66] Ossenkoppele R, et al. The behavioural/dysexecutive variant of Alzheimer's disease: clinical, neuroimaging and pathological features. Brain 2015;138:2732−49.

[67] Salloway S, et al. Two phase 3 trials of bapineuzumab in mild-to-moderate Alzheimer's disease. N Engl J Med 2014;370:322−33. [Accessed 15 July 2014].

[68] Aisen PS, Cummings J, Doody R, Kramer L, Salloway S, Selkoe DJ, Sims J, Sperling RA, Vellas B. The future of anti-amyloid trials. J Prev Alzheimers Dis 2020;7:146−51.

[69] Motter R, Vigo-Pelfrey C, Kholodenko D, Barbour R, Johnson-Wood K, Galasko D, Chang L, Miller B, Clark C, Green R, Olson D, Southwick P, Wolfert R, Munroe B, Lieberburg I, Seubert P, Schenk D. Reduction of β-amyloid peptide42 in the cerebrospinal fluid of patients with Alzheimer's disease. Ann Neurol 1995;38:643−8. https://onlinelibrary.wiley.com/doi/full/10.1002/ana.410380413. [Accessed 4 December 2020].

[70] Andreasen N, Hesse C, Davidsson P, Minthon L, Wallin A, Winblad B, Vanderstichele H, Vanmechelen E, Blennow K. Cerebrospinal fluid β-amyloid((1-42)) in Alzheimer disease: differences between early- and late-onset Alzheimer disease and stability during the course of disease. Arch Neurol 1999;56:673−80. [Accessed 4 December 2020].

[71] Blennow K, Wallin A, Ågren H, Spenger C, Siegfried J, Vanmechelen E. Tau protein in cerebrospinal fluid - a biochemical marker for axonal degeneration in Alzheimer disease? Mol Chem Neuropathol 1995;26:231−45. https://link.springer.com/article/10.1007/BF02815140. [Accessed 4 December 2020].

[72] Vanmechelen E, Vanderstichele H, Davidsson P, Van Kerschaver E, Van Der Perre B, Sjögren M, Andreasen N, Blennow K. Quantification of tau phosphorylated at threonine 181 in human cerebrospinal fluid: a sandwich ELISA with a synthetic phosphopeptide for standardization. Neurosci Lett 2000;285:49−52.

[73] Sunderland T, Linker G, Mirza N, Putnam KT, Friedman DL, Kimmel LH, Bergeson J, Manetti GJ, Zimmermann M, Tang B, Bartko JJ, Cohen RM. Decreased β-amyloid1-42 and increased tau levels in cerebrospinal fluid of patients with Alzheimer disease. J Am Med Assoc 2003;289:2094−103. [Accessed 7 December 2020].

[74] Bittner T, Zetterberg H, Teunissen CE, Ostlund RE, Militello M, Andreasson U, Hubeek I, Gibson D, Chu DC, Eichenlaub U, Heiss P, Kobold U, Leinenbach A, Madin K, Manuilova E, Rabe C, Blennow K. Technical performance of a novel, fully automated electrochemiluminescence immunoassay for the quantitation of β-amyloid (1-42) in human cerebrospinal fluid. Alzheimer's Dementia 2016;12:517−26. https://doi.org/10.1016/j.jalz.2015.09.009. [Accessed 4 December 2020].

[75] Hansson O, Seibyl J, Stomrud E, Zetterberg H, Trojanowski JQ, Bittner T, Lifke V, Corradini V, Eichenlaub U, Batrla R, Buck K, Zink K, Rabe C, Blennow K, Shaw LM. CSF biomarkers of Alzheimer's disease concord with amyloid-β PET and predict clinical progression: a study of fully automated immunoassays in BioFINDER and ADNI cohorts. Alzheimer's Dementia 2018;14:1470−81. https://doi.org/10.1016/j.jalz.2018.01.010. [Accessed 4 December 2020].

[76] Klunk WE, et al. Imaging brain amyloid in Alzheimer's disease with Pittsburgh Compound-B. Ann Neurol 2004;55:306−19. [Accessed 24 July 2014].

[77] Ikonomovic MD, Klunk WE, Abrahamson EE, Mathis CA, Price JC, Tsopelas ND, Lopresti BJ, Ziolko S, Bi W, Paljug WR, Debnath ML, Hope CE, Isanski BA, Hamilton RL, DeKosky ST. Post-mortem correlates of in vivo PiB-PET amyloid imaging in a typical case of Alzheimer's disease. Brain 2008;131:1630−45.

[78] Clark CM, Pontecorvo MJ, Beach TG, Bedell BJ, Coleman RE, Doraiswamy PM, Fleisher AS, Reiman EM, Sabbagh MN, Sadowsky CH, Schneider JA, Arora A, Carpenter AP, Flitter ML, Joshi AD, Krautkramer MJ, Lu M, Mintun MA, Skovronsky DM. Cerebral PET with florbetapir compared with neuropathology at autopsy for detection of neuritic amyloid-β plaques: a prospective cohort study. Lancet Neurol 2012;11:669−78.

[79] Curtis C, et al. Phase 3 trial of flutemetamol labeled with radioactive fluorine 18 imaging and neuritic plaque density. JAMA Neurol 2015;72:287−94. [Accessed 8 December 2020].

[80] Sabri O, et al. Florbetaben PET imaging to detect amyloid beta plaques in Alzheimer's disease: phase 3 study. Alzheimer's Dementia 2015;11:964−74. https://doi.org/10.1016/j.jalz.2015.02.004. [Accessed 8 December 2020].

[81] Xia C-F, et al. [18F]T807, a novel tau positron emission tomography imaging agent for Alzheimer's disease. Alzheimer's Dementia 2013;9:666−76. [Accessed 17 July 2018].

[82] Nelson PT, et al. Correlation of Alzheimer disease neuropathologic changes with cognitive status: a review of the literature. J Neuropathol Exp Neurol 2012;71:362−81. http://jnen.oxfordjournals.org/lookup/doi/10.1097/NEN.0b013e31825018f7. [Accessed 1 November 2016].

[83] Cho H, Choi JY, Hwang MS, Lee JH, Kim YJ, Lee HM, Lyoo CH, Ryu YH, Lee MS. Tau PET in Alzheimer disease and mild cognitive impairment. Neurology 2016;87:375−83. [Accessed 2 September 2016].

[84] Ossenkoppele R, et al. Tau PET patterns mirror clinical and neuroanatomical variability in Alzheimer's disease. Brain 2016;139:1551−67. [Accessed 6 August 2018].

[85] Schöll M, Lockhart SN, Schonhaut DR, O'Neil JP, Janabi M, Ossenkoppele R, Baker SL, Vogel JW, Faria J, Schwimmer HD, Rabinovici GD, Jagust WJ. PET imaging of tau deposition in the aging human brain. Neuron 2016;89:971−82. [Accessed 2 September 2016].

[86] Maass A, Lockhart SN, Harrison TM, Bell RK, Mellinger T, Swinnerton K, Baker SL, Rabinovici GD, Jagust WJ. Entorhinal tau pathology, episodic memory decline, and neurodegeneration in aging. J Neurosci 2018;38.

[87] Whitwell JL, Graff-Radford J, Tosakulwong N, Weigand SD, Machulda MM, Senjem ML, Spychalla AJ, Vemuri P, Jones DT, Drubach DA, Knopman DS, Boeve BF, Ertekin-Taner N, Petersen RC, Lowe VJ, Jack CR, Josephs KA. Imaging correlations of tau, amyloid, metabolism, and atrophy in typical and atypical Alzheimer's disease. Alzheimer's Dementia 2018;14:1005−14.

[88] La Joie R, et al. Prospective longitudinal atrophy in Alzheimer's disease correlates with the intensity and topography of baseline tau-PET. Sci Transl Med 2020;12. https://pubmed.ncbi.nlm.nih.gov/31894103/. [Accessed 23 June 2020].

[89] Lowe VJ, Curran G, Fang P, Liesinger AM, Josephs KA, Parisi JE, Kantarci K, Boeve BF, Pandey MK, Bruinsma T, Knopman DS, Jones DT, Petrucelli L, Cook CN, Graff-Radford NR, Dickson DW, Petersen RC, Jack CR, Murray ME. An autoradiographic evaluation of AV-1451 Tau PET in dementia. Acta Neuropathol Commun 2016;4:1−19. https://link.springer.com/articles/10.1186/s40478-016-0315-6. [Accessed 8 December 2020].

[90] Baker SL, Maass A, Jagust WJ. Considerations and code for partial volume correcting [18F]-AV-1451 tau PET data. Data Brief 2017;15:648−57. [Accessed 7 March 2018].

[91] Baker SL, Harrison TM, Maass A, Joie R La, Jagust WJ. Effect of off-target binding on 18F-flortaucipir variability in healthy controls across the life span. J Nucl Med 2019;60:1444−51. [Accessed 4 December 2020].

[92] Ng KP, Pascoal TA, Mathotaarachchi S, Therriault J, Kang MS, Shin M, Guiot M-C, Guo Q, Harada R, Comley RA, Massarweh G, Soucy J-P, Okamura N, Gauthier S, Rosa-Neto P. Monoamine oxidase B inhibitor, selegiline, reduces 18F-THK5351 uptake in the human brain. Alzheimer's Res Ther 2017;9:25. http://alzres.biomedcentral.com/articles/10.1186/s13195-017-0253-y. [Accessed 10 August 2018].

[93] Josephs KA, Martin PR, Botha H, Schwarz CG, Duffy JR, Clark HM, Machulda MM, Graff-Radford J, Weigand SD, Senjem ML, Utianski RL, Drubach DA, Boeve BF, Jones DT, Knopman DS, Petersen RC, Jack CR, Lowe VJ, Whitwell JL. [18 F]AV-1451 tau-PET and primary progressive aphasia. Ann Neurol 2018;83:599−611.

[94] Leuzy A, Chiotis K, Lemoine L, Gillberg PG, Almkvist O, Rodriguez-Vieitez E, Nordberg A. Tau PET imaging in neurodegenerative tauopathies—still a challenge. Mol Psychiatr 2019;24:1112−34.

[95] van Dyck CH, Swanson CJ, Aisen P, Bateman RJ, Chen C, Gee M, Kanekiyo M, Li D, Reyderman L, Cohen S, Froelich L, Katayama S, Sabbagh M, Vellas B, Watson D, Dhadda S, Irizarry M, Kramer LD, Iwatsubo T. Lecanemab in early Alzheimer's disease. N Engl J Med 2023;388:9−21.

[96] Sims JR, et al. Donanemab in early symptomatic Alzheimer disease. JAMA 2023;330:512−27. https://doi.org/10.1001/jama.2023.13239.

[97] Landau SM, Lu M, Joshi AD, Pontecorvo M, Mintun MA, Trojanowski JQ, Shaw LM, Jagust WJ. Comparing positron emission tomography imaging and cerebrospinal fluid measurements of β-amyloid. Ann Neurol 2013;74:826−36. [Accessed 8 December 2020].

[98] Mattsson N, Insel PS, Landau S, Jagust W, Donohue M, Shaw LM, Trojanowski JQ, Zetterberg H, Blennow K, Weiner M. Diagnostic accuracy of CSF Ab42 and florbetapir PET for Alzheimer's disease. Ann Clin Transl Neurol 2014;1:534−43. https://onlinelibrary.wiley.com/doi/10.1002/acn3.81. [Accessed 8 December 2020].

[99] Palmqvist S, et al. Discriminative accuracy of plasma phospho-tau217 for Alzheimer disease vs other neurodegenerative disorders. JAMA, J Am Med Assoc 2020;324:772−81. [Accessed 4 December 2020].

[100] Janelidze S, Berron D, Smith R, Strandberg O, Proctor NK, Dage JL, Stomrud E, Palmqvist S, Mattsson-Carlgren N, Hansson O. Associations of plasma phospho-tau217 levels with tau positron emission tomography in early Alzheimer disease. JAMA Neurol 2021;78:149−56. [Accessed 27 February 2023].

[101] Li Y, Schindler SE, Bollinger JG, Ovod V, Mawuenyega KG, Weiner MW, Shaw LM, Masters CL, Fowler CJ, Trojanowski JQ, Korecka M, Martins RN, Janelidze S, Hansson O, Bateman RJ. Validation of plasma amyloid-β 42/40 for detecting Alzheimer disease amyloid plaques. Neurology 2022;98:e688−99. [Accessed 27 February 2023].

[102] Dubois B, Feldman HH, Jacova C, Dekosky ST, Barberger-Gateau P, Cummings J, Delacourte A, Galasko D, Gauthier S, Jicha G, Meguro K, O'brien J, Pasquier F, Robert P, Rossor M, Salloway S, Stern Y, Visser PJ, Scheltens P. Research criteria for the diagnosis of Alzheimer's disease: revising the NINCDS-ADRDA criteria. Lancet Neurol 2007;6:734−46. [Accessed 10 July 2014].

[103] Albert MS, DeKosky ST, Dickson D, Dubois B, Feldman HH, Fox NC, Gamst A, Holtzman DM, Jagust WJ, Petersen RC, Snyder PJ, Carrillo MC, Thies B, Phelps CH. The diagnosis of mild cognitive impairment due to Alzheimer's disease: recommendations from the National Institute on Aging-Alzheimer's Association workgroups on diagnostic guidelines for Alzheimer's disease. Alzheimers Dement 2011;7:270−9. [Accessed 9 July 2014].

[104] McKhann GM, Knopman DS, Chertkow H, Hyman BT, Jack CR, Kawas CH, Klunk WE, Koroshetz WJ, Manly JJ, Mayeux R, Mohs RC, Morris JC, Rossor MN, Scheltens P, Carrillo MC, Thies B, Weintraub S, Phelps CH. The diagnosis of dementia due to Alzheimer's disease: recommendations from the National Institute on Aging-Alzheimer's Association workgroups on diagnostic guidelines for Alzheimer's disease. Alzheimer's Dementia 2011;7:263−9. [Accessed 8 August 2018].

[105] Sperling RA, et al. Toward defining the preclinical stages of Alzheimer's disease: recommendations from the National Institute on Aging-Alzheimer's Association workgroups on diagnostic guidelines for Alzheimer's disease. Alzheimers Dement 2011;7:280−92. [Accessed 10 July 2014].

[106] Jagust W. Imaging the evolution and pathophysiology of Alzheimer disease. Nat Rev Neurosci 2018;19:687−700.

[107] Jack CR, Bennett DA, Blennow K, Carrillo MC, Feldman HH, Frisoni GB, Hampel H, Jagust WJ, Johnson KA, Knopman DS, Petersen RC, Scheltens P, Sperling RA, Dubois B. A/T/N: an unbiased descriptive classification scheme for Alzheimer disease biomarkers. Neurology 2016;87:539−47.

[108] Klunk WE, Koeppe RA, Price JC, Benzinger TL, Devous MD, Jagust WJ, Johnson KA, Mathis CA, Minhas D, Pontecorvo MJ, Rowe CC, Skovronsky DM, Mintun MA. The Centiloid project: standardizing quantitative amyloid plaque estimation by PET. Alzheimer's Dementia 2015;11:1—15.

[109] Schöll M, Ossenkoppele R, Strandberg O, Palmqvist S, Jögi J, Ohlsson T, Smith R, Hansson O. Distinct 18F-AV-1451 tau PET retention patterns in early- and late-onset Alzheimer's disease. Brain 2017;140:2286—94. [Accessed 8 December 2020].

[110] Whitwell JL, Martin P, Graff-Radford J, Machulda MM, Senjem ML, Schwarz CG, Weigand SD, Spychalla AJ, Drubach DA, Jack CR, Lowe VJ, Josephs KA. The role of age on tau PET uptake and gray matter atrophy in atypical Alzheimer's disease. Alzheimer's Dementia 2019;15:675—85.

[111] Dubois B, et al. Clinical diagnosis of Alzheimer's disease: recommendations of the international working group. Lancet Neurol 2021;20:484—96. [Accessed 27 February 2023].

[112] Rabinovici GD, Gatsonis C, Apgar C, Chaudhary K, Gareen I, Hanna L, Hendrix J, Hillner BE, Olson C, Lesman-Segev OH, Romanoff J, Siegel BA, Whitmer RA, Carrillo MC. Association of amyloid positron emission tomography with subsequent change in clinical management among medicare beneficiaries with mild cognitive impairment or dementia. J Am Med Assoc 2019;321:1286—94. [Accessed 11 December 2020].

[113] Ossenkoppele R, et al. Amyloid and tau PET-positive cognitively unimpaired individuals are at high risk for future cognitive decline. Nat Med 2022;28:2381—7. [Accessed 27 February 2023].

[114] Strikwerda-Brown C, et al. Association of elevated amyloid and tau positron emission tomography signal with near-term development of Alzheimer disease symptoms in older adults without cognitive impairment. JAMA Neurol 2022;79:975—85. [Accessed 27 February 2023].

[115] Lingler JH, Klunk WE. Disclosure of amyloid imaging results to research participants: has the time come? Alzheimers Dement 2013;9:741—4. [Accessed 8 June 2017].

[116] Jack CR, et al. Revised criteria for diagnosis ans staging of Alzheimer's disease: Alzheimer's Association Workgroup. Alzheimers Dement. 2024;20:5143—69. https://doi.org/10.1002/alz.13859.

Chapter 3

Early Alzheimer's disease: Clinical overview

James Selwood[a] and Elizabeth Coulthard[b]
[a]*University of Bristol, Devon Partnership Trust, Exeter, United Kingdom;* [b]*University of Bristol, North Bristol NHS Trust, Bristol, United Kingdom*

Alzheimer's disease pathology

Early Alzheimer's disease pathology

Post-mortem histopathology defines AD and is central to the search for treatments. In AD, amyloid-beta (Aβ) plaques and tau tangles accumulate in the brain, each with a distinct characteristic pattern [1]. Both Aβ plaque deposition and tau are associated with synaptic loss and may be the trigger for neuronal degeneration that causes marked atrophy seen macroscopically at post-mortem. Extracellular Aβ plaques and intracellular hyperphosphorylated tau tangles are still the gold standard for confirming a diagnosis of AD at post-mortem [1].

Amyloid-beta

Aβ is derived from the processing of Aβ precursor protein (APP), a single-pass transmembrane protein which, together with Aβ peptides, may play a key role in regulating neuronal activity and synaptic vesicle release [2]. Two APP processing pathways exist - one of which produces the form most associated with AD - $Aβ_{42}$ - and the more abundant $Aβ_{40}$ [3]. Compared to $Aβ_{40}$, the $Aβ_{42}$ isoform is more hydrophobic and toxic, making it a core component of Aβ plaques. Oligomerization leads to the formation of Aβ fibrils and then plaques [4].

The Aβ cascade hypothesis was first proposed in 1992 and states that the generation of Aβ plaques is the primary event in AD pathogenesis, triggering the formation of tau tangles and cell death [5]. In vivo, Aβ and tau positron emission tomography (PET) imaging reveal that Aβ plaque deposition precedes tangle formation [6], in keeping with a key role for Aβ in AD progression.

Tau

Tau is a microtubule-associated protein predominantly found in neurons coded by the microtubule-associated protein tau (MAPT) gene located on chromosome 17q21. Normal tau stabilizes the microtubules within neurons, regulates axonal transport and promotes neurite formation [7]. Abnormal soluble tau within axons from the locus coeruleus is probably the earliest AD-related change in the brain, but does not necessarily progress to AD [1].

Pathological spread of tau is an intercellular process that occurs via multiple mechanisms referred to as pathological seeding [8]. In progressive AD, tau hyperphosphorylation leads to cytoskeleton disassembly, disruption to axonal transport and aggregation of tau into filamentous deposits, or neurofibrillary tangles (NFTs). Progressive accumulation of NFTs follows a characteristic pattern across the brain defined by Braak stages. Aβ plaque deposition follows a different pattern according to Thal staging (see Fig. 3.1) [9]. Unlike Aβ plaque distribution, NFT pathology shows a higher level of correlation with early cognitive symptoms [10].

Aβ plaques alone are a common post-mortem finding without evidence of dementia [11]. This suggests that sporadic Aβ plaques alone do not always trigger the cascade that leads to AD dementia. The interplay between Aβ and tau may be important for propagation through the brain [12]. This synergistic relationship between Aβ and tau is illustrated in Fig. 3.2. However, genome wide association studies (GWAS) reveal over 40 susceptibility loci for AD, most of which do not relate to Aβ or tau, suggesting a "multiplex model" of AD development [13].

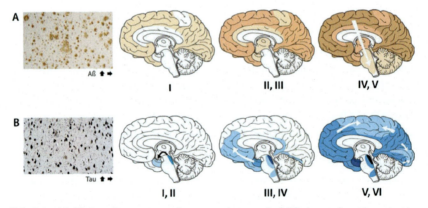

FIG. 3.1 (A) Thal staging represents the progressive spread of Aβ plaques from the neocortex (I) to the entorhinal cortex and hippocampus (II and III) and finally the brainstem and cerebellum (IV and V). (B) Braak staging represents the progressive spread of NFTs from the entorhinal region (I and II) to the hippocampal region (III and IV) and finally the neocortex (V and VI) [9].

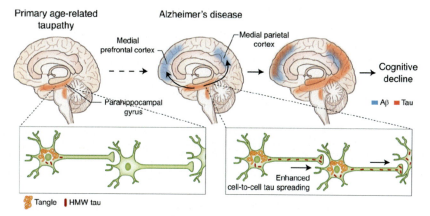

FIG. 3.2 In primary age-related tauopathy, tau tangles may be found in the brainstem nuclei and entorhinal cortex in cognitively healthy older individuals with no Aβ plaque deposition. In AD, the deposition of Aβ plaques is associated with tau propagation from the parahippocampus to the neocortex and with increased intercellular spreading of high molecular weight (HMW) hyper-phosphorylated tau [14].

Co-pathology

Although Aβ is typically associated with plaque formation, it may also be deposited within blood vessel walls, causing cerebral amyloid angiopathy (CAA). Microhaemorrhages occur as Aβ accumulates within the vessel walls. The resulting vascular dysfunction contributes to the accumulation of extracellular Aβ plaques, exemplifying the interaction between different disease processes. AD vascular disease is not limited to CAA. Atherosclerosis is also likely to be present and may lead to gross ischemia and infarcts [15].

Post-mortem examination of brains from patients with AD shows other significant co-pathologies alongside Aβ and tau, including alpha-synuclein (~50% of AD cases), TAR DNA-binding protein 43, or TDP-43, (~35% AD cases) and vascular pathology. Alpha-synuclein and vascular co-pathologies have been associated with more rapid decline [11], strongly implicating pathology over and above plaques and tangles in the expression of AD, at least in the later stages. Changes in cerebral perfusion [16], breakdown in the blood-brain barrier [17], neuroinflammatory/microglial changes [18], mitochondrial abnormalities [19] and iron deposition [20] are other neuropathological changes that occur early, often before AD symptoms occur. In summary, Aβ and tau are accompanied by multiple pathologies evident even in the early stages of AD. Clinical use of biomarkers of Aβ, tau and copathologies detects these pathologies and facilitates characterization of AD at an early stage, but such tests always require clinical context for interpretation [21].

The syndrome of mild cognitive impairment

Historical attempts to classify cognitive change less marked than, and often preceding, full-blown dementia have included benign senescent forgetfulness, age-associated memory impairment, age-associated cognitive decline and cognitive impairment not dementia [22]. MCI has come into common use and is broadly defined as an objective cognitive decline with no significant impact on activities of daily living [23,24].

The latest edition of the Diagnostic and Statistical Manual of Mental Disorders (DSM-V) has used the term mild neurocognitive disorder [22]. This is also defined by both subjective and objective evidence of modest cognitive impairments without significant impact on activities of daily living. However, DSM-V has abandoned strict cut-offs in favor of a more individual assessment [22]. Critically, early definitions of MCI and similar diagnostic labels did not confer any pathological etiology and these remain in common clinical use. However, the National Institute on Aging-Alzheimer's Association (NIA-AA) criteria now define MCI due to AD pathology (see Box 3.1) using biological staging alongside symptomatic/clinical disease staging [21,25]. NIA-AA criteria propose a more fine-grained clinical staging with levels 0—6 ranging through asymptomatic to severe dementia with previous definitions of MCI most closely aligning with Stage 3.

BOX 3.1 Diagnostic criteria for MCI

Petersen criteria [26]	NIA-AA criteria [25]
• Subjective memory complaint • Objective evidence of memory impairment • Preserved activities of daily living • Not demented	• Evidence of concern about a change in cognition • Impairment in one or more cognitive domains • Preservation of independence in functional abilities • Not demented • Absence of systemic or brain disease that could account for cognitive decline Biomarkers that support a diagnosis of MCI due to AD include: • Positive CSF Aβ42/40. Phosphorylated-tau181/Aβ42, t-tau/Aβ42 or Aβ PET • Positive CSF total tau or tau PET • Hippocampal/medial temporal lobe atrophy on MRI • Temporoparietal/precuneus hypometabolism or hypoperfusion on blood-flow PET

The reported incidence and prevalence of MCI vary considerably in the literature. A recent systematic review looking specifically at the incidence of MCI reported 22.5 new cases per 1000 person-years for ages 75–79. This increased to 60.1 for those aged 85 and over [27]. The COSMIC collaboration reported a range of published MCI prevalence estimates in people over 60 years of age between 5% and 36.7% [28].

Other mild syndromes that can be considered as dementia prodromes or risk factors have been defined including subjective cognitive decline (SCD) and mild behavioral impairment (MBI).

Subjective cognitive decline

SCD is defined by self-reported cognitive symptoms, but without evidence of a decline or lower than normal scores on cognitive testing. Some patients may initially present with SCD that later develops into MCI, particularly in those with neurodegeneration but a high level of premorbid functioning. SCD may therefore be considered a pre-MCI state in these individuals – although rates of progression to MCI vary according to the population and criteria applied [29]. In this model, patients initially transition from being cognitively healthy to SCD. As neurodegeneration progresses, the individual develops more overt symptoms that are detectable on cognitive testing. It is at this stage that MCI may be identified. A diagnosis of dementia is later made when the individual shows significant functional impairment. Therefore, SCD may be another stage of cognitive decline in incipient AD, falling between healthy cognition and MCI [30].

Mild behavioral impairment

MBI describes neurobehavioral symptoms that emerge later in life including apathy, emotional dysregulation, impulsivity, inappropriate social behavior and psychosis. MBI is a predementia state in some individuals, although its symptoms have been identified in cognitively normal older populations [31]. The MBI Checklist (MBI-C) is a 34-question rating scale that may be used to identify this syndrome [32]. Much like the MCI concept, MBI gives no indication of the etiology and simply recognizes that a neuropsychiatric disturbance exists.

Sub-classification mild cognitive impairment

The syndrome of MCI can be fractionated according to the number of affected cognitive domains: single or multi-domain. More precise classifications focus on which cognitive domains are impaired: amnestic MCI refers to a deficit in memory, while non-amnestic MCI refers to deficits in other cognitive domains, such as language or executive function. Amnestic and multi-domain MCI syndromes carry the greatest risk of converting to Alzheimer's dementia [33,34].

Conversion to dementia

The risk of conversion varies in the literature, ranging from 5% to 15% per year [26]. This range is likely to reflect the heterogenous nature of MCI, which will include many patients who do not have degenerative conditions and therefore "never" convert to dementia. At 5 years follow-up, approximately 50% of patients with MCI have not declined [35]. These patients are often described as having a stable MCI.

Risk of conversion depends critically on the etiology of MCI, which can be broadly divided into degenerative and non-degenerative causes. The degenerative group includes all the dementias, though they may not first present at the MCI stage. AD is the commonest dementia to emerge from MCI — explaining why criteria largely focus on early features of AD, such as amnesia. Non-degenerative causes vary widely and include vitamin B12 deficiency or hypothyroidism. Psychiatric disturbances, such as depression, are also important causes of MCI.

Neuropsychology is a standard clinical tool that is helpful for predicting conversion. In a systematic review of 28 studies with 2365 participants, both verbal memory and language tests were highly predictive of conversion over a mean follow-up period of 31 months [36]. Structural brain imaging may also be used to refine the diagnosis. For example, hippocampal or medial lobe atrophy is included in the NIA-AA recommendations on diagnostic guidelines for MCI due to AD [25]. Access to molecular biomarkers is not available everywhere, but growing evidence supports their use. In a multi-center study of 250 patients with MCI, the use of CSF biomarkers in addition to standard measures (demographic, clinical and imaging) significantly improved the prediction of conversion risk [37]. Prognostic algorithms are emerging that can be used clinically to predict risk of developing dementia from MCI, taking into account neuroimaging and molecular biomarkers [38].

We are now in a position clinically to offer prognostic information to patients. But questions remain about how we assess psychological readiness to receive prognostic information and the ethical framework that should underpin predictive AD dementia diagnosis at the MCI stage. Even when early post-diagnostic support is available, receiving an MCI diagnosis may cause anxiety about the future [39]. Further complications arise from use of relatively new diagnostic techniques with, for example, changes made to normal ranges within CSF testing as experience grows and a proportion of biomarker test results being in the equivocal range. Furthermore, even when an underlying degenerative cause is identified, the lack of specific treatments can cause frustration [40]. As more effective treatments emerge and certainty around use of predictive tests grows, the clinical imperative to diagnose AD early, at the MCI or SCD stages, will grow.

Management of mild cognitive impairment
Clinical assessment

An approach to clinical assessment is presented in Box 3.2 and investigations that may help accurate diagnosis are discussed below.

Neuropsychological assessment commonly uses test such as the Montreal Cognitive Assessment (MoCA) [44], the Addenbrooke's Cognitive Examination (ACE-R) [45] and/or the Repeatable Battery for the Assessment of

---

BOX 3.2 Clinical assessment

Clinical assessment includes history from the patient and a caregiver/witness exploring multiple cognitive domains (including memory, language, attention, executive and visuospatial function), the impact of symptoms on daily activities (such as driving), and neurological examination. Drug and alcohol history should include prescribed and over-the-counter medications to identify drugs with a high anticholinergic burden or other cognitive side effects [41]. Primary sensory impairment, particularly hearing loss, is highlighted as both a risk factor and exacerbator of cognitive decline and should be explored [42,43]. Vascular risk factors, such as hypertension, as well as other risk factors for dementia listed below can be helpful as part of the consultation:
- Education level
- Traumatic brain injury
- Mid-life hypertension
- Obesity
- Smoking
- Hearing loss
- Visual impairment
- Alcohol intake
- Depression/anxiety or other psychiatric symtpoms
- Social isolation
- Physical inactivity
- Diabetes
- Cholesterol level - LDL
- Menopause/HRT status
- Sleep apnea symptoms – daytime sleepiness, heavy snoring with pauses etc
- Insomnia

The examination can uncover functional deficits not always obvious from the history, for example, extrapyramidal signs (including gait) might suggest alpha-synucleinopathy. Pyramidal signs and/or lower motor neuron signs may suggest frontotemporal dementia associated with motor neuron disease. Impaired praxis (e.g., difficulty with buttons or shoe laces during examination) and other visuospatial deficits such as simultanagnosia and optic apraxia may suggest posterior cortical atrophy.

Neuropsychological Status (RBANS) [46]. If cognitive deficits are detected, the pattern of impairments may be used to sub-classify the MCI as amnestic, non-amnestic or multi-domain. For example, impairments detected on verbal and/or visual memory tests but normal scores on other cognitive tests would suggest an amnestic picture. In contrast, impairments in executive function alone would suggest a non-amnestic MCI. In cases where impairments are detected on two or more cognitive domains, the MCI is classified as multi-domain.

General medical screen includes vitamin B12, thyroid function, diabetes screening and possibly cholesterol levels [41]. Other clues to reversible cause of cognitive impairment include acute/subacute onset, seizure activity and/or unexplained hyponatraemia, which should trigger an encephalitis screen such as LGI1 (voltage-gated potassium channel) or NMDA receptor antibodies. Epilepsy occasionally presents as late-onset cognitive impairment and seizures may accelerate cognitive decline in AD [47].

Structural brain imaging serves two purposes. Firstly, it is used to identify intra-cranial pathology, such as tumors, vascular disease, and normal pressure hydrocephalus (NPH). Secondly, the imaging may identify degenerative changes, such as hippocampal atrophy [48]. However, the absence of focal atrophy does not exclude a degenerative process. A Cochrane review concluded that hippocampal or medial temporal lobe atrophy is neither sensitive nor specific to identify AD in patients presenting at the MCI stage [49]. Furthermore, some findings may be constitutional while generalized involutional changes are often age-related. Serial imaging can be helpful to identify change over time [50].

Other imaging modalities can increase diagnostic certainty. ^{18}F fluorodeoxyglucose (FDG) PET imaging quantifies brain glucose metabolism. Glucose hypometabolism is suggestive of neurodegeneration [51]. Single-photon emission-computed tomography (SPECT) is now rarely used to assess regional cerebral blood flow using an intravenous radiolabeled tracer. Uptake of the radiotracer is proportional to cerebral blood flow. A reduced uptake indicates hypoperfusion and reduced neural activity. Hypoperfusion of the temperoparietal regions is suggestive of AD [52]. A systematic review comparing both imaging modalities reported sensitivities of diagnosing AD in the range of 75%−99% for PET and 65%−85% for SPECT. Specificities were in the range of 71%−93% for PET and 72%−87% for SPECT. This favored the use of PET imaging but was based on limited evidence [53]. A later Cochrane review of the accuracy of ^{18}F-FDG PET in predicting the development of AD concluded it is of limited clinical value. This was due to the wide variation in specificities and unclear cut-offs for positive scans [54]. Overall, these imaging studies may support an AD diagnosis, but normal results do not exclude it.

Molecular biomarker testing of AD can be achieved through molecular PET imaging or cerebrospinal fluid sampling (CSF) using lumbar puncture. A radiolabeed ligand, capable of binding to Aβ, is injected into the patient's bloodstream. The ligand crosses the blood-brain barrier and binds to extracellular Aβ deposits. PET imaging allows the clinician to determine whether the amount of Aβ binding is physiological or abnormal. A positive result suggests significant plaque deposition that supports an AD diagnosis [50]. Aβ PET imaging is increasingly used in dementia research to identify subjects eligible for AD drug trials, but is often not available in routine clinical practice.

CSF sampling simultaneously measures both Aβ and tau species as well as other markers such as neurofilament light chain. Reduced CSF Aβ$_{42}$ suggests significant intra-cerebral Aβ deposition [55] as Aβ plaques absorb newly generated Aβ, thereby preventing its normal clearance into the CSF. Expressing Aβ as Aβ$_{42}$/Aβ$_{40}$ ratio compensates for individual differences in Aβ production and pre-analysis variations [56]. Measurement of total tau provides a measure of the degree of neuronal loss, but is not specific to AD. For example, a high total tau can be seen in the encephalitides and stroke. A very high total tau suggests rapid neurodegeneration, as seen in Creutzfeld-Jakob Disease [57]. Phosphorylated tau is a more specific measure of AD [58]. Due to expense and the technical challenges of CSF collection and PET, blood biomarkers could have far broader clinical utility [59].

To identify those patients who are converting to dementia, patients with MCI can be offered regular long-term monitoring. Determining whether there has been a significant decline in social or occupational functioning may be captured objectively through assessment tools, such as the functional activities questionnaire (FAQ) [60]. Repeat neuropsychology assessments may be used to provide an objective measure of any further cognitive decline.

Treatment
Pharmacological
Monoclonal antibodies targeting Aβ (including aducanumab, lecanemab and donanemab) are the first licensed therapies for MCI due to Alzheimer's disease, and are entering routine clinical practice around the world [61,62]. There are no other licensed pharmacological treatments for MCI due to AD, but several are in development. According to a drug development pipeline report in 2024, 164 agents were in clinical trials [63] — 75% of agents are aiming for disease-modification. An increasing number of trials are now targeting pathways beyond Aβ and tau [64].

The cholinesterase inhibitors (donepezil, galantamine and rivastigmine) are licensed for AD dementia, increasing the synaptic availability of acetylcholine to compensate for the degenerative loss of cholinergic neurons. For moderate to severe AD, the non-competitive *N*-methyl-D-aspartate (NMDA) receptor antagonist memantine reduces neurotoxicity induced by the excitatory effects of glutamate. All four medications have only modest benefit and are primarily used to improve symptoms and quality of life. They do not delay AD progression [65].

Unfortunately, a systematic review concluded that the cholinesterase inhibitors and memantine were ineffective in MCI [66]. Whether the initiation of these medications improve symptoms and quality of life in patients with MCI and positive AD biomarkers requires further study.

Interest in the therapeutic potential of anti-inflammatories was sparked by the emerging role of inflammation in AD pathology. Aspirin and other non-steroidal anti-inflammatory drugs (NSAIDs) have been investigated in healthy individuals and patients with MCI. In a recent Cochrane review of four large randomized controlled trials (RCTs) with 23,187 participants, aspirin and other NSAIDs (celecoxib, rofecoxib and naproxen) were all ineffective in preventing dementia but did cause harm [67]. Immunomodulatory anti-inflammatory approaches are emerging and not yet in the clinical sphere [68].

Psychiatric disturbances are common in cognitive clinics. The relationship between affective disturbance and neurodegeneration is complex. Mid-life depression is a recognised risk factor for dementia [42]. The underlying neurobiology is not fully understood but depression has been associated with pathological changes including low CSF $A\beta_{42}$, entorhinal cortical thinning and hippocampal atrophy [69]. Affective disturbances may therefore be an early manifestation of neurodegeneration. Anxiety may also arise as a natural result of a patient's insight into their own cognitive decline. Affective disturbances can also occur independently of any degenerative process and may present as a functional cognitive syndrome [70]. In these cases, the identification and treatment of the affective disorder may reverse the cognitive decline.

Non-pharmacological

Useful frameworks to consider when developing a clinical service that offers non-pharmacological interventions include the Lancet Commission [43] and the successful implementation of a lifestyle regime in, for example, the Finnish Geriatric Intervention Study to Prevent Cognitive Impairment and Disability (FINGER) study [71].

The 2024 Lancet Commission on dementia prevention, intervention and care identified 14 potentially modifiable risk factors for dementia (see Box 3.2) [42]. However, these factors are generally based on association rather than proven potential to modify trajectory of decline. Nevertheless, they provide a framework for Brain Health Clinics that are emerging to people with early cognitive symptoms reduce their risk of dementia. The UK-based PROTECT study is just one example of active research that aims to better understand why people develop dementia, particularly in relation to lifestyle factors, and how people can be encouraged to adopt changes [72]. Studies such as PROTECT also provide the infrastructure to pilot interventions to help delay AD dementia.

Promoting a healthy diet is of benefit to all aspects of general health and may specifically improve cognitive performance. In the FINGER study, 1260 participants at risk of cognitive decline received either intensive diet counseling or regular health advice over 2 years. Dietary changes during this period improved executive function while long-term diet was shown to affect more global cognition [73]. On another level, Souvenaid is a nutritional product developed for people with AD. This yoghurt is a complex of various nutrients involved in synaptic structure and function. They include B vitamins, choline, vitamin E, nucleotides, omega-3 fatty acids and selenium. A systematic review of three studies with 1011 participants concluded that Souvenaid did not improve function or behavior but did improve verbal recall in patients with mild AD [74]. Patients keen to try Souvenaid should be informed that there is limited evidence of cognitive benefit. A Cochrane review in 2020 concluded that Souvenaid does not reduce risk of conversion to dementia and does not benefit other outcomes in patients with MCI or mild-moderate AD dementia [75].

Physical exercise appears to be of specific benefit to cognition in patients with MCI. A systematic review of 27 studies including 2077 participants showed that physical exercise improved global cognition, executive function and delayed recall with benefits also seen in verbal fluency and attention. Moderately intense exercise and mind-body exercises had the most notable effects [76].

Other clinically appropriate lifestyle interventions include addressing excess alcohol consumption and smoking with appropriate referral to local alcohol services, nicotine replacement therapy and/or access to smoking cessation services where available.

Finally, cognitive training refers to the repeated practice of stimulating cognitive exercises. This may include traditional puzzles, such as crosswords, but also refers to other more novel tasks. One of the proposed benefits of cognitive training relates to the concept of cognitive reserve. People with a higher cognitive reserve are thought to be less susceptible and more resilient to degenerative change that might compromise function. Cognitive reserve is typically linked to earlier life experiences including educational and occupational achievement [77]. Some studies suggest that cognitive training improves brain activity in the short term [78].

Technological advances have concomitantly led to the development of an overwhelming number of available "apps" to deliver cognitive training. A Cochrane review of eight clinical trials with 660 participants did not find evidence to support the use of computerized cognitive training in preventing dementia. Similarly, the evidence did not suggest improved or maintained cognition in people with established MCI [79]. While brain training activities are unlikely to cause harm, they do not appear to lead to global cognitive improvements.

Overall, diagnosis and management of MCI due to AD is rapidly evolving and will continue to do so as new diagnostic tools and treatments are developed. While there are multiple research avenues for reducing dementia risk in people with MCI due to AD, relatively few intervention have proven benefit in this population.

References

[1] Braak H, Del Tredici K. The preclinical phase of the pathological process underlying sporadic Alzheimer's disease. Brain 2015;138(Pt 10):2814−33.

[2] O'Brien RJ, Wong PC. Amyloid precursor protein processing and Alzheimer's disease. Annu Rev Neurosci 2011;34:185−204.

[3] Agrawal N, Skelton AA. Structure and function of Alzheimer's amyloid betaeta proteins from monomer to fibrils: a mini review. Protein J 2019;38(4):425−34.

[4] Fandrich M. Oligomeric intermediates in amyloid formation: structure determination and mechanisms of toxicity. J Mol Biol 2012;421(4−5):427−40.

[5] Hardy JA, Higgins GA. Alzheimer's disease: the amyloid cascade hypothesis. Science 1992;256(5054):184−5.

[6] Bateman RJ, Xiong C, Benzinger TL, Fagan AM, Goate A, Fox NC, et al. Clinical and biomarker changes in dominantly inherited Alzheimer's disease. N Engl J Med 2012;367(9):795−804.

[7] Wang Y, Mandelkow E. Tau in physiology and pathology. Nat Rev Neurosci 2016;17(1):5−21.

[8] Gibbons GS, Lee VMY, Trojanowski JQ. Mechanisms of cell-to-cell transmission of pathological tau: a review. JAMA Neurol 2019;76(1):101−8.

[9] Jouanne M, Rault S, Voisin-Chiret AS. Tau protein aggregation in Alzheimer's disease: an attractive target for the development of novel therapeutic agents. Eur J Med Chem 2017;139:153−67.

[10] Guillozet AL, Weintraub S, Mash DC, Mesulam MM. Neurofibrillary tangles, amyloid, and memory in aging and mild cognitive impairment. Arch Neurol 2003;60(5):729−36.

[11] Robinson JL, Lee EB, Xie SX, Rennert L, Suh E, Bredenberg C, et al. Neurodegenerative disease concomitant proteinopathies are prevalent, age-related and APOE4-associated. Brain 2018;141(7):2181−93.

[12] Nisbet RM, Polanco JC, Ittner LM, Götz J. Tau aggregation and its interplay with amyloid-β. Acta Neuropathol 2015;129(2):207−20.

[13] Sims R, Hill M, Williams J. The multiplex model of the genetics of Alzheimer's disease. Nat Neurosci 2020;23(3):311−22.

[14] Busche MA, Hyman BT. Synergy between amyloid-β and tau in Alzheimer's disease. Nat Neurosci 2020;23(10):1183−93.

[15] Greenberg SM, Bacskai BJ, Hernandez-Guillamon M, Pruzin J, Sperling R, van Veluw SJ. Cerebral amyloid angiopathy and Alzheimer disease - one peptide, two pathways. Nat Rev Neurol 2020;16(1):30−42.

[16] Sierra-Marcos A. Regional cerebral blood flow in mild cognitive impairment and Alzheimer's disease measured with arterial spin labeling magnetic resonance imaging. Int J Alzheimers Dis 2017;2017:5479597.

[17] Sweeney MD, Sagare AP, Zlokovic BV. Blood-brain barrier breakdown in Alzheimer disease and other neurodegenerative disorders. Nat Rev Neurol 2018;14(3):133−50.

[18] Cameron B, Landreth GE. Inflammation, microglia, and Alzheimer's disease. Neurobiol Dis 2010;37(3):503−9.

[19] Swerdlow RH, Burns JM, Khan SM. The Alzheimer's disease mitochondrial cascade hypothesis: progress and perspectives. Biochim Biophys Acta 2014;1842(8):1219−31.

[20] Liu JL, Fan YG, Yang ZS, Wang ZY, Guo C. Iron and Alzheimer's disease: from pathogenesis to therapeutic implications. Front Neurosci 2018;12:632.

[21] Jack Jr. CR, Andrews JS, Beach TG, Buracchio T, Dunn B, Graf A, et al. Revised criteria for diagnosis and staging of Alzheimer's disease: Alzheimer's Association Workgroup. Alzheimers Dementia.

[22] Bradfield NI, Ames D. Mild cognitive impairment: narrative review of taxonomies and systematic review of their prediction of incident Alzheimer's disease dementia. BJPsych Bull 2020;44(2):67−74.

[23] Winblad B, Palmer K, Kivipelto M, Jelic V, Fratiglioni L, Wahlund LO, et al. Mild cognitive impairment–beyond controversies, towards a consensus: report of the International Working Group on Mild Cognitive Impairment. J Intern Med 2004;256(3):240−6.

[24] Dunne RA, Aarsland D, O'Brien JT, Ballard C, Banerjee S, Fox NC, et al. Mild cognitive impairment: the Manchester consensus. Age Ageing 2021;50(1):72−80.

[25] Albert MS, DeKosky ST, Dickson D, Dubois B, Feldman HH, Fox NC, et al. The diagnosis of mild cognitive impairment due to Alzheimer's disease: recommendations from the National Institute on Aging-Alzheimer's Association workgroups on diagnostic guidelines for Alzheimer's disease. Alzheimer Dementia J Alzheimer Assoc 2011;7(3):270−9.

[26] Petersen RC, Roberts RO, Knopman DS, Boeve BF, Geda YE, Ivnik RJ, et al. Mild cognitive impairment: ten years later. Arch Neurol 2009;66(12):1447−55.

[27] Gillis C, Mirzaei F, Potashman M, Ikram MA, Maserejian N. The incidence of mild cognitive impairment: a systematic review and data synthesis. Alzheimers Dement (Amst) 2019;11:248−56.

[28] Sachdev PS, Lipnicki DM, Kochan NA, Crawford JD, Thalamuthu A, Andrews G, et al. The prevalence of mild cognitive impairment in diverse geographical and ethnocultural regions: the COSMIC collaboration. PLoS One 2015;10(11):e0142388.

[29] Archer HA, Newson MA, Coulthard EJ. Subjective memory complaints: symptoms and outcome in different research settings. J Alzheimers Dis 2015;48(Suppl. 1):S109−14.

[30] Jessen F, Amariglio RE, van Boxtel M, Breteler M, Ceccaldi M, Chetelat G, et al. A conceptual framework for research on subjective cognitive decline in preclinical Alzheimer's disease. Alzheimers Dement 2014;10(6):844−52.

[31] Creese B, Griffiths A, Brooker H, Corbett A, Aarsland D, Ballard C, Ismail Z. Profile of mild behavioral impairment and factor structure of the Mild Behavioral Impairment Checklist in cognitively normal older adults. Int Psychogeriatr 2020;32(6):705−17.

[32] Ismail Z, Agüera-Ortiz L, Brodaty H, Cieslak A, Cummings J, Fischer CE, et al. The mild behavioral impairment checklist (MBI-C): a rating scale for neuropsychiatric symptoms in pre-dementia populations. J Alzheimers Dis 2017;56(3):929−38.

[33] Bradfield NI, Ellis KA, Savage G, Maruff P, Burnham S, Darby D, et al. Baseline amnestic severity predicts progression from amnestic mild cognitive impairment to Alzheimer disease dementia at 3 years. Alzheimer Dis Assoc Disord 2018;32(3):190−6.

[34] Tabert MH, Manly JJ, Liu X, Pelton GH, Rosenblum S, Jacobs M, et al. Neuropsychological prediction of conversion to Alzheimer disease in patients with mild cognitive impairment. Arch Gen Psychiatr 2006;63(8):916−24.

[35] Rossini PM, Miraglia F, Alu F, Cotelli M, Ferreri F, Iorio RD, et al. Neurophysiological hallmarks of neurodegenerative cognitive decline: the study of brain connectivity as a biomarker of early dementia. J Personalized Med 2020;10(2).

[36] Belleville S, Fouquet C, Hudon C, Zomahoun HTV, Croteau J. Consortium for the early identification of Alzheimer's d-Q. Neuropsychological measures that predict progression from mild cognitive impairment to Alzheimer's type dementia in older adults: a systematic review and meta-analysis. Neuropsychol Rev 2017;27(4):328−53.

[37] Handels RLH, Vos SJB, Kramberger MG, Jelic V, Blennow K, van Buchem M, et al. Predicting progression to dementia in persons with mild cognitive impairment using cerebrospinal fluid markers. Alzheimers Dement 2017;13(8):903−12.

[38] van Maurik IS, Zwan MD, Tijms BM, Bouwman FH, Teunissen CE, Scheltens P, et al. Interpreting biomarker results in individual patients with mild cognitive impairment in the Alzheimer's biomarkers in daily practice (ABIDE) project. JAMA Neurol 2017;74(12):1481−91.

[39] Gomersall T, Astell A, Nygard L, Sixsmith A, Mihailidis A, Hwang A. Living with ambiguity: a metasynthesis of qualitative research on mild cognitive impairment. Gerontologist 2015;55(5):892−912.

[40] Gomersall T, Smith SK, Blewett C, Astell A. 'It's definitely not Alzheimer's': perceived benefits and drawbacks of a mild cognitive impairment diagnosis. Br J Health Psychol 2017;22(4):786−804.

[41] Arvanitakis Z, Shah RC, Bennett DA. Diagnosis and management of dementia: review. JAMA 2019;322(16):1589−99.

[42] Livingston G, Huntley J, Sommerlad A, Ames D, Ballard C, Banerjee S, et al. Dementia prevention, intervention, and care: 2020 report of the Lancet Commission. Lancet 2020;396(10248):413−46.

[43] Livingston G, Huntley J, Liu KY, Costafreda SG, Selbæk G, Alladi S, et al. Dementia prevention, intervention, and care: 2024 report of the Lancet standing Commission. Lancet 2024.

[44] Nasreddine ZS, Phillips NA, Bédirian V, Charbonneau S, Whitehead V, Collin I, et al. The Montreal Cognitive Assessment, MoCA: a brief screening tool for mild cognitive impairment. J Am Geriatr Soc 2005;53(4):695−9.

[45] Mioshi E, Dawson K, Mitchell J, Arnold R, Hodges JR. The Addenbrooke's Cognitive Examination Revised (ACE-R): a brief cognitive test battery for dementia screening. Int J Geriatr Psychiatr 2006;21(11):1078−85.

[46] Randolph C, Tierney MC, Mohr E, Chase TN. The repeatable battery for the assessment of neuropsychological status (RBANS): preliminary clinical validity. J Clin Exp Neuropsychol 1998;20(3):310−9.

[47] Vossel KA, Tartaglia MC, Nygaard HB, Zeman AZ, Miller BL. Epileptic activity in Alzheimer's disease: causes and clinical relevance. Lancet Neurol 2017;16(4):311−22.

[48] Veitch DP, Weiner MW, Aisen PS, Beckett LA, Cairns NJ, Green RC, et al. Understanding disease progression and improving Alzheimer's disease clinical trials: recent highlights from the Alzheimer's Disease Neuroimaging Initiative. Alzheimers Dement 2019;15(1):106−52.

[49] Lombardi G, Crescioli G, Cavedo E, Lucenteforte E, Casazza G, Bellatorre AG, et al. Structural magnetic resonance imaging for the early diagnosis of dementia due to Alzheimer's disease in people with mild cognitive impairment. Cochrane Database Syst Rev 2020;3:CD009628.

[50] Filippi M, Agosta F, Barkhof F, Dubois B, Fox NC, Frisoni GB, et al. EFNS task force: the use of neuroimaging in the diagnosis of dementia. Eur J Neurol 2012;19(12). e131-e140, 1487-e140.

[51] Foster NL, Heidebrink JL, Clark CM, Jagust WJ, Arnold SE, Barbas NR, et al. FDG-PET improves accuracy in distinguishing frontotemporal dementia and Alzheimer's disease. Brain 2007;130(Pt 10):2616—35.

[52] Catafau AM. Brain SPECT in clinical practice. Part I: perfusion. J Nucl Med 2001;42(2):259—71.

[53] Davison CM, O'Brien JT. A comparison of FDG-PET and blood flow SPECT in the diagnosis of neurodegenerative dementias: a systematic review. Int J Geriatr Psychiatr 2014;29(6):551—61.

[54] Smailagic N, Vacante M, Hyde C, Martin S, Ukoumunne O, Sachpekidis C. (1)(8)F-FDG PET for the early diagnosis of Alzheimer's disease dementia and other dementias in people with mild cognitive impairment (MCI). Cochrane Database Syst Rev 2015;1:CD010632.

[55] Schipke CG, Koglin N, Bullich S, Joachim LK, Haas B, Seibyl J, et al. Correlation of florbetaben PET imaging and the amyloid peptide Ass42 in cerebrospinal fluid. Psychiatry Res Neuroimaging 2017;265:98—101.

[56] Hansson O, Lehmann S, Otto M, Zetterberg H, Lewczuk P. Advantages and disadvantages of the use of the CSF Amyloid beta (Abeta) 42/40 ratio in the diagnosis of Alzheimer's Disease. Alzheimers Res Ther 2019;11(1):34.

[57] Skillback T, Rosen C, Asztely F, Mattsson N, Blennow K, Zetterberg H. Diagnostic performance of cerebrospinal fluid total tau and phosphorylated tau in Creutzfeldt-Jakob disease: results from the Swedish Mortality Registry. JAMA Neurol 2014;71(4):476—83.

[58] Blennow K, Zetterberg H. Biomarkers for Alzheimer's disease: current status and prospects for the future. J Intern Med 2018;284(6):643—63.

[59] Zetterberg H, Blennow K. Moving fluid biomarkers for Alzheimer's disease from research tools to routine clinical diagnostics. Mol Neurodegener 2021;16(1):10.

[60] Pfeffer RI, Kurosaki TT, Harrah Jr CH, Chance JM, Filos S. Measurement of functional activities in older adults in the community. J Gerontol 1982;37(3):323—9.

[61] van Dyck CH, Swanson CJ, Aisen P, Bateman RJ, Chen C, Gee M, et al. Lecanemab in early Alzheimer's disease. N Engl J Med 2022.

[62] Sims JR, Zimmer JA, Evans CD, Lu M, Ardayfio P, Sparks J, et al. Donanemab in early symptomatic Alzheimer disease: the TRAILBLAZER-ALZ 2 randomized clinical trial. JAMA 2023;330(6):512—27.

[63] Cummings J, Zhou Y, Lee G, Zhong K, Fonseca J, Cheng F. Alzheimer's disease drug development pipeline: 2024. Alzheimers Dementia Transl Res Clin Intervent 2024;10(2):e12465.

[64] Cummings J, Lee G, Ritter A, Sabbagh M, Zhong K. Alzheimer's disease drug development pipeline: 2020. Alzheimers Dement (N Y) 2020;6(1):e12050.

[65] Vaz M, Silvestre S. Alzheimer's disease: recent treatment strategies. Eur J Pharmacol 2020;887:173554.

[66] Cooper C, Li R, Lyketsos C, Livingston G. Treatment for mild cognitive impairment: systematic review. Br J Psychiatry 2013;203(3):255—64.

[67] Jordan F, Quinn TJ, McGuinness B, Passmore P, Kelly JP, Tudur Smith C, et al. Aspirin and other non-steroidal anti-inflammatory drugs for the prevention of dementia. Cochrane Database Syst Rev 2020;4:CD011459.

[68] Kunkle BW, Grenier-Boley B, Sims R, Bis JC, Damotte V, Naj AC, et al. Genetic meta-analysis of diagnosed Alzheimer's disease identifies new risk loci and implicates Aβ, tau, immunity and lipid processing. Nat Genet 2019;51(3):414—30.

[69] Ismail Z, Gatchel J, Bateman DR, Barcelos-Ferreira R, Cantillon M, Jaeger J, et al. Affective and emotional dysregulation as pre-dementia risk markers: exploring the mild behavioral impairment symptoms of depression, anxiety, irritability, and euphoria. Int Psychogeriatr 2018;30(2):185—96.

[70] Ball HA, McWhirter L, Ballard C, Bhome R, Blackburn DJ, Edwards MJ, et al. Functional cognitive disorder: dementia's blind spot. Brain 2020;143(10):2895—903.

[71] Ngandu T, Lehtisalo J, Solomon A, Levälahti E, Ahtiluoto S, Antikainen R, et al. A 2 year multidomain intervention of diet, exercise, cognitive training, and vascular risk monitoring versus control to prevent cognitive decline in at-risk elderly people (FINGER): a randomised controlled trial. Lancet 2015;385(9984):2255—63.

[72] Huntley J, Corbett A, Wesnes K, Brooker H, Stenton R, Hampshire A, Ballard C. Online assessment of risk factors for dementia and cognitive function in healthy adults. Int J Geriatr Psychiatry 2018;33(2):e286—93.

[73] Lehtisalo J, Levalahti E, Lindstrom J, Hanninen T, Paajanen T, Peltonen M, et al. Dietary changes and cognition over 2 years within a multidomain intervention trial-the Finnish geriatric intervention study to prevent cognitive impairment and disability (FINGER). Alzheimers Dement 2019;15(3):410—7.

[74] Onakpoya IJ, Heneghan CJ. The efficacy of supplementation with the novel medical food, Souvenaid, in patients with Alzheimer's disease: a systematic review and meta-analysis of randomized clinical trials. Nutr Neurosci 2017;20(4):219—27.

[75] Burckhardt M, Watzke S, Wienke A, Langer G, Fink A. Souvenaid for Alzheimer's disease. Cochrane Database Syst Rev 2020;12:CD011679.

[76] Biazus-Sehn LF, Schuch FB, Firth J, Stigger FS. Effects of physical exercise on cognitive function of older adults with mild cognitive impairment: a systematic review and meta-analysis. Arch Gerontol Geriatr 2020;89:104048.

[77] Stern Y. Cognitive reserve in ageing and Alzheimer's disease. Lancet Neurol 2012;11(11):1006—12.

[78] Grady C. The cognitive neuroscience of ageing. Nat Rev Neurosci 2012;13(7):491—505.

[79] Gates NJ, Vernooij RW, Di Nisio M, Karim S, March E, Martinez G, Rutjes AW. Computerised cognitive training for preventing dementia in people with mild cognitive impairment. Cochrane Database Syst Rev 2019;3:CD012279.

Chapter 4

The need for early diagnosis—Clinical, societal and health economic drivers

Ingrid van Maurik[a,b,c] and Wiesje M. van der Flier[a,b,d]
[a]Alzheimer Center Amsterdam, Neurology, Vrije Universiteit Amsterdam, Amsterdam UMC, Amsterdam, The Netherlands; [b]Epidemiology and Data Science, Vrije Universiteit Amsterdam, Amsterdam UMC, Amsterdam, The Netherlands; [c]Northwest Academy, Northwest Clinics Alkmaar, Alkmaar, The Netherlands; [d]Amsterdam Neuroscience, Neurodegeneration, Amsterdam, The Netherlands

Epidemiology of dementia

The aging population – prevalence and incidence of dementia and dementia subtypes

Dementia is one of the most disabling diseases and ranks in the top 10 diseases causing most deaths worldwide [1]. The worldwide prevalence of dementia disorders is estimated at 50 million with nearly 10 million new cases of dementia each year [1–4]. The world-wide age-standardized prevalence of dementia ≥60 years of age is approximately 5%. The proportion of people living with (subtypes of) dementia increases with age and doubles every 5 years between the ages of 50–80 years, up to over 30% of patients above 80 years of age [2].

The majority, up to 70%, of these dementia cases are due to Alzheimer's disease (AD) [5]. Other common causes are dementia with Lewy Bodies (DLB), accounting for approximately 24% of cases [6–8], and vascular dementia (VaD), accounting for 15%–20% [9,10]. Frontotemporal dementia (FTD) accounts for a smaller proportion of 5% of dementia cases, but is a common cause of young-onset dementia [11]. Mixed pathology is common; pathological studies show that concomitant AD was present in 66% of DLB and 77% of VaD patients [12]. The other way around, concomitant pathology of DLB and/or VaD was also common in AD, as 43% of AD patients had pure AD and 57% had coexisting DLB and/or VaD pathology [13]. Pathological features co-exist and as a result, so do the clinical features of dementia subtypes. All these dementia subtypes are diagnosed on clinical characteristics, but are these characteristics are not specific enough to discriminate between subtypes. Advances in diagnostic testing, biomarkers and knowledge on the genetic components of dementia subtypes, gives the opportunity to diagnose dementia subtypes more accurately. Within the dementia field, efforts are being made to move form a clinical to a biological diagnosis. The National Institute on Aging — Alzheimer's Associations (NIA-AA) have published a research framework proposing a biological definition of Alzheimer's Disease. Instead of a clinical diagnoses, the diagnosis of Alzheimer's disease is based on the presence of amyloid-beta, tauopathy and neurodegeneration. This has two consequences. First, more people could receive a diagnosis of Alzheimer's disease, as the definition is broader; according to the criteria only evidence of amyloid pathology is required to receive a diagnosis within the Alzheimer continuum. This could also be the case in a stage where objectified impairment is still mild or even absent. But second, it also results in a more narrow definition of the diagnosis; if a case seems like a typical AD patient, but there is no evidence of underlying AD pathology, than this patient will not receive the diagnosis AD.

Only one study reports on the prevalence of this biological defined AD diagnosis [14]. They showed the A-T-N- subgroup was most prevalent from age 50 to the late 70s, while the A+T+N+ subgroup becomes most prevalent from the early 80s onward (Fig. 4.4 [14]). Other prevalence and incidence studies on this biological defined AD diagnosis are currently lacking, but such a biomarker driven diagnosis will help to better describe the epidemiology of dementia subtypes (Fig. 4.1).

A biomarker driven diagnosis can help to accurately diagnose patients even in the earliest phases. There is a wealth of data that shows that the pathological changes related to these neurodegenerative disorders occur decades prior to the onset of dementia [5]. Advances in diagnostic testing for Alzheimer's disease, caused a shift toward a diagnosis in the stage of mild cognitive impairment (MCI). The prevalence of MCI, as the pre-dementia stage which is also more common in higher age, increases from 7% in people aged 60%—65% to 25% in people aged 80—85 years [15].

Global time trends and inequalities in dementia prevalence and incidence

The Alzheimer Disease International report in 2015 estimated that the global prevalence will increase up to 131.5 million cases in 2050 [1,16]. While the prevalence of dementia is increasing, the incidence has declined by 13% per decade over the past 25 years [17]. This decline in incidence is most prominent in HIC. Most studies suggest that this decline is subject to a changing risk profile for dementia. A reduction in cardiovascular and cerebrovascular diseases, in particular stroke rates, have been well documented since the mid-20th

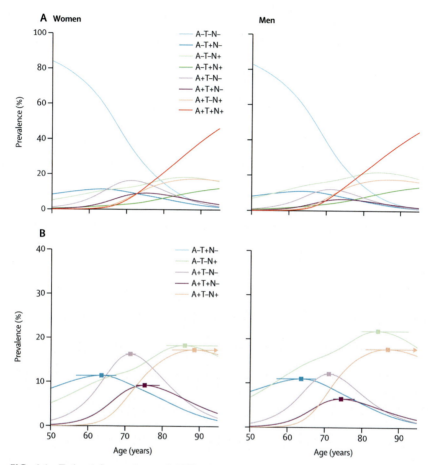

FIG. 4.1 Estimated prevalence of ATN subgroups by age and sex. Estimated prevalence curves by age for all ATN groups on an enlarged scale with the estimated peak for each curve shown with a square and a 95% confidence interval. *Reprinted with permission from Jack C.R. Jr., Wiste H.J., Weigand S.D., et al. Age-specific and sex-specific prevalence of cerebral beta-amyloidosis, tauopathy, and neurodegeneration in cognitively unimpaired individuals aged 50-95 years: a cross-sectional study. Lancet Neurol 2017;16(6):435—444. http://doi.org/10.1016/S1474-4422(17)30077-7.*

century [18—21]. Moreover, the management of vascular risk factors seems much more efficient as more antihypertensive and lipid-lowering drugs are used, leading to fewer dementia cases [19,20]. In addition, a positive influence of improvements in lifestyle factors, such as less smoking and a healthier diet, has been observed [22,23]. A delay in the onset of dementia seems to be an important factor in the decreasing dementia incidence too. Higher education helps to build cognitive reserve, leading to a delay in dementia onset [22,24]. Cognitive reserve refers to the hypothetical construct that explains differences

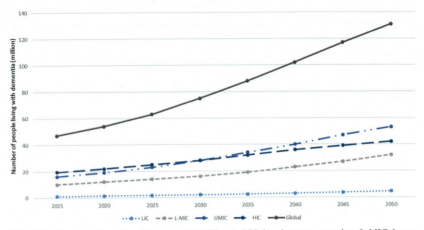

FIG. 4.2 Number of people with dementia (millions). *LIC*, low income countries; *L-MIC*, low to middle income countries; *HIC*, high income countries, *UMIC*, upcoming middle income countries. *Adapted from Prince M, Wimo A, Guerchet M, et al. World Alzheimer report 2015 - the global impact of dementia. London: 2015*

between individuals in the ability to maintain cognitive function in the presence of neurodegenerative pathology [25].

The projected increase in prevalence is largely attributable to low and middle income countries (LMIC, Fig. 4.2) [1]. Almost 60% of all dementia cases live in low to middle income countries (LMIC) [1,2,26]. This notion is supported by a recent systematic analysis that reported on the prevalence and incidence of dementia on a regional level of 195 countries and territories, showing the prevalence of dementia increased and will further increase in middle income countries, while declining in Western, high income countries (HIC, Fig. 4.3 [2]).

While these numbers are alarming, we also know that these number are likely still an underestimation, as dementia diagnosis is often made in a late stage and underdiagnosis of dementia is a common phenomenon. Even in HIC, like the UK, the national under detection rate was approximately 50%. With data available from Brazil and India, the under detection rate in LMIC countries is expected to be much higher (77% and 90% respectively) [27,28]. This under detection rate is specifically high in primary care, due to lack or training, time and resources [29—31]. Moreover, the fear of damaging the doctor-patient relationship by disclosing dementia without treatment options, has been found to be an important barrier for general practitioners (GPs) [32,33].

Risk factors

The strongest risk factors for dementia, specifically Alzheimer's disease, are age, female sex and Apolipoprotein E (APOE)-ε4 genotype. The risk for AD

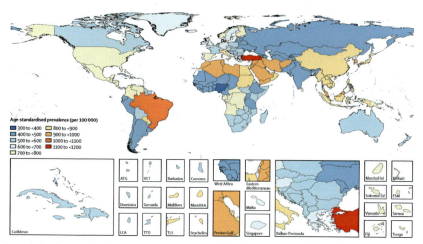

FIG. 4.3 Age-standardised prevalence for Alzheimer's disease and other dementias per 1,00,000 population by location for both sexes. *ATG*, Antigua and Barbuda; *FSM*, Federated States of Micronesia; *Isl*, Islands; *LCA*, Saint Lucia; *TLS*, Timor-Leste; *TTO*, Trinidad and Tobago; *VCT*, Saint Vincent and the Grenadines. *From Collaborators GBDD. Global, regional, and national burden of Alzheimer's disease and other dementias, 1990−2016: a systematic analysis for the Global Burden of Disease Study 2016. Lancet Neurol 2019;18(1):88−106. CC-BY.*

was found to depend for 60%−80% on heritable factors [34], in which the APOE-ε4 allele explains a substantial part. Per *APOE-ε4* allele the risk of AD dementia increases by threefold [35]. APOE-ε4 is thus not a causative gene, like *PSEN1, PSEN2* and *APP*. APOE-ε4 is not used in clinical practice, but is widely used for risk stratification purposes in research [36]. In the last decade, genome-wide association studies (GWAS) have been performed to search for novel genetic variants. These GWAS studies are performed on AD case-control datasets and by-proxy AD case-control studies and have identified over 40 genomic loci that modify the risk of AD [37−39]. Often, contrary to *APOE-ε4*, novel genes found in GWAS studies contribute only for a very small proportion to the total risk of AD and their clinical relevance is therefore unclear. But by combining these small in so-called polygenic risk score (PRS) are associated with neuropathological hallmarks of AD [40,41], conversion of MCI to AD [42−44], and the age of onset of disease in both APOE- ε4 carriers and non-carriers [40,45]. The discriminative accuracy between AD patients and controls of these PRS including APOE is estimated to be 75%−85% and potentially useful to increase power and reduce the duration of trials in pre-symptomatic patients at high genetic risk [40,41,45,46].

A significant proportion of dementia risk is explained by modifiable factors and dementia cases could be prevented by eliminating or reducing modifiable risk factors [47−50]. Estimates of this proportion vary between a third and 40% [47−50]. Of note, these modifiable risk factors are for dementia in general, but it is currently unclear how eliminating modifiable risk factors

56 The Management of Early Alzheimer's Disease

affect risk on AD. Most evidence is available for cardiovascular risk factors and suboptimal life-style. In these two categories, well-known risk factors are hypertension, diabetes, obesity, smoking and physical inactivity. The latest report of the Lancet Commission on dementia prevention, intervention, and care, summarized new evidence of modifiable risk factors for dementia in a 12 risk factor life-course model (Fig. 4.4 [51]). In this model, three new risk factors were included being traumatic brain injury, alcohol consumption, and air pollution. The model shows in which life phase (early life (<45 years),

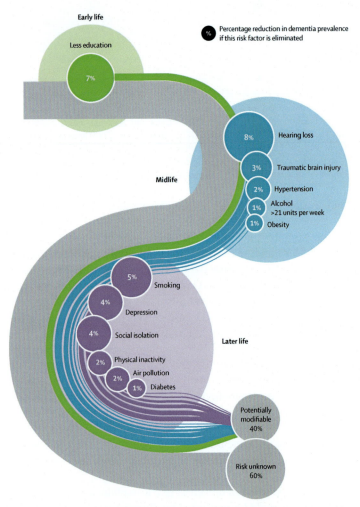

FIG. 4.4 Population attributable fraction of potentially modifiable risk factors for dementia. *From Livingston G, Huntley J, Sommerlad A, et al. Dementia prevention, intervention, and care: 2020 report of the Lancet Commission. Lancet 2020;396(10248):413–446. Reprinted with permission.*

midlife (45–65 years) and late life (>65 years)) the risk factors contribute to an increased dementia risk and it is in these phases where there is the highest potential to intervene. For example, policy makers should prioritize childhood education as this positively affects cognitive reserve. Mid- and late life risk factors either reduce neuropathological damage or help to maintain or increase cognitive reserve. The potential of preventing dementia is high, especially in LMIC countries as changing these risk factors will have the highest benefit in individuals that are most deprived. In LMIC, not everyone has access to secondary education. Moreover, globally the rates of hypertension, obesity, hearing loss, and diabetes are growing, and thus an even greater proportion of dementia could potentially be prevented.

Dementia costs

Along with the increasing prevalence of people living with dementia, the economic costs related with the disease are rapidly increasing. Alzheimer's Disease International World Alzheimer Report 2010 by Alzheimer's Disease International published that the worldwide costs of dementia were estimated at US$604 billion [52]. To put this number in perspective; this is 1% of the global domestic product which is a significant economic impact of just one disorder. In 2015 this number was updated to be US$818 billion, being a 35% increase compared with 2010 and was expected to further increase up to US$1 trillion in 2018 and US$2 trillion near 2030 (Fig. 4.5 [1,16]). Dementia disorders have been found to coexist with other chronic conditions like cardiovascular diseases and depression. Having multiple chronic diseases has been known to even further drive the economic burden [26].

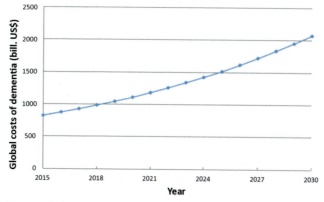

FIG. 4.5 Forecasted global costs of dementia 2015–30. *From Wimo A, Guerchet M, Ali GC, et al. The worldwide costs of dementia 2015 and comparisons with 2010. Alzheimers Dement 2017;13(1):1–7. CC-BY.*

There is an enormous imbalance between the global prevalence and distribution of costs for dementia [53]. HIC account for 86% of the global costs for dementia, while prevalence is highest in the LMIC countries [16]. While this estimate of global costs in HIC remained stable compared to 2010, the costs in the upper middle-income countries had doubled (from 5% to 10%) [16,54].

Dementia related costs are driven by informal care and direct social care costs [54]. Both account for approximately 40% of the global costs with little variation between countries. The impact of informal care for dementia seems highest in LMIC, as the resources for social care are limited. While in HIC the costs are driven by long-term institutionalization. A delay of 1 month in institutionalization could reduce costs with approximately $US2000 per patient [55—57]. Nonetheless, the greatest challenge for all is to effectively organize patient management, and ultimately treatment, which all begins with a diagnostic process [54].

In the context of a disease where to date, no curative treatment options are available, demonstrating health-economic benefits of a diagnostic process, clinical utility of a diagnostic test or an early diagnosis is challenging. In this situation, health benefits are likely to take place later in the disease process. While in the early phases of the disease, costs are primarily made in the healthcare system, the social care sector accounts for most of the costs later in the disease [54]. Against this background, costs are occurring in one sector and the benefits in another [53]. This imbalance might particularly problematic when a disease-modifying therapy becomes available, as the healthcare sector will bear the costs for the diagnostic process and treatment but will not gain from the benefits. This is already the case for cost-effectiveness of an early diagnosis.

Even without treatment or interventions, an early diagnosis has important benefits as it offers opportunities for care management and cost savings [58,59]. An overview of benefits and challenges regarding an early diagnosis is shown in Table 4.1. Importantly, an early and accurate diagnosis has important benefits for patient empowerment as patients may be better equipped to participate in future (care) planning and safety of patients can be maximized [59]. Labeling the condition of patient helps to educate caregivers on the disease and encourages the development of coping strategies to handle future changes in patient's function [60,61]. Ultimately, such an early diagnosis helps to prevent crisis later on in the disease and could lead to better (and less costly) disease trajectories which has great opportunities for cost savings.

Most literature on the benefits and challenges of an early diagnosis is based on expert opinions, rather than scientific evidence. There is a particular scarcity of studies addressing the patient perspective. While health care professionals (HCP) may indicate in survey studies that they communicate early dementia diagnoses to their patients and that they engage patients and caregivers in the process of decision making, this still seems to be suboptimal in

TABLE 4.1 Benefits and challenges of an early diagnosis.

Benefits		Challenges	
Provides answers to concerns about cognitive and functional impairment of the relative, which may help to reduce anxiety	P, F	Stigma, fear and anxiety, social isolation	P, F
Can promote shared management	P,F,H	Belief that memory problems are a normal part of aging causes delays in seeking help	A
Encourages planning of future care, including design of a specific, individualized treatment and management plan	P,F,H	Lack of awareness about signs and symptoms of dementia	H
Allows prompt evaluation and treatment of reversible causes of impairment	P,F,H	Financial constraints or negative effects of reimbursement systems	A
Allows potential management of symptoms with medications or other interventions	P,F,H	Cultural beliefs impact degree of perceived stigma and support alternative explanations of symptoms	A
Helps facilitate treatment or management of coexisting medical conditions that may worsen cognitive function and prevents prescription of medications for co-existing conditions that may impair cognitive function	P,F,H	Limited preventative or treatment options	P,F,H
Enables the inclusion of patients in clinical trials for researching new treatments	P,F,H	Shortage of specialist diagnostic services	H,S
Aids management of possible behavioral symptoms	P,F,H	Diagnostic uncertainty	H
Encourages the development of coping strategies to handle future changes in patient's function	P,F,H	Reluctance to disclose diagnosis	H
Could reduce the overall costs of dementia	S		
Could postpone institutionalization into residences and nursing homes	A		
Could reduce dangerous and challenging behavior (e.g., traffic accidents, etc.)	A		

A, all; F, family/caregiver level; H, healthcare provider level; P, patient level; S, societal level.
Adapted from Dubois B, Padovani A, Scheltens P, Rossi A, Dell'Agnello G. Timely diagnosis for Alzheimer's disease: a literature review on benefits and challenges. J Alzheimers Dis 2016;49(3):617-631.

clinical practice [62–64]. To make use of the benefits of an early diagnosis, clear communication of this diagnosis is pivotal. Evidence from a recent audio-tape study showed that there is definitely room for improvement, as clinicians showed limited behavior to facilitate engagement of patients or promoted understanding of the information provided [64–66]. Patients and caregivers seldom showed initiative to express their preferences, yet they did express an unmet information need after the consultation [64]. Especially in the stage of MCI, clinicians' current approach may not match with the patients' and caregivers' needs. Clinicians often addressed the cause of complaints in a tentative manner, showed a tendency to foster hope and did not meaningfully convey results for diagnostic testing. Patients with MCI do not have dementia at the moment of the MCI diagnosis, but they do have an increased risk of progression. Therefore a diagnosis of MCI rather implies a prognosis. Clearly discussing an early diagnosis and its implications with patients and caregivers, may cause them relief by enabling them to label the patient's condition. However, this discussion was only rarely observed.

Access to diagnosis

The diagnostic chain for people with cognitive complaints is complex and long. The duration from first complaints to diagnosis ranges on average from one to 5 years in HIC [67–69]. A diagnosis of young-onset dementia takes almost 2 years longer than late-onset dementia [68]. Often, the diagnostic process starts with the primary care physician, where dementia is often under-recognized especially in early stages [70]. As a result, patients and caregivers live with uncertainty for a long time, sometimes in denial of symptoms which has adverse effects on coping and may delay treatment.

In out-patient clinics, or memory clinics, specialized health care professionals have more resources at their disposal. Although clinical guidelines differ in their details, they all include a clinical and neuropsychological examination [71–73]. How this is carried out, differs from clinic to clinic – some only perform a clinical examination, others additionally administer a screening test for global cognition or a complete neuropsychological test battery. Imaging of the brain, either with CT or MRI, helps to identify reversible causes of cognitive decline and specific features of brain atrophy can be analyzed [73,74]. More recently, techniques like the quantification of amyloid-beta and tau proteins in cerebrospinal fluid, or via positron emission tomography (PET) scanning have been introduced [75,76]. These techniques help to determine the presence or absence of any molecular pathology that may underpin the cognitive disorder and is of particular diagnostic value in early stages of disease when neurodegenerative disorders are harder to distinguish from non-degenerative conditions (such as age-associated memory decline, anxiety and depression) on clinical grounds alone and when structural brain imaging has lower diagnostic sensitivity and specificity [77]. However, current molecular biomarker-based tests are unsuited to widespread application given their high cost, invasiveness and infrastructural requirements. Obtaining CSF via a lumbar puncture requires training of clinical staff and access to specialized medical facilities such as dedicated clinic suites and emergency medical cover, while ligand-PET requires nuclear medicine facilities and rapid access to radiochemistry labs where ligands are synthesized, given the relatively short half-life of ^{18}F and ^{11}C compounds on which current ligands are based.

Diagnostic routines for dementia differ between countries, but also within individual countries, with as more advanced techniques or memory clinics being less accessible in rural areas [78,79].

The diagnostic process in LMIC is even more challenging and resources are poor [80–82]. One can distinguish three main factors that contribute to the limited number of dementia diagnoses. First, access to health care is limited. Geriatric care is only limitedly available, and if available patients have to travel long distances to receive a diagnosis or treatment [83,84]. Furthermore, HCPs have very limited personnel and resources at their disposal to perform accurate diagnostics. Secondly, health literacy is low and people are unaware that appropriate care is available [84]. Overall, there is a high tolerance for health problems in elderly. The characteristics complaints, such as forgetfulness, are often believed to be the result of normal aging. Also the interplay of behavioral, cognitive and physical aspects of the disease are often ignored or not connected. This is also the case at the level of HCP [85], where their ability to recognize early signs of neurodegeneration is affected by a lack of knowledge. Lastly, the social stigma in LMIC is a barrier for diagnosing dementia [85]. The diagnostic label may affect the social value of the patient. But also in the absence of a diagnosis, there is a negative attitude toward the signs and complaints of dementia. Especially if behavioral changes are involved. Table 4.2 provides an overview of additional factors that contribute to the limited number of dementia diagnoses in resource poor areas.

TABLE 4.2 Likely causes of limited dementia diagnoses in resource-poor areas.

- Initial symptoms are subtle and fluctuating
- Inability to recognize cognitive, behavioral, or functional impairment as a consequence of dementing illness
- Accommodation, denial, and rationalization of symptoms as due to aging and adverse circumstances
- High tolerance for health problems in elderly individuals
- Person with early symptoms does not feel the need to seek treatment

Continued

TABLE 4.2 Likely causes of limited dementia diagnoses in resource-poor areas.—cont'd

- Relatives do not believe symptoms are treatable
- Demoralization regarding medical assistance, due to ignorance of available treatments
- Do not know where to go for diagnosis
- Family is afraid of psychiatric stigma (demented is often equated to being crazy)
- Physician underemphasizes the importance of symptoms
- Perception that physicians will miss the diagnosis and will make the family waste time and money in apparently unnecessary tests
- Costs of diagnosis in terms of effort, time, economic costs

From Maestre GE. Assessing dementia in resource-poor regions. Curr Neurol Neurosci Rep 2012;12(5):511-519. Springer nature - Creative Commons Attribution 2.0 International License.

Future perspectives

The field moves toward a biological definition of AD, and in time the same will apply to other causes of dementia. To this end, the use of biomarkers in the clinic will take a prominent place and recent years have witnessed an accumulation of studies not just on ways to combine biomarker evidence, but also on how to convey this meaningfully in the clinical setting. The imminent arrival of disease modifying treatments (DMTs) will not only bring solutions, but also poses new challenges for the healthcare sector, as large numbers of people are expected to seek early diagnosis and treatment, and triaging processes for DMT eligibility will have to be introduced, such as a combination of blood based biomarkers with a cognitive test [86]. The introduction of blood based biomarkers into clinical practice will also be of great value to LMIC, that have less access to imaging modalities or lumbar puncture facilities.

Even in the absence of curative treatment, a timely diagnosis is important as it helps to educate patients and caregivers and to arrange proper care. Future use of decision tools and risk calculating applications, encompassing data from blood biomarkers and polygenic risk scores, could help move the field toward personalized diagnostics and care for AD [87,88]. Genetic information will identify biological pathways, such as immune response to pathology, cholesterol mechanism and dysregulation of the endo-lysosomal trafficking [38,39], defective in each person, leading to individually- tailored interventions and management.

Conclusion

Globally the number of people living with dementia is increasing and this increase is largely attributable to low and middle income countries. Along with the increasing prevalence of people living with dementia, the economic costs related with the disease are rapidly increasing. An accurate and timely diagnosis is necessary for prompt implementation of dementia care, empowering patients and their caregivers to initiate management strategies. Patients may be better equipped to participate in future (care) planning and by being able to label their condition it helps to also educate caregivers on the disease and encourages the development of coping strategies to handle future changes in patient's function. Ultimately, an early diagnosis helps to deploy prevention strategies and improved disease trajectories.

References

[1] Prince M, Wimo A, Guerchet M, et al. World Alzheimer report 2015 - the global impact of dementia. 2015. London.

[2] Collaborators GBDD. Global, regional, and national burden of Alzheimer's disease and other dementias, 1990-2016: a systematic analysis for the Global Burden of Disease Study 2016. Lancet Neurol 2019;18(1):88—106.

[3] Prince M, Ali GC, Guerchet M, et al. Recent global trends in the prevalence and incidence of dementia, and survival with dementia. Alzheimer's Res Ther 2016;8(1):23.

[4] Prince M, Bryce R, Albanese E, et al. The global prevalence of dementia: a systematic review and metaanalysis. Alzheimers Dement 2013;9(1):63—75.

[5] Scheltens P, Blennow K, Breteler MM, et al. Alzheimer's disease. Lancet 2016;388(10043):505—17.

[6] Savica R, Grossardt BR, Bower JH, et al. Incidence of dementia with Lewy bodies and Parkinson disease dementia. JAMA Neurol 2013;70(11):1396—402.

[7] Hogan DB, Fiest KM, Roberts JI, et al. The prevalence and incidence of dementia with Lewy bodies: a systematic review. Can J Neurol Sci 2016;43(Suppl. 1):S83—95.

[8] Rocha Cabrero F, Morrison EH. Lewy bodies. Treasure Island (FL): StatPearls; 2020.

[9] Rizzi L, Rosset I, Roriz-Cruz M. Global epidemiology of dementia: Alzheimer's and vascular types. BioMed Res Int 2014;2014:908915.

[10] Lobo A, Launer LJ, Fratiglioni L, et al. Prevalence of dementia and major subtypes in Europe: a collaborative study of population-based cohorts. Neurologic Diseases in the Elderly Research Group. Neurology 2000;54(11 Suppl. 5):S4—9.

[11] Onyike CU, Diehl-Schmid J. The epidemiology of frontotemporal dementia. Int Rev Psychiatr 2013;25(2):130—7.

[12] Barker WW, Luis CA, Kashuba A, et al. Relative frequencies of Alzheimer disease, Lewy body, vascular and frontotemporal dementia, and hippocampal sclerosis in the State of Florida Brain Bank. Alzheimer Dis Assoc Disord 2002;16(4):203—12.

[13] Rabinovici GD, Carrillo MC, Forman M, et al. Multiple comorbid neuropathologies in the setting of Alzheimer's disease neuropathology and implications for drug development. Alzheimers Dement 2017;3(1):83—91.

[14] Jack Jr CR, Wiste HJ, Weigand SD, et al. Age-specific and sex-specific prevalence of cerebral beta-amyloidosis, tauopathy, and neurodegeneration in cognitively unimpaired individuals aged 50-95 years: a cross-sectional study. Lancet Neurol 2017;16(6):435—44.

[15] Petersen RC, Lopez O, Armstrong MJ, et al. Practice guideline update summary: mild cognitive impairment: report of the guideline development, dissemination, and implementation subcommittee of the American academy of Neurology. Neurology 2018;90(3):126—35.

[16] Wimo A, Guerchet M, Ali GC, et al. The worldwide costs of dementia 2015 and comparisons with 2010. Alzheimers Dement 2017;13(1):1—7.

[17] Wolters FJ, Chibnik LB, Waziry R, et al. Twenty-seven-year time trends in dementia incidence in Europe and the United States: the Alzheimer cohorts consortium. Neurology 2020;95(5):e519—31.

[18] Koton S, Schneider AL, Rosamond WD, et al. Stroke incidence and mortality trends in US communities, 1987 to 2011. JAMA 2014;312(3):259—68.

[19] Satizabal CL, Beiser AS, Chouraki V, et al. Incidence of dementia over three decades in the framingham heart study. N Engl J Med 2016;374(6):523—32.

[20] Grasset L, Brayne C, Joly P, et al. Trends in dementia incidence: evolution over a 10-year period in France. Alzheimers Dement 2016;12(3):272—80.

[21] Taudorf L, Norgaard A, Islamoska S, et al. Declining incidence of dementia: a national registry-based study over 20 years. Alzheimers Dement 2019;15(11):1383—91.

[22] Roehr S, Pabst A, Luck T, Riedel-Heller SG. Is dementia incidence declining in high-income countries? A systematic review and meta-analysis. Clin Epidemiol 2018;10:1233—47.

[23] Kivipelto M, Helkala EL, Laakso MP, et al. Midlife vascular risk factors and Alzheimer's disease in later life: longitudinal, population based study. BMJ 2001;322(7300):1447—51.

[24] Matthews FE, Stephan BC, Robinson L, et al. A two decade dementia incidence comparison from the Cognitive Function and Ageing Studies I and II. Nat Commun 2016;7:11398.

[25] Stern Y. Cognitive reserve in ageing and Alzheimer's disease. Lancet Neurol 2012;11(11):1006—12.

[26] Hajat C, Stein E. The global burden of multiple chronic conditions: a narrative review. Prev Med Rep 2018;12:284—93.

[27] Dias A, Patel V. Closing the treatment gap for dementia in India. Indian J Psychiatr 2009;51(Suppl. 1):S93—7.

[28] Nakamura AE, Opaleye D, Tani G, Ferri CP. Dementia underdiagnosis in Brazil. Lancet 2015;385(9966):418—9.

[29] Bamford C, Eccles M, Steen N, Robinson L. Can primary care record review facilitate earlier diagnosis of dementia? Fam Pract 2007;24(2):108—16.

[30] Connolly A, Gaehl E, Martin H, Morris J, Purandare N. Underdiagnosis of dementia in primary care: variations in the observed prevalence and comparisons to the expected prevalence. Aging Ment Health 2011;15(8):978—84.

[31] Eichler T, Thyrian JR, Hertel J, et al. Rates of formal diagnosis in people screened positive for dementia in primary care: results of the DelpHi-Trial. J Alzheimers Dis 2014;42(2):451—8.

[32] Giezendanner S, Monsch AU, Kressig RW, et al. General practitioners' attitudes towards early diagnosis of dementia: a cross-sectional survey. BMC Fam Pract 2019;20(1):65.

[33] Koch T, Iliffe S, project E-E. Rapid appraisal of barriers to the diagnosis and management of patients with dementia in primary care: a systematic review. BMC Fam Pract 2010;11:52.

[34] Gatz M, Reynolds CA, Fratiglioni L, et al. Role of genes and environments for explaining Alzheimer disease. Arch Gen Psychiatr 2006;63(2):168—74.

[35] Farrer LA, Cupples LA, Haines JL, et al. Effects of age, sex, and ethnicity on the association between apolipoprotein E genotype and Alzheimer disease. A meta-analysis. APOE and Alzheimer Disease Meta Analysis Consortium. JAMA 1997;278(16):1349—56.

[36] Green RC, Roberts JS, Cupples LA, et al. Disclosure of APOE genotype for risk of Alzheimer's disease. N Engl J Med 2009;361(3):245—54.

[37] Lambert JC, Ibrahim-Verbaas CA, Harold D, et al. Meta-analysis of 74,046 individuals identifies 11 new susceptibility loci for Alzheimer's disease. Nat Genet 2013;45(12):1452—8.

[38] Jansen IE, Savage JE, Watanabe K, et al. Genome-wide meta-analysis identifies new loci and functional pathways influencing Alzheimer's disease risk. Nat Genet 2019;51(3):404—13.

[39] Kunkle BW, Grenier-Boley B, Sims R, et al. Genetic meta-analysis of diagnosed Alzheimer's disease identifies new risk loci and implicates Abeta, tau, immunity and lipid processing. Nat Genet 2019;51(3):414—30.

[40] Desikan RS, Fan CC, Wang Y, et al. Genetic assessment of age-associated Alzheimer disease risk: development and validation of a polygenic hazard score. PLoS Med 2017;14(3):e1002258.

[41] Escott-Price V, Myers AJ, Huentelman M, Hardy J. Polygenic risk score analysis of pathologically confirmed Alzheimer disease. Ann Neurol 2017;82(2):311—4.

[42] Rodriguez-Rodriguez E, Sanchez-Juan P, Vazquez-Higuera JL, et al. Genetic risk score predicting accelerated progression from mild cognitive impairment to Alzheimer's disease. J Neural Transm 2013;120(5):807—12.

[43] Lacour A, Espinosa A, Louwersheimer E, et al. Genome-wide significant risk factors for Alzheimer's disease: role in progression to dementia due to Alzheimer's disease among subjects with mild cognitive impairment. Mol Psychiatr 2017;22(1):153—60.

[44] Adams HH, de Bruijn RF, Hofman A, et al. Genetic risk of neurodegenerative diseases is associated with mild cognitive impairment and conversion to dementia. Alzheimers Dement 2015;11(11):1277—85.

[45] van der Lee SJ, Wolters FJ, Ikram MK, et al. The effect of APOE and other common genetic variants on the onset of Alzheimer's disease and dementia: a community-based cohort study. Lancet Neurol 2018;17(5):434—44.

[46] Escott-Price V, Sims R, Bannister C, et al. Common polygenic variation enhances risk prediction for Alzheimer's disease. Brain 2015;138(Pt 12):3673—84.

[47] Edwards Iii GA, Gamez N, Escobedo Jr G, Calderon O, Moreno-Gonzalez I. Modifiable risk factors for Alzheimer's disease. Front Aging Neurosci 2019;11:146.

[48] Licher S, Ahmad S, Karamujic-Comic H, et al. Genetic predisposition, modifiable-risk-factor profile and long-term dementia risk in the general population. Nat Med 2019;25(9):1364—9.

[49] Peters R, Booth A, Rockwood K, et al. Combining modifiable risk factors and risk of dementia: a systematic review and meta-analysis. BMJ Open 2019;9(1):e022846.

[50] Yaffe K. Modifiable risk factors and prevention of dementia: what is the latest evidence? JAMA Intern Med 2018;178(2):281—2.

[51] Livingston G, Huntley J, Sommerlad A, et al. Dementia prevention, intervention, and care: 2020 report of the Lancet Commission. Lancet 2020;396(10248):413—46.

[52] Wimo A, Jonsson L, Bond J, et al. The worldwide economic impact of dementia 2010. Alzheimers Dement 2013;9(1):1–11 e13.
[53] Skoldunger A, Johnell K, Winblad B, Wimo A. Mortality and treatment costs have a great impact on the cost-effectiveness of disease modifying treatment in Alzheimer's disease–a simulation study. Curr Alzheimer Res 2013;10(2):207–16.
[54] Wimo A, Ballard C, Brayne C, et al. Health economic evaluation of treatments for Alzheimer's disease: impact of new diagnostic criteria. J Intern Med 2014;275(3):304–16.
[55] Costa N, Derumeaux H, Rapp T, et al. Methodological considerations in cost of illness studies on Alzheimer disease. Health Econ Rev 2012;2(1):18.
[56] Davidson M, Schnaider Beeri M. Cost of Alzheimer's disease. Dialogues Clin Neurosci 2000;2(2):157–61.
[57] Zhu CW, Sano M. Economic considerations in the management of Alzheimer's disease. Clin Interv Aging 2006;1(2):143–54.
[58] Leifer BP. Early diagnosis of Alzheimer's disease: clinical and economic benefits. J Am Geriatr Soc 2003;51(5 Suppl. Dementia):S281–8.
[59] Dubois B, Padovani A, Scheltens P, Rossi A, Dell'Agnello G. Timely diagnosis for Alzheimer's disease: a literature review on benefits and challenges. J Alzheimers Dis 2016;49(3):617–31.
[60] Gomersall T, Astell A, Nygard L, et al. Living with ambiguity: a metasynthesis of qualitative research on mild cognitive impairment. Gerontol 2015;55(5):892–912.
[61] Gomersall T, Smith SK, Blewett C, Astell A. 'It's definitely not Alzheimer's': perceived benefits and drawbacks of a mild cognitive impairment diagnosis. Br J Health Psychol 2017;22(4):786–804.
[62] Kunneman M, Smets EMA, Bouwman FH, et al. Clinicians' views on conversations and shared decision making in diagnostic testing for Alzheimer's disease: the ABIDE project. Alzheimers Dement 2017;3(3):305–13.
[63] van der Flier WM, Kunneman M, Bouwman FH, Petersen RC, Smets EMA. Diagnostic dilemmas in Alzheimer's disease: room for shared decision making. Alzheimers Dement 2017;3(3):301–4.
[64] Visser LNC, Kunneman M, Murugesu L, et al. Clinician-patient communication during the diagnostic workup: the ABIDE project. Alzheimers Dement 2019;11:520–8.
[65] Visser LNC, Pelt SAR, Kunneman M, et al. Communicating uncertainties when disclosing diagnostic test results for (Alzheimer's) dementia in the memory clinic: the ABIDE project. Health Expect 2020;23(1):52–62.
[66] Visser LNC, van Maurik IS, Bouwman FH, et al. Clinicians' communication with patients receiving a MCI diagnosis: the ABIDE project. PLoS One 2020;15(1):e0227282.
[67] Salloway S, Correia S. Alzheimer disease: time to improve its diagnosis and treatment. Cleve Clin J Med 2009;76(1):49–58.
[68] van Vliet D, de Vugt ME, Bakker C, et al. Time to diagnosis in young-onset dementia as compared with late-onset dementia. Psychol Med 2013;43(2):423–32.
[69] Wilkinson D, Stave C, Keohane D, Vincenzino O. The role of general practitioners in the diagnosis and treatment of Alzheimer's disease: a multinational survey. J Int Med Res 2004;32(2):149–59.
[70] Bradford A, Kunik ME, Schulz P, Williams SP, Singh H. Missed and delayed diagnosis of dementia in primary care: prevalence and contributing factors. Alzheimer Dis Assoc Disord 2009;23(4):306–14.
[71] Mental and Behavioural Disorders (F00-F99). The international classification of diseases, 10th rev: ICD-10. Geneva: World Health Organization; 1992. p. 311–88.

[72] Association AP. Diagnostic and statistical manual of mental disorders. 5th ed. Arlington, VA: American Psychiatric Association; 2013.
[73] McKhann GM, Knopman DS, Chertkow H, et al. The diagnosis of dementia due to Alzheimer's disease: recommendations from the National Institute on Aging-Alzheimer's Association workgroups on diagnostic guidelines for Alzheimer's disease. Alzheimers Dement 2011;7(3):263–9.
[74] Ten Kate M, Barkhof F, Boccardi M, et al. Clinical validity of medial temporal atrophy as a biomarker for Alzheimer's disease in the context of a structured 5-phase development framework. Neurobiol Aging 2017;52:167–82.
[75] Chiotis K, Saint-Aubert L, Boccardi M, et al. Clinical validity of increased cortical uptake of amyloid ligands on PET as a biomarker for Alzheimer's disease in the context of a structured 5-phase development framework. Neurobiol Aging 2017;52:214–27.
[76] Mattsson N, Lonneborg A, Boccardi M, et al. Clinical validity of cerebrospinal fluid Abeta42, tau, and phospho-tau as biomarkers for Alzheimer's disease in the context of a structured 5-phase development framework. Neurobiol Aging 2017;52:196–213.
[77] Lombardi G, Crescioli G, Cavedo E, et al. Structural magnetic resonance imaging for the early diagnosis of dementia due to Alzheimer's disease in people with mild cognitive impairment. Cochrane Database Syst Rev 2020;3:CD009628.
[78] Abner EL, Jicha GA, Christian WJ, Schreurs BG. Rural-urban differences in Alzheimer's disease and related disorders diagnostic prevalence in Kentucky and West Virginia. J Rural Health 2016;32(3):314–20.
[79] Szymczynska P, Innes A, Mason A, Stark C. A review of diagnostic process and post-diagnostic support for people with dementia in rural areas. J Prim Care Community Health 2011;2(4):262–76.
[80] Boggatz T, Dassen T. Ageing, care dependency, and care for older people in Egypt: a review of the literature. J Clin Nurs 2005;14(8B):56–63.
[81] Chu LW, Chi I. Nursing homes in China. J Am Med Dir Assoc 2008;9(4):237–43.
[82] Ogunniyi A, Hall KS, Baiyewu O, et al. Caring for individuals with dementia: the Nigerian experience. W Afr J Med 2005;24(3):259–62.
[83] Perry B, Gesler W. Physical access to primary health care in Andean Bolivia. Soc Sci Med 2000;50(9):1177–88.
[84] Maestre GE. Assessing dementia in resource-poor regions. Curr Neurol Neurosci Rep 2012;12(5):511–9.
[85] Patel V, Prince M. Ageing and mental health in a developing country: who cares? Qualitative studies from Goa, India. Psychol Med 2001;31(1):29–38.
[86] Mattke S, Cho SK, Bittner T, Hlavka J, Hanson M. Blood-based biomarkers for Alzheimer's pathology and the diagnostic process for a disease-modifying treatment: projecting the impact on the cost and wait times. Alzheimers Dement 2020;12(1):e12081.
[87] van Maurik IS, Visser LN, Pel-Littel RE, et al. Development and usability of ADappt: web-based tool to support clinicians, patients, and caregivers in the diagnosis of mild cognitive impairment and Alzheimer disease. JMIR Form Res 2019;3(3):e13417.
[88] Tolonen A, Rhodius-Meester HFM, Bruun M, et al. Data-driven differential diagnosis of dementia using multiclass disease state index classifier. Front Aging Neurosci 2018;10:111.

Chapter 5

An overview of current diagnostic strategies

Stefano F. Cappa[a,b] **and Chiara Cerami**[a,b]
[a]*University Institute for Advanced Studies (IUSS), Pavia, Italy;* [b]*IRCCS Mondino Foundation, Pavia, Italy*

The need for an early diagnosis of Alzheimer's disease (AD) is widely acknowledged among researchers, sharing the opinion that early detection of individuals at risk of progression to dementia is associated with significant benefits at the individual and societal level [1]. The positions of other stakeholders, such as clinical practitioners, patient associations and the lay public are more nuanced. An interesting example of current discussion is provided, by example, by a focus group consisting of healthy elderly subjects, informal caregivers, nursing staff, researchers, and clinicians, about crucial issues linked to early diagnosis, such as their willingness to know the results of their positron emission tomography (PET) amyloid scan, a biomarker of AD pathology discussed here in the Diagnostic tool section [2]. While researchers and informal caregivers in general wanted to know the results, the three other groups opted for not knowing, or were undecided. The main reason for a positive response was the possibility to make plans for the future, while the lack of a disease-modifying treatment was the most frequently reported negative factor. It is remarkable that emotional impact, listed by most respondents as a negative factor, was considered by some as a possible advantage (going out of a condition of doubt). The concept of early diagnosis needs to be further qualified, by considering the crucial distinction between preclinical and prodromal diagnosis. The original notion of preclinical AD was derived from epidemiological studies on aging populations, reporting subtle cognitive decline, mostly of information processing speed and executive functions, years before the diagnosis of dementia [3]. The definition of AD as a continuum, which can be diagnosed on the basis of the positivity of biomarkers before its clinical expression, has subsequently played a major role in promoting the concept of preclinical detection of at-risk individuals [1] and in "shifting definition of AD in living people from a syndromal to a biological

The Management of Early Alzheimer's Disease. https://doi.org/10.1016/B978-0-12-822240-9.00007-7
Copyright © 2025 Elsevier Inc. All rights reserved, including those for text and data mining, AI training, and similar technologies.

construct" [4]. While this framework is intended for research purposes, its application in everyday clinical practice is a topic of current debate (see Ref. [5,6] for a discussion). A diagnosis of preclinical AD is possible in carriers of pathogenic mutations and in longitudinal studies of other at risk populations [7] and gives rise to a number of ethical challenges [8], as covered elsewhere in this book. There is accumulating evidence that cognitively unimpaired individuals with biomarker positivity may never develop clinical manifestations in their lifetime [9]. In contrast, a diagnosis of prodromal AD is now widely accepted in clinical practice, and is the focus of the present chapter. The history of prodromal AD is closely linked to the development of the concept of mild cognitive impairment (MCI) [10,11], defined as "a transitional state between the cognitive changes of normal aging and AD" [12]. The present definitions of "prodromal AD" [13,14], "MCI due to AD" [12] and "mild neurocognitive disorder due to AD" [15] are crucially linked to the definition of specific clinical criteria and to the use of diagnostic biomarkers [16].

Several recent guidelines have been proposed for the diagnosis of this clinical condition (see, for example, [17]) [18,19], representing the current diagnostic strategies in US and Europe [20]. One of the main aspects of the process is clinical assessment and neuropsychological testing as the "gateway" to biomarker assessment (imaging and biological markers).

Clinical assessment

According to all criteria, the entry to the diagnostic process is when "the patient or a close contact voices concern about memory or impaired cognition" [17]. While subjective complaints alone are not sufficient to diagnose MCI, they are often considered as a risk factor for AD. The Subjective Cognitive Decline (SCD) Initiative criteria include two major features [21]. First, a self-experienced persistent decline in cognitive capacity, compared with a previously normal cognitive status, unrelated to an acute event. This is a personal perspective, as observation of such a decline by others is not required. The second criterion is normal performance on standardized cognitive tests adjusted for age, sex, and education. While SCD per se does not predict progression to objective decline in most individuals, there is now considerable evidence indicating that its association with biomarker positivity (preclinical AD in the Jack et al. [4] classification scheme) may predict disease progression [21]. This is confirmed by the observation of longitudinal changes of test scores within the normal range, approaching pathological levels over the course of several years [22]. The presence of SCD and subtle cognitive changes in association with biomarker positivity may thus be considered as an at-risk stage for disease progression, sometimes labeled as pre-MCI [23]. The situation in which the concern about cognitive decline is voiced by a close contact person, but is not endorsed by the patient is also worth considering. In this case, the discrepancy of judgment between the subject and the informant may indicate a decreased self-awareness of cognitive status. The possibility that informants' ratings may predict the progression to dementia better than self-reported complaints has been suggested by results coming from the INSIGHT-PreAD study [24]. The score of an "awareness of cognitive decline index (ACDI)", based on the discrepancy between subject and informant in an SCD questionnaire indicated a closer direct relationship with both biological and topographical AD biomarkers (i.e., the presence of a higher amyloid burden and lower cortical glucose metabolism) in the low awareness than in the high awareness group. In a follow-up study, an "anosognosia index" showed an interesting quadratic relation with measures of amyloid load, characterized by increasing awareness up to the positivity level, followed by decline with increasing amyloid load. This finding suggests that reduction of awareness may be a better marker for AD pathology than complaints about cognitive decline [25]. Further studies are needed to test this hypothesis [26].

Many tools are available to support clinicians and systematically collect information about subjective cognitive complaints and awareness. The subjective perception of memory changes can be evaluated with the Memory Function Questionnaire [27]. The Everyday Cognition scale [28] is informant-based, questionnaire, developed for the assessment of cognitively-mediated functional abilities, which can also be used for self-assessment of ability changes in several domains (memory, language, attention, planning and organization), comparing the present status with 10 years before. The Cognitive Function Instrument (CFI) [29] is a brief questionnaire, focused on the evaluation of changes in memory and cognitive functioning by the subject and an informant. Anosognosia in dementia can be assessed with a specific questionnaire [30] or by measuring the discrepancy between subjective perception and objective performance with the Anosognosia Index [31].

In addition, the Washington University Clinical Dementia Rating Scale (see https://knightadrc.wustl.edu/cdr/cdr.htm for updated information, including scoring and training) offers an overview of cognitive skills in daily life. This is a 5-point scale used to characterize six domains of cognition and function: Memory, Orientation, Judgment & Problem Solving, Community Affairs, Home & Hobbies, and Personal Care. The rating is obtained through a semi-structured interview of the patient and a reliable informant, and provides through an algorithm a Global Score, which is widely used in both clinical and research setting for the staging of AD [32].

It may appear trivial to mention the neurological examination as a component of clinical assessment, but it is unfortunately true that in many situations time pressure may have a negative impact on the need for a comprehensive evaluation of neurological status. For example, a survey

TABLE 5.1 Positive neurological signs and/or symptoms in cognitive syndromes.

Neurological signs and/or symptoms	Present in
Hypokinetic syndrome	Vascular dementia; Lewy body dementia; Parkinson's disease dementia; progressive supranuclear palsy; corticobasal syndrome
Hyperkinetic syndrome	Huntington's disease; Wilson's disease; corticobasal syndrome
Upper motor neuron or motor cortex involvement	Vascular dementia; motor neuron disease; frontotemporal dementia; corticobasal syndrome
Cerebellar involvement	Paraneoplastic syndromes; Creutzfeldt-Jakob disease
Myoclonus and/or seizures	Alzheimer's disease; Creutzfeldt-Jakob disease; post-anoxic encephalopathy; Autoimmune encephalitis
Lower motor neuron involvement	Motor neuron disease; frontotemporal dementia with motor neuron disease
Visual field defect	Posterior cortical atrophy
Eye movement disorders	Progressive supranuclear palsy; Wernicke-Korsakoff disease
Ipo- or Anosmia	Alzheimer's disease; Parkinson's disease; Huntington's disease

documented that neurologists and psychiatrists working in memory clinics do not always assess motor functions [33]. This neglect can have serious consequences on the process of differential diagnosis of cognitive syndromes. Table 5.1 (inspired by Cooper and Greene [34]) reports a list of neurological signs, whose association with cognitive complaints may have diagnostic usefulness. It is worth mentioning that the mental status evaluation is part of the neurological examination, and is not replaceable with cognitive screening tests, or vice versa. In particular, the clinical evaluation of language, praxis and visuo-spatial function is a source of information, which is not always collected during the formal neuropsychological examination. For example, disorders of speech production and apraxia are crucial for early diagnosis of conditions such as progressive supranuclear palsy [35] and corticobasal syndrome (CBS) [36], which may present with cognitive, rather than motor complaints; unilateral neglect is an unusual feature of posterior cortical atrophy (PCA), usually an atypical AD presentation [37].

Neuropsychological testing

According to current diagnostic guidelines [16,38], neuropsychological tests can be considered as a "gateway biomarker" in the AD diagnostic process [39]. In the case of typical presentations of AD, the performance on episodic memory tests is crucial for early diagnosis and is the basis for the definition of MCI or prodromal AD according to current diagnostic criteria.

Screening tools

The first stage of cognitive assessment is the administration of a screening test for the global assessment of cognitive function. These tests vary considerably in length and detail. The most widely used short tests (less than 15 min), typically administered by clinicians during the screening visit, are the Mini Mental State Examination (MMSE) [40] and the Montreal Cognitive Assessment [41]. Examples of longer screening tests are the AD Assessment Scale-Cognition (ADAS-CoG) [42] and the Addenbrooke's Cognitive Examination (ACE-R) [43]. While the presence of objective cognitive impairment in a screening test is often considered sufficient for a diagnosis of MCI in everyday clinical practice, current diagnostic criteria recommend formal neuropsychological testing of multiple cognitive domains.

There is however no consensus on the most appropriate tests to be employed. Surveys of assessment tools for AD used across Europe have been published by the European AD consortium [44] and by a task force of the European Federation of Neurological Societies [45], demonstrating a wide variety in assessment tools. More recently, a project funded by the Joint Program for Neurodegenerative Diseases (JPND) and by the Italian Ministry of Health aimed at providing a consensus framework for the harmonization of assessment tools to be applied to research in neurodegenerative disorders affecting cognition across Europe [46]. A panel of European experts reviewed the current methods of neuropsychological assessment and identified pending issues. The conclusion was that the possibility to provide firm recommendations suffers from the limited information about the psychometric properties of many of the tools in current clinical use, and from the absence of good quality, up to date normative data (i.e., appropriate age groups and education levels) for many neuropsychological tools. A survey of current clinical cognitive testing in US by the American Academy of Neurology Behavioral Neurology Section Workgroup led to similar conclusions [47]. An important point to be considered is that many of the tests used in clinical settings were developed several decades ago, with the aim of testing different patient populations (developmental disorders and stroke). The enormous progress of cognitive neurosciences has provided novel, translationally relevant information about the organization of cognitive functions in the normal brain, which needs to be exploited in proposing new measures with increased sensitivity and specificity to early stages of neurodegenerative diseases.

It must be underlined that, while the screening tests are effective in detecting a deviation from normal performance [48,49], as stand-alone measurements their predictive value of progression to dementia is limited. In the case of the most widely used screening test, the MMSE, a Cochrane systematic review [50] failed to find evidence supporting a substantial role in the identification of MCI patients who may develop dementia. The analysis of multiple cognitive domains provides a detailed phenotyping, required for differential diagnosis among the multiple possible causes of cognitive impairment and, combining measures with high sensitivity and high specificity, has a better predictive value than single measures of global cognition [51].

Cognitive domains testing

Impairment of episodic memory is the key element for the diagnosis of typical (amnestic) AD. Widely used tests require the recall of verbal and non verbal information after an interval (delayed recall). The recall of a prose passage (Logical Memory test in the Wechsler Memory Scale [52]) originates from the "Babcock story" [53], a verbal memory measure in which participants are read a brief story and then asked to recall it immediately. The story is then read a second time, and delayed recall is collected after an interval of 20 min. Performance on prose rather than isolated word recall benefits from the semantic organization of the information, reducing the load on the self-generation of recall strategies requiring executive resources [54]. Other widely used verbal memory measures are based on the recall of lists of unrelated [55,56] or related [57,58] word lists. Among the non verbal tests, the delayed recall of Rey-Osterrieth complex figure is probably one of the best known [59].

Delayed recall tests are sensitive, but not specific for AD. A decline in performance is found also in healthy aging [60] and in depression [61], and is not helpful for the differential diagnosis between AD and other causes of dementia. An impaired recall performance is also found in subcortical vascular dementia [62] and frontotemporal dementia (FTD) [63]. In addition, delayed recall tests are not suitable for measuring disease severity and progression as they reach floor levels early in the disease course [64]. Tests controlling for effective memory encoding and retrieval may be particularly suitable to identify the hippocampal amnestic syndrome, which is typical of AD. This is characterized by the presence of a specific episodic memory involvement, with a reduced free recall performance that is marginally improved, or not improved at all, by cueing. The Free and Cued Selective Reminding Test (FCSRT) has been used with the aim of maximizing the differentiation between the genuine hippocampal deficit of AD and age-associated memory dysfunctions, due to impaired attention, inefficient information processing, and ineffective retrieval [65]. The FCSRT as well as the "bedside" 5-Word cued recall test [66] increases the specificity for AD [67]. In the case of FTD, FCSRT performance detects a subgroup of patients whose performance is normalized by cueing [68]. There is also evidence supporting the value of the FCSRT to predict progression toward dementia in at risk populations [14,69–71].

Besides delayed free recall, other aspects of memory function are sensitive to the early involvement of the hippocampal region by AD pathology. The temporary maintenance of associations in working memory, assessed with the visual short-term memory binding test [72], is specifically affected in early AD [73] and in asymptomatic carriers of pathogenetic AD mutations [74]. Allocentric spatial working memory, another hippocampal function, can be assessed with the Four Mountains test and is affected in biomarker-positive MCI subjects [75]. The assessment of language abilities in a standard neuropsychological assessment is usually limited to word-finding, with a picture-naming task (such as the Boston Naming test- [76]) and to "fluency" or "controlled associations" tasks, requiring the generation of words beginning with a specified letter (phonemic fluency) or belonging to a semantic category (semantic fluency) [77]. The fluency task is also considered among the "executive" tests (see below), because of its requirement for selection of information and continuous monitoring of performance. Semantic fluency, in particular, is characterized by an extremely high sensitivity to cognitive decline [78]. Impairment in semantic memory tasks, in general, have been suggested to represent an early marker of perirhinal cortex dysfunction [79], a pathological feature which is considered to reflect tau deposition in the early stages of AD [80,81].

A more in depth assessment of language abilities is required in the case of atypical (language-based) presentations, suggesting a possible diagnosis of primary progressive aphasia (PPA). The classification according to the current diagnostic criteria [82] requires a qualitative and quantitative assessment of the patient's connected speech production, tasks of picture naming and word-picture matching to assess single word comprehension, a repetition test allowing an assessment of phonological and auditory verbal short-term memory abilities, and a sentence-picture matching task to assess comprehension of syntactically complex sentences. A dedicated test is the Mini Standard Language Examination [83]. Looking for the presence of apraxia is useful for differential diagnosis among neurodegenerative dementias [84]. Many standardized tests for limb apraxia and oral apraxia [85], developed for the assessment of stroke patients, are available.

Visuo-spatial and visuo-constructional function is also assessed with neuropsychological tests used for the evaluation of focal brain damage. The function of ventral and dorsal visual processing pathways is related, respectively, to object and spatial perceptual abilities [86]. The widely used Clock Drawing Test, which includes both these aspects as well as executive function, is better considered as a global screening tool than a visuo-spatial function test.

A specific assessment can be performed with the Benton tests of matching unfamiliar faces and the spatial orientation of lines [87]. A standard task of visuo-constructional abilities is the copy of Rey's figure, which provides the baseline for the already-mentioned delayed recall test [55]. The ability to perceive multiple objects at the same time (impaired in simultanagnosia) can be evaluated with the Poppelreuter overlapping pictures test [88]. A brief but comprehensive screening battery, which allows to evaluate the function of both ventral and dorsal visual processing pathways, is the Visual Object and Space Perception Battery [89].

The neuropsychological examination is completed by the assessment of attention and executive functions. The "dysexecutive syndrome" includes both cognitive and behavioral aspects and has been traditionally associated with frontal lobe function [90]. Cognitive neuroscience research has significantly changed this perspective. Many of the traditional "frontal lobe" tests, such as the Wisconsin Card Sorting, have been shown to involve multiple neural networks including non-frontal areas [91] and are highly related to measures of fluid intelligence [92]. In addition, the key role of orbitofrontal and ventromedial frontal cortex in social cognition is now recognized as a central determinant of the clinical features of many neurodegenerative disorders [93]. A classical, brief executive function test is the Trail Making [94]. Part A, connecting encircled numbers irregularly distributed on a sheet, is a basic test of processing speed; part B, in which the subject has to alternate between letter and numbers, engages resources required for planning, sequencing, set-shifting, and inhibition of response. Another widely used attentional/executive family of tasks is based on the Stroop effect, i.e., the cognitive interference occurring when the processing of a specific stimulus feature affects the simultaneous processing of a second stimulus attribute [95]. The most widely used clinical version is the Color-Word Stroop, in which the subjects are required to read three different tables as fast as possible. Two of them represent the "congruent condition", in which participants are reading color names printed in black ink and naming color patches. In the third, incongruent condition color-words are printed in an inconsistent color (for instance "red" is printed in green ink), and the participants are required to name the color of the ink. The interference effect in the latter condition is measured on the basis of the number of errors and/or the decline in processing speed in the last (incongruous) condition [96]). An increased interference is considered as a sign of executive dysfunction. A useful screening test for frontal function is the Frontal Assessment Battery [97]. An assessment of social cognition performance [98] is a useful complement to the investigation of behavioral disorders associated with orbitofrontal dysfunction [99]. Emotion recognition can be assessed with the classical Ekman 60-faces test [100]. A brief task assessing the ability to attribute mental and affective states to others is the Story-based Empathy task

(SET- [101]); the mini-Social Cognition and Emotional Assessment (SEA- [102]) is a short battery exploring different facets of social cognition.

Anxiety and depression are common in subjects with a diagnosis of MCI and may increase the risk of progression to AD [103]. Apathy, with or without depression, has also been found to increase the progression risk in MCI subjects [104]. Anxiety and depression may represent a predictor of cognitive decline also in elderly subjects without cognitive complaints [105,106]. The latter observations have led to the concept of Mild Behavioral Impairment (MBI) [107], defined as a condition in which the emergence of sustained and impactful neuropsychiatric symptoms after age 50 predicts cognitive decline. Several rating scales have been specifically developed for the assessment of neuropsychiatric disturbances in dementia, such as the Neuropsychiatric Inventory (NPI) [108] and the Frontal Behavioral Inventory (FBI) [109]. Apathy can also be assessed with specialized tools [110]. In contrast, depression and anxiety are often evaluated with traditional psychiatric instruments, such as the Hamilton scales [111,112], the Geriatric Depression Scale (GDS) [113].

A complete neuropsychological and behavioral examination provides invaluable information supporting diagnostic orientation at the MCI stage. In an individual presenting with memory complaints, an isolated disorder of episodic memory, with features suggesting the presence of selective hippocampal dysfunction, suggests the prodromal stage of typical AD. Language, visuospatial and executive/behavioral presentations, associated with defective performance in the respective neuropsychological domains, evoke the possibility of early stages of atypical AD as defined by the IGW-2 criteria [114], or of a non-AD dementia. While the use of instrumental diagnostic tools (see below) is generally required for diagnostic prediction, the clinical and neuropsychological features allow the formulation of a first hypothesis.

A language presentation is compatible with the early stages of PPA, a syndrome which can be due to AD or to a pathology belonging to the FTD spectrum. The logopenic/phonological variant of primary progressive aphasia (lv-PPA) is by far the most common language presentation of AD, and needs to be differentiated from the other variants, i.e., non-fluent/agrammatic and semantic PPA, which are due to AD pathology only in a minority of cases [115]. A probabilistic diagnosis of the underlying pathology is suggested by the neurolinguistic features [82]. The typical logopenic profile includes hesitant but well-articulated speech, anomia, sometimes phonological errors, preserved word comprehension, defective repetition and impaired comprehension of syntactically complex sentences. The presence of speech apraxia and agrammatism, on the other hand, suggest a diagnosis of non-fluent/agrammatic variant, while a prominent impairment in word use, including paraphasia in production and impairment in word comprehension, in the context of fluent speech are typical of the semantic variant. Both syndromes are in most cases associated with pathology belonging to the fronto-temporal spectrum [115].

A presentation with disorders of visual and spatial perception is usually associated with another atypical AD variant, the PCA syndrome. This clinical picture is characterized at the onset by prominent visuo-spatial cognitive features, such as deficits in space and object perception, simultanagnosia, constructional dyspraxia, prosopagnosia, oculomotor apraxia, optic ataxia and alexia. A prominent pattern of visuoperceptual dysfunction can be also a prodromal feature of Lewy body dementia (LBD) [116], and of some cases of CBS [117]. In both cases, careful history taking and neurological examination (including apraxia evaluation) are the key to a correct diagnostic orientation [36,118].

Prominent behavioral manifestations suggest the early phase of the behavioral variant of FTD (bvFTD) [119]. Within the FTD spectrum, the right sided variant of semantic dementia must also be considered as a syndromic variant, in particular if associated with visuo-perceptual impairment [120]. Frontal AD (fvAD) is also a possibility to be considered [121]. While biomarker evidence is necessary for a differential diagnosis with FTD, a comprehensive neuropsychological assessment may provide useful information. In the dementia stage, cognitive impairment is more widespread in fvAD than in bvFTD. In particular, classical working memory and executive function test are more severely affected in fvAD than in bvFTD [122].

Functional assessment instruments

The assessment of functional abilities is the fourth pillar of the definition of mild cognitive impairment. In the original formulation, "normal activities of daily living" were a required criterion for MCI [11]. The NIA criteria [12] accept that mild problems performing complex functional tasks, such as paying bills, preparing a meal, or shopping, may be present, but in the context of maintenance of independence of function in daily life, with minimal aids or assistance. An excellent tool for the assessment of the Instrumental Activities of Daily Living (IADL) is the Amsterdam IADL questionnaire [123].

Diagnostic tools to support clinical diagnosis

Although AD diagnosis is based on clinical judgment, many clinical studies in the last two decades have consistently shown that instrumental tools provide relevant information to support a correct early diagnosis [16]. Among the tools available in clinical settings to support AD pathology detection, neuroimaging techniques and cerebrospinal fluid (CSF) assessment have gained increasing importance in research studies, reaching a key role in guiding early and differential diagnosis. Technical advances and encouraging developments have made it possible to measure such disease markers using fully automated assays, with high precision and stability among centers. This remarkable progress is pushing toward a widespread use of these diagnostic tests in clinical research settings. Disease markers reflecting core AD pathophysiology are going to play an increasingly important role in trials and clinical settings in general. Nonetheless, their assessment is expensive and often invasive, and can be inefficient without a specific correlation with clinical data. Therefore, a stepwise approach for the use of biomarkers is to recommend in suspected AD and non-AD cases, starting from the clinical and neuropsychological information as gateway for their correct use.

Cerebrospinal fluid

The core CSF biomarkers for the in vivo detection of AD pathology are amyloid beta (Aβ42 and Aβ40), total tau (t-tau), and phosphorylated tau (p-tau181 and p-tau217), respectively assessing the presence of amyloid pathology, axonal degeneration and tau pathology [124]. Such biomarkers have been extensively validated in clinical settings, showing very high diagnostic performance for the detection of AD pathology even in the early disease phases [125] with similar concentrations in different ethnic groups [126]. However, current literature on direct comparison of fluid biomarkers between ethnic groups is very limited and the relevance of sociodemographic determinants and/or comorbidities should be further elucidated.

Fully automated assays and standardization of preanalytical procedures have been developed and implemented in clinical research to reduce variability in measurements across laboratories and minimize patient misclassification rates [127]. Reduced Aβ42 and increased t-tau and p-tau levels are highly correlated with AD neuropathology in MCI subjects [125] and formally incorporated into the research diagnostic criteria [12,13]. Level of Aβ42 decreases early in the pathological cascade of AD, remaining stably low thereafter [128]. According to this latter study, Aβ42 levels are fully decreased at least 5–10 years before conversion to AD dementia. It is thus a valuable marker for prodromal or even preclinical diagnosis in at risk subjects (e.g., asymptomatic familiar AD cases). T-tau and p-tau levels, which are usually two or three time higher in AD patients in comparison to healthy elderly people, are useful for diagnosis but not for conversion to dementia, since they become pathological later in the course of AD [128]. Two decades of research in the field indicated a poor specificity of t-tau levels, whose increase is frequently observed in other medical conditions associated with neurodegeneration, such as acute cerebrovascular disorders or brain trauma [129]. Elevated p-tau is a better biomarker for discriminating AD from other dementias, as its levels are not increased in non-AD pathologies (e.g., FTD and LBD) [130–132]. Incremental value in predicting AD has been shown for CSF markers in combination with other biomarkers [133]. Additionally, CSF biomarker ratios and composite markers have been proved superior to single CSF biomarkers [134,135].

CSF biomarkers provide useful information for in vivo detection of AD pathology in cognitive patients with unusual presentations, providing relevant data for a better management and treatment options. None of the abovementioned CSF biomarkers is however helpful to make distinctions between typical and atypical variants of AD (e.g., frontal, logopenic or posterior variants) [114]. CSF biomarkers showed the same performances in identifying AD pathology, irrespective of the cognitive phenotype [134]. Nonetheless, some differences can be detected in the atypical AD group due to its heterogeneous balance between amyloid processing and neurodegeneration. As reported by Paterson and co-workers [136], posterior variant individuals may present lower Tau levels compared to frontal variant ones, showing the highest Tau levels. Clinical-neuropsychological data with the support of topographical distribution of neuronal loss and/or brain synaptic dysfunction is certainly mandatory in atypical AD cases to provide relevant information for the differential diagnosis with non-AD syndromes.

Neurofilament light chain (NfL) levels have been recently proved to be promising biomarker of neurodegeneration useful to detect AD and non-AD neurodegenerative conditions [137]. Elevated CSF NfL concentration is useful to distinguish neurodegenerative patients from healthy controls with stronger discriminatory power for FTD compared to AD group [137]. Nonetheless, clinical interpretation of NfL level measurement should consider the overlap between AD and non-AD pathologies mimicking AD, in which elevation in NfL concentration may be even higher than the fold-change observed for AD cases [137]. Clinical validity of NfL concentrations in prodromal dementia phases is still to be defined. A better understanding of the performance of this nonspecific neurodegeneration marker in comorbid phenotypes is certainly needed before evaluating its implementation in the diagnostic roadmap of dementia and in monitoring treatment efficacy. Blood-based biomarkers, covered in greater detail elsewhere in this book in the chapter by Henrik Zetterberg, hold great promise to revolutionize the diagnostic and prognostic work-up of AD in clinical practice [130] reducing the need for more costly and invasive investigations (CSF and PET imaging). Longitudinal measurements of such biomarkers will certainly improve the detection of effects of drugs such as anti-amyloid-β (Aβ) immunotherapies or other relevant disease-modifying treatments.

Magnetic Resonance Imaging

Standard structural brain imaging investigations (i.e., Computed Tomography — CT — and Magnetic Resonance Imaging — MRI) play a preliminary critical role in cognitive neurology by helping in detecting non-degenerative causes (e.g., hydrocephalus, brain tumors, or vascular lesions) of cognitive decline. MRI is useful to assess the presence and degree of comorbidities (e.g., cerebrovascular disease) in suspected neurodegenerative patients. Even if it is widely accepted that structural MR imaging provides limited information about the AD pathological process, especially in the early disease stages [138], its exclusionary role represents a cornerstone of the AD diagnostic workup in clinical settings. Structural imaging can aid in discriminating non-AD from AD cases and is often the first step in the diagnostic roadmap after the clinical-neuropsychological assessment.

The topographical patterns of atrophy assessed by structural MRI provide useful information. Medial temporal atrophy and hippocampal regional volume reduction are the most validated MRI markers for typical AD in the dementia and predementia stages [139,140]. Nonetheless, this atrophy pattern has low sensitivity and specificity and does not qualify structural MRI as a stand-alone add-on test for early detection of AD pathology in prodromal phase [139]. Positive finding of hippocampal or medial temporal lobe atrophy in MRI scans of MCI subjects must alert the clinician and lead to a close follow-up of the case, because of the possibility of progression to typical AD dementia. However, a negative MRI finding does not help in excluding AD pathology. The use of visual assessment of regional atrophy rather than semi- or fully-automated segmentation tools may account for the limited accuracy of structural MRI data at the single subject level [141]. MRI plays also a crucial role in anti-amyloid clinical trial safety monitoring, particularly for the identification of amyloid-related imaging abnormalities (ARIA).

Atypical AD cases present specific signature of brain vulnerability on MRI. Cross sectional voxel-based morphometry studies in posterior variants highlight a widespread pattern of bilateral gray matter loss [142]. Occipital (fusiform gyrus, middle occipital gyrus, and lingual gyrus) and parietal (cuneus and precuneus) regions are involved together with the inferior temporal gyrus [139,143]. Comparison with typical AD reveals significantly greater bilateral atrophy in primary and associative visual cortex and in the right posterior parietal lobe in PCA subjects [140,142,144,145]. However, large areas of overlapping atrophy in parietal, occipital, posterior cingulate and temporal cortices have been reported in both atypical and typical AD cases [146]. The imaging marker of the logopenic variant is a predominant damage of the left temporo-parietal cortex involving the phonologic short-term loop [82]. Widespread temporo-parietal involvement is usually present in CBS individuals with AD pathology while non-AD CBS cases show more predominant asymmetric fronto-insular (premotor, supplemental motor and insula) damage [147], parallelly to what found with glucose PET imaging [148]. Finally, macroscopic MRI changes of brain volume are not able to improve the classification between AD and non-AD cases (e.g., FTD) [149]. Additional information on microstructural white matter integrity has been suggested to add complementary data improving differential diagnosis [149]. More advanced MRI techniques, such as Diffusion tensor imaging (DTI) and functional MRI (fMRI) at the resting state, are thus more promising MRI tools but are not yet included in the clinical roadmap [140].

Positron emission tomography imaging

PET imaging techniques allow the early detection and quantification of molecular and functional brain changes years before regional volumetric MRI changes. They have thus a key role in the diagnostic algorithm of AD dementia, particularly in early phases and in case of early onset or atypical presentations, when AD diagnosis is very challenging due to mild symptomatology or overlapping symptoms with non-AD dementias.

PET imaging with amyloid ligands allows the detection of amyloidosis in vivo with standardized tracer-specific procedures that guarantees high-reproducibility among centers and high negative predictive value toward AD pathology [150,151]. The commercialization of ^{18}F-labeled amyloid tracers (florbetapir, florbetaben, flutemetamol), obviating the need for on-site cyclotron and radiochemistry facilities, which is required for ^{11}C-labeled compound production (e.g., ^{11}C-PiB), made available amyloid PET imaging to secondary nuclear medicine centers. Practical issues such as training status of clinicians, availability of and distance to PET scanners and cyclotrons, adverse effects and reimbursement issues according to priorities of the health-care system, however, limit the use of this imaging technique making clinical practice different in each country [151,152]. Diffuse brain amyloid pathology measured with PET represents an established prognostic marker in subjects with MCI, distinguishing stable subjects from those who have a higher risk to develop AD dementia [141]. It is not sufficient, however, to stage disease, or to predict if, when and how the individual subject may progress to dementia. Amyloid plaques may follow consistent deposition patterns in different regions of the brain, making it possible to stage in vivo amyloid pathology [153]. Amyloid PET imaging may crucially influence diagnosis, prescriptions, and patient management, as proved by a recent survey on clinicians [154].

The complex relationship between amyloid positivity and neurodegeneration is far from well-understood, and the triggers of AD pathology progression remain unknown [155,156]. Several studies have demonstrated good diagnostic agreement between CSF Aβ42 levels and amyloid-PET measures (see for an example [157]). However, some recent evidence reports that CSF β-amyloid levels may become abnormal even prior to an amyloid PET signal [158–160], suggesting that CSF may be a more sensitive tool for the detection of the earliest phases of amyloid accumulation. Such finding can be influenced by the accuracy of the technique used for the in vivo quantification of amyloid-PET tracer deposition and by CSF biomarkers levels. Appropriate criteria for amyloid-PET use have been summarized, defining three populations most likely to benefit from the procedure: MCI patients with unclear pathological substrate; atypical dementia presentations compatible with AD and early onset progressive cognitive decline [161]. Notably, the regional pattern of amyloid load in atypical AD subjects is diffuse and comparable to what is observed in typical subjects, supporting once more the idea that amyloid markers do not differentiate between typical and atypical AD, and non-AD presentations (see for example [162]).

PET imaging with ^{18}F-fluorodeoxyglucose (^{18}F-FDG) adds crucial information in the early assessment of early onset and atypical AD cases [141,163]. In the diagnostic roadmap of AD, FDG PET plays a key role in the definition of the topographical downstream neurodegeneration. It is particularly useful for the early differential diagnosis of MCI patients, as it is able to predict MCI conversion to AD and non-AD pathologies [164] and can show specific hypometabolic signatures suggestive of the different clinical presentations of AD pathology [165–167]. FDG PET has also proven to be useful to predict clinical outcome at the individual level, in MCI who underwent an amyloid PET scan. While a negative FDG PET scan predicts clinical stability during follow-up of many years (even in amyloid-positive cases), a positive FDG PET scan is associated with increased risk of progressive cognitive decline, also in amyloid-negative cases [150].

As in the case of amyloid PET, there is a consistent body of evidence regarding the importance of semi-quantitative analysis in the definition of specific FDG PET patterns (see Ref. [141] for a review). Characteristic hypometabolic patterns include posterior cingulate and temporo-parietal involvement in typical AD cases, and differential functional metabolic signatures in atypical AD [164]. FDG PET imaging shows a selective left-sided pattern of cortical hypometabolism in the logopenic variant [162] and parieto-occipital in PCA, involving frontal eye field structures in some cases [143]. CBS patients with AD biomarkers usually show typical AD-like hypometabolic pattern involving posterior cingulate cortex, precuneus and temporo-parietal cortex, whereas CBS cases with non-AD pathology have predominant bilateral hypometabolism in fronto-insular cortex and basal ganglia [148]. Combined PET imaging is needed in case of frontal variant, in which brain structural imaging findings mimic those of FTD [168].

More recently other tracers targeting tau aggregates and neuro-inflammatory response have been developed. Tau PET imaging has shown a very high diagnostic accuracy for distinguishing AD dementia from other non-AD neurodegenerative syndromes [169] and one PET tracer (flortaucipir) has been approved for clinical use in US [130]. Tau PET imaging has shown a strong clinico-anatomical correlation in typical AD, with more heterogeneous distribution in patients with frontal variant [170]. Recent evidence suggests that Tau PET may be the best stand-alone marker predicting progression to dementia in MCI subjects [171].

Conclusions

In this chapter, we have aimed to offer the reader an overview of current strategies in the clinical diagnosis of AD. Although cost-effective roadmaps for the use of biomarkers and their strategic implementation in clinical contexts have been prospected (see for example the Geneva task force for the roadmap of Alzheimer's biomarkers [16]), there is a strong need for further efforts in a more systematic validation of clinical guidelines for the use of AD biomarkers. Clinical examination and screening measures of global cognitive assessment represent mandatory information that should be always collected in the first clinical assessment of an individual with suspected cognitive impairment. If the screening tool indicates the presence of cognitive impairment, a comprehensive neuropsychological assessment, the choice of which depends in part on local expertise and availability of resources, should always be considered to aid differential diagnosis of AD, supplemented by imaging data and biomarkers whenever possible.

This perspective, based on the complementary role of the multiple available biomarkers, will undergo major changes in the coming years by the updating of currently available methodologies and by the introduction of new tools. The arrival of disease-modifying therapies targeting specific aspects of pathophysiology underscores the importance of accurate diagnosis through a stepwise approach, starting with detailed characterization of the clinical-cognitive phenotype and individual risk factors and culminating in the collection of fine-grained biomarker data (Fig. 5.1).

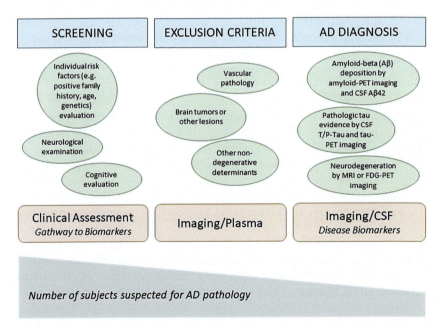

FIG. 5.1 Stepwise approach for subjects suspected for AD pathology. Individual risk factors, neurological examination and cognitive assessment should be included for the screening of subjects (i.e., clinical assessment — gateway to biomarker). Basic information form imaging and plasma should be collected in order to rule out non-AD determinants for cognitive deficits. Amyloid and tau pathology, as well as neurodegeneration biomarkers should be part of the AD disease biomarker roadmap.

References

[1] Dubois B, Padovani A, Scheltens P, Rossi A, Dell'Agnello G. Timely diagnosis for Alzheimer's disease: a literature review on benefits and challenges. J Alzheimer's Dis 2016;49:617−31.

[2] Vanderschaeghe G, Vandenberghe R, Dierickx K. Stakeholders' views on early diagnosis for Alzheimer's Disease, clinical trial participation and amyloid PET disclosure: A focus group study. J Bioeth Inq 2019;16:45−59.

[3] Epelbaum S, Genthon R, Cavedo E, Habert MO, Lamari F, Gagliardi G, et al. Preclinical Alzheimer's disease: a systematic review of the cohorts underlying the concept. Alzheimer's Dement 2017;13:454−67.

[4] Jack Jr CR, Bennett DA, Blennow K, et al. NIA-AA research framework: toward a biological definition of Alzheimer's disease. Alzheimers Dementia 2018;14:535−62.

[5] Dubois B, Villain N, Frisoni GB, Rabinovici GD, Sabbagh M, Cappa S, Feldman HH. Clinical diagnosis of Alzheimer's disease: recommendations of the international working group. Lancet Neurol 2021;20(6):484−96.

[6] Dubois B, Villain N, Schneider L, Fox N, Campbell N, Galasko D, et al. Alzheimer disease as a clinical-biological construct—an International Working Group recommendation. JAMA Neurol 2024.

[7] Ritchie K, Ropacki M, Albala B, Harrison J, Kaye J, Kramer J, et al. Recommended cognitive outcomes in preclinical Alzheimer's disease: Consensus statement from the European Prevention of Alzheimer's Dementia project. Alzheimer's Dement 2017;13:186−95.

[8] Molinuevo JL, Cami J, Carné X, Carrillo MC, Georges J, Isaac MB, et al. Ethical challenges in preclinical Alzheimer's disease observational studies and trials: Results of the Barcelona summit. Alzheimer's Dement 2016;12:614−22.

[9] Brookmeyer R, Abdalla N. Estimation of lifetime risks of Alzheimer's disease dementia using biomarkers for preclinical disease. Alzheimers Dementia 2018;14:981−8.

[10] Flicker C, Ferris SH, Reisberg B. Mild cognitive impairment in the elderly: predictors of dementia. Neurology 1991;41:1006−1006.

[11] Petersen RC, Smith GE, Waring SC, Ivnik RJ, Tangalos EG, Kokmen E. Mild cognitive impairment: clinical characterization and outcome. Arch Neurol 1999;56:303−8.

[12] Albert MS, DeKosky ST, Dickson D, et al. The diagnosis of mild cognitive impairment due to Alzheimer's disease: recommendations from the National Institute on Aging-Alzheimer's Association workgroups on diagnostic guidelines for Alzheimer's disease. Alzheimers Dement 2011;7:270−9.

[13] Dubois B, Feldman HH, Jacova C, DeKosky ST, Barberger-Gateau P, Cummings J, Delacourte A, Galasko D, Gauthier S, Jicha G, Meguro K, O'Brien J, Pasquier F, Robert P, Rossor M, Salloway S, Stern Y, Visser PJ, Scheltens P. Research criteria for the diagnosis of Alzheimer's disease: revising the NINCDS−ADRDA criteria. Lancet Neurol 2007;6(8):734−46. https://doi.org/10.1016/S1474-4422(07)70178-3.

[14] Sarazin M, Berr C, De Rotrou J, Fabrigoule C, Pasquier F, Legrain S, et al. Amnestic syndrome of the medial temporal type identifies prodromal AD: a longitudinal study. Neurology 2007;69:1859−67.

[15] Sachdev PS, Blacker D, Blazer DG, Ganguli M, Jeste DV, Paulsen JS, et al. Classifying neurocognitive disorders: the DSM-5 approach. Nat Rev Neurol 2014;10:634.

[16] Frisoni GB, Boccardi M, Barkhof F, Blennow K, Cappa S, Chiotis K, Démonet J-F, Garibotto V, Giannakopoulos P, Gietl A, Hansson O, Herholz K, Jack CR, Nobili F,

Nordberg A, Snyder HM, Ten Kate M, Varrone A, Albanese E, Winblad B. Strategic roadmap for an early diagnosis of Alzheimer's disease based on biomarkers. Lancet Neurol 2017;16(8):661−76. https://doi.org/10.1016/S1474-4422(17)30159-X.
[17] Petersen RC, Lopez O, Armstrong MJ, Getchius TS, Ganguli M, Gloss D, et al. Practice guideline update summary: Mild cognitive impairment: Report of the Guideline Development, Dissemination, and Implementation Subcommittee of the American Academy of Neurology. Neurology 2018;90:126−35.
[18] Boccardi M, Nicolosi V, Festari C, et al. Italian consensus recommendations for a biomarker-based aetiological diagnosis in mild cognitive impairment patients. Eur J Neurol 2019.
[19] Frisoni GB, Festari C, Massa F, Ramusino MC, Orini S, Aarsland D, Nobili F. European intersocietal recommendations for the biomarker-based diagnosis of neurocognitive disorders. Lancet Neurol 2024;23(3):302−12.
[20] Bertens D, Vos S, Kehoe P, et al. Use of mild cognitive impairment and prodromal AD/MCI due to AD in clinical care: a European survey. Alzheimers Res Ther 2019;11:74.
[21] Jessen, F., Amariglio, R.E., Buckley, R.F., van der Flier, W.M., Han, Y., Molinuevo, J.L., et al., "The characterisation of subjective cognitive decline," The Lancet neurology, vol. 19, pp. 271−278, 2020.
[22] Insel PS, Weiner M, Mackin RS, Mormino E, Lim YY, Stomrud E, et al. Determining clinically meaningful decline in preclinical Alzheimer disease. Neurology 2019;93:e322−33.
[23] Loewenstein DA, Greig MT, Schinka JA, Barker W, Shen Q, Potter E, et al. An investigation of PreMCI: subtypes and longitudinal outcomes. Alzheimer's Dement 2012;8:172−9.
[24] Cacciamani F, Tandetnik C, Gagliardi G, et al. Low cognitive awareness, but not complaint, is a good marker of preclinical Alzheimer's disease. J Alzheimer Dis 2017;59:753−62.
[25] Gagliardi G., Houot M., Cacciamani F., Habert M.-O., Dubois B., Epelbaum S. "The metamemory ratio: a new cohort-independent way to measure cognitive awareness in asymptomatic individuals at risk for Alzheimer's disease," Alzheimer's research & therapy, vol. 12, pp. 1−11, 2020
[26] Cappa SF, Ribaldi F, Chicherio C, Frisoni GB. Subjective cognitive decline: memory complaints, cognitive awareness, and metacognition. Alzheimers Dementia 2024. https://doi.org/10.1002/alz.13905.
[27] Gilewski MJ, Zelinski EM, Schaie KW. The Memory Functioning Questionnaire for assessment of memory complaints in adulthood and old age. Psychol Aging 1990;5:482.
[28] Farias ST, Mungas D, Reed BR, Cahn-Weiner D, Jagust W, Baynes K, et al. The measurement of everyday cognition (ECog): scale development and psychometric properties. Neuropsychology 2008;22:531.
[29] Amariglio RE, Donohue MC, Marshall GA, et al. Tracking early decline in cognitive function in older individuals at risk for Alzheimer disease dementia: the Alzheimer's Disease Cooperative Study Cognitive Function Instrument. JAMA Neurol 2015;72:446−54.
[30] Migliorelli R, Tesón A, Sabe L, Petracca G, Petracchi M, Leiguarda R, et al. Anosognosia in Alzheimer's disease: a study of associated factors. J Neuropsychiatry Clin Neurosci 1995;7(3):338−44.
[31] Dalla Barba G, Parlato V, Iavarone A, Boller F. Anosognosia, intrusions and 'frontal' functions in Alzheimer's disease and depression. Neuropsychologia 1995;33:247−59.

[32] Morris JC. The Clinical Dementia Rating (CDR): Current version and scoring rules. Neurology 1993;43:2412–4.

[33] Elamin M., Bennett G., Symonds A., Pal S., Abrahams S., Parra M., et al. Introducing a brief screening a tool for motor signs in patients with dementia (P6. 203). Neurology 2015;84(14_supplement):P6-203.

[34] Cooper S, Greene JDW. The clinical assessment of the patient with early dementia. J Neurol Neurosurg Psychiatry 2005;76(suppl 5):v15–24.

[35] Höglinger GU, Respondek G, Stamelou M, et al. Clinical diagnosis of progressive supranuclear palsy: the movement disorder society criteria. Mov Disord 2017;32:853–64.

[36] Armstrong MJ, Litvan I, Lang AE, et al. Criteria for the diagnosis of corticobasal degeneration. Neurology 2013;80:496–503.

[37] Crutch SJ, Schott JM, Rabinovici GD, et al. Consensus classification of posterior cortical atrophy. Alzheimer's Dement 2017.

[38] McKhann GM, Knopman DS, Chertkow H, Hyman BT, Jack Jr CR, Kawas CH, et al. The diagnosis of dementia due to Alzheimer's disease: recommendations from the National Institute on Aging-Alzheimer's Association workgroups on diagnostic guidelines for Alzheimer's disease. Alzheimer's Dement 2011;7:263–9.

[39] Cerami C, Dubois B, Boccardi M, Monsch AU, Demonet JF, Cappa SF, et al. Clinical validity of delayed recall tests as a gateway biomarker for Alzheimer's disease in the context of a structured 5-phase development framework. Neurobiol Aging 2017;52:153–66.

[40] Folstein MF, Folstein SE, McHugh PR. Mini-mental state": a practical method for grading the cognitive state of patients for the clinician. J Psychiat Res 1975;12:189–98.

[41] Nasreddine ZS, Phillips NA, Bédirian V, Charbonneau S, Whitehead V, Collin I, et al. The Montreal Cognitive Assessment, MoCA: a brief screening tool for mild cognitive impairment. J Am Geriat Soc 2005;53:695–9.

[42] Doraiswamy P, Krishen A, Stallone F, Martin W, Potts N, Metz A, et al. Cognitive performance on the Alzheimer's Disease Assessment Scale: effect of education. Neurology 1995;45:1980–4.

[43] Mioshi E, Dawson K, Mitchell J, Arnold R, Hodges JR. The Addenbrooke's Cognitive Examination Revised (ACE-R): a brief cognitive test battery for dementia screening. Int J Geriatr Psychiatry 2006;21:1078–85.

[44] Paulino Ramirez Diaz S, Gil Gregório P, Manuel Ribera Casado J, Reynish E, Jean Ousset P, Vellas B, et al. The need for a consensus in the use of assessment tools for Alzheimer's disease: the Feasibility Study (assessment tools for dementia in Alzheimer Centres across Europe), a European Alzheimer's Disease Consortium's (EADC) survey. Int J Geriatr Psychiatry 2005;20:744–8.

[45] Maruta C, Guerreiro M, De Mendonça A, Hort J, Scheltens P. The use of neuropsychological tests across Europe: the need for a consensus in the use of assessment tools for dementia. Eur J Neurol 2011;18:279–85.

[46] Costa A, Bak T, Caffarra P, Caltagirone C, Ceccaldi M, Collette F. et al.The need for harmonisation and innovation of neuropsychological assessment in neurodegenerative dementias in Europe: consensus document of the Joint Program for Neurodegenerative Diseases Working Group. Alzheimer's Res Ther 2017;9:27.

[47] Daffner KR, Gale SA, Barrett A, et al. Improving clinical cognitive testing: report of the AAN Behavioral Neurology Section Workgroup. Neurology 2015;85:910–8.

[48] Breton A, Casey D, Arnaoutoglou NA. Cognitive tests for the detection of mild cognitive impairment (MCI), the prodromal stage of dementia: meta-analysis of diagnostic accuracy studies. Int J Geriatr Psychiatr 2019;34:233—42.
[49] Tsoi KK, Chan JY, Hirai HW, et al. Cognitive tests to detect dementia: a systematic review and meta-analysis. JAMA Int Med 2015;175:1450—8.
[50] Arevalo-Rodriguez I, Smailagic N, i Figuls MR, et al. Mini-Mental State Examination (MMSE) for the detection of Alzheimer's disease and other dementias in people with mild cognitive impairment (MCI). Cochrane Database Syst Rev 2015.
[51] Belleville S, Fouquet C, Hudon C, et al. Neuropsychological measures that predict progression from mild cognitive impairment to Alzheimer's type dementia in older adults: a systematic review and meta-analysis. Neuropsychol Rev 2017;27:328—53.
[52] Wechsler D. Wechsler Memory Scale. American Psychological Association; 1945.
[53] Babcock H, Levy L. Test and manual of directions; the revised examination for the measurement of efficiency of mental functioning. 1940.
[54] Tremont G, Halpert S, Javorsky DJ, et al. Differential impact of executive dysfunction on verbal list learning and story recall. Clin Neuropsychol 2000;14:295—302.
[55] Rey A. L'examen Clinique En Psychologie. Paris: Presses Universitaires de France; 1964.
[56] Tierney M, Szalai J, Snow W, et al. Prediction of probable Alzheimer's disease in memory-impaired patients: A prospective longitudinal study. Neurology 1996;46:661—5.
[57] Brandt J. The Hopkins Verbal Learning Test: development of a new memory test with six equivalent forms. Clin Neuropsychol 1991;5:125—42.
[58] Delis DC, Freeland J, Kramer JH, et al. Integrating clinical assessment with cognitive neuroscience: construct validation of the California Verbal Learning Test. J Consult Clin Psychol 1988;56:123.
[59] Rey A. L'examen psychologique dans les cas d'encéphalopathie traumatique. (Les problems.). Archives de psychologie 1941.
[60] Squire LR, Wixted JT. The cognitive neuroscience of human memory since HM. Annu Rev Neurosci 2011;34:259—88.
[61] Burt DB, Zembar MJ, Niederehe G. Depression and memory impairment: a meta-analysis of the association, its pattern, and specificity. Psychol Bull 1995;117:285.
[62] Graham N, Emery T, Hodges J. Distinctive cognitive profiles in Alzheimer's disease and subcortical vascular dementia. J Neurol Neurosurg Psychiatry 2004;75:61—71.
[63] Hornberger M, Piguet O, Graham A, et al. How preserved is episodic memory in behavioral variant frontotemporal dementia? Neurology 2010;74:472—9.
[64] Locascio JJ, Growdon JH, Corkin S. Cognitive test performance in detecting, staging, and tracking Alzheimer's disease. Arch Neurol 1995;52:1087—99.
[65] Grober E, Buschke H. Genuine memory deficits in dementia. Dev Neuropsychol 1987;3:13—36.
[66] Dubois B., Touchon J., Portet F., Ousset P., Vellas B., Michel B. "The 5 words": a simple and sensitive test for the diagnosis of Alzheimer's disease. Presse Med (Paris, France: 1983) 2002;31:1696—1699
[67] Economou A, Routsis C, Papageorgiou SG. Episodic memory in Alzheimer disease, frontotemporal dementia, and dementia with Lewy bodies/Parkinson disease dementia: disentangling retrieval from consolidation. Alzheimer's Dis Assoc Disord 2016;30:47—52.
[68] Bertoux M, de Souza LC, Corlier F, et al. Two distinct amnesic profiles in behavioral variant frontotemporal dementia. Biol Psychiatr 2014a;75:582—8.

[69] Wagner M, Wolf S, Reischies F, Daerr M, Wolfsgruber S, Jessen F, et al. Biomarker validation of a cued recall memory deficit in prodromal Alzheimer disease. Neurology 2012:379−86.

[70] Dierckx E, Engelborghs S, De Raedt R, Van Buggenhout M, De Deyn P, Verté D, et al. Verbal cued recall as a predictor of conversion to Alzheimer's disease in Mild Cognitive Impairment. Int J Geriatr Psychiatry 2009;24:1094−100.

[71] Grande G., Vanacore N., Vetrano D. L., Cova I., Rizzuto D., et al. Free and Cued Selective Reminding Test predicts progression to Alzheimer's disease in people with Mild Cognitive Impairment. Neurological sciences 39:1867−1875.

[72] Parra MA, Calia C, García AF, Olazarán-Rodríguez J, Hernandez-Tamames JA, Alvarez-Linera J, et al. Refining memory assessment of elderly people with cognitive impairment: insights from the short-term memory binding test. Arch Gerontol Geriatr 2019;83:114−20.

[73] Della Sala S, Parra MA, Fabi K, et al. Short-term memory binding is impaired in AD but not in non-AD dementias. Neuropsychologia 2012;50:833−40.

[74] Parra MA, Abrahams S, Logie RH, et al. Visual short-term memory binding deficits in familial Alzheimer's disease. Brain 2010;133:2702−13.

[75] Moodley K, Minati L, Contarino V, et al. Diagnostic differentiation of mild cognitive impairment due to Alzheimer's disease using a hippocampus-dependent test of spatial memory. Hippocampus 2015;25:939−51.

[76] Goodglass H, Kaplan E, Weintraub S. Boston Naming Test. Lea & Febiger; 1983.

[77] Lezak MD, Howieson D, Loring D, et al. Neuropsychological Assessment. New York: Oxford University Press; 2004. p. 429−98.

[78] Monsch AU, Bondi MW, Butters N, et al. Comparisons of verbal fluency tasks in the detection of dementia of the Alzheimer type. Arch Neurol 1992;49:1253−8.

[79] Davies R, Graham KS, Xuereb JH, et al. The human perirhinal cortex and semantic memory. Eur J Neurosci 2004;20:2441−6.

[80] Hirni DI, Kivisaari SL, Monsch AU, et al. Distinct neuroanatomical bases of episodic and semantic memory performance in Alzheimer's disease. Neuropsychologia 2013;51:930−7.

[81] Maass A, Berron D, Harrison TM, Adams JN, La Joie R, Baker S, et al. Alzheimer's pathology targets distinct memory networks in the ageing brain. Brain 2019;142:2492−509.

[82] Gorno-Tempini ML, Hillis AE, Weintraub S, Kertesz A, Mendez M, Cappa SF, Ogar JM, Rohrer JD, Black S, Boeve BF, Manes F, Dronkers NF, Vandenberghe R, Rascovsky K, Patterson K, Miller BL, Knopman DS, Hodges JR, Mesulam MM, Grossman M. Classification of primary progressive aphasia and its variants. Neurology 2011;76(11):1006−14. https://doi.org/10.1212/WNL.0b013e31821103e6.

[83] Patel N, Peterson KA, Ingram RU, et al. A 'Mini Linguistic State Examination'to classify primary progressive aphasia. Brain Commun 2022;4:fcab299.

[84] Ahmed S, Baker I, Thompson S, et al. Utility of testing for apraxia and associated features in dementia. J Neurol Neurosurg Psychiatr 2016;87:1158−62.

[85] De Renzi E, Pieczuro A, Vignolo L. Oral apraxia and aphasia. Cortex 1966;2:50−73.

[86] Mishkin M, Ungerleider LG, Macko KA. Object vision and spatial vision: two cortical pathways. Trends Neurosci 1983;6:414−7.

[87] Benton AL, Abigail B, Sivan AB, et al. Contributions to neuropsychological assessment: a clinical manual. USA: Oxford University Press; 1994.

[88] Della Sala S, Laiacona M, Trivelli C, et al. Poppelreuter-Ghent's overlapping figures test: its sensitivity to age, and its clinical use. Arch Clin Neuropsychol 1995;10:511−34.

[89] Warrington EK, James M. A new test of object decision: 2D silhouettes featuring a minimal view. Cortex 1991;27(3):377–83.
[90] Godefroy O, Martinaud O, Narme P, et al. Dysexecutive disorders and their diagnosis: A position paper. Cortex 2018;109:322–35.
[91] Nyhus E, Barceló F. The Wisconsin Card Sorting Test and the cognitive assessment of prefrontal executive functions: a critical update. Brain Cogn 2009;71:437–51.
[92] Smith V, Pinasco C, Achterberg J, et al. Fluid intelligence and naturalistic task impairments after focal brain lesions. cortex 2022;146:106–15.
[93] Elamin M, Pender N, Hardiman O, et al. Social cognition in neurodegenerative disorders: a systematic review. J Neurol Neurosurg Psychiatry 2012;83:1071–9.
[94] Battery AIT. Manual of directions and scoring. Washington, DC: War Department, Adjutant General's Office; 1944.
[95] Stroop JR. Factors affecting speed in serial verbal reactions. Psychol Monogr 1938;50:38.
[96] Scarpina F, Tagini S. The stroop color and word test. Front Psychol 2017;8:557.
[97] Dubois B, Slachevsky A, Litvan I, Pillon B. The FAB: a frontal assessment battery at bedside. Neurology 2000;55:1621–6.
[98] Torralva T, Roca M, Gleichgerrcht E, et al. A neuropsychological battery to detect specific executive and social cognitive impairments in early frontotemporal dementia. Brain 2009;132:1299–309.
[99] Dodich A, Crespi C, Santi GC, et al. Evaluation of Discriminative Detection Abilities of Social Cognition Measures for the Diagnosis of the Behavioral Variant of Frontotemporal Dementia: a Systematic Review. Neuropsychol Rev 2020:1–16.
[100] Ekman P, Friesen W. Pictures of Facial Affect. Palo Alto, CA: Consulting Psychologists Press; 1976.
[101] Dodich A, Cerami C, Canessa N, et al. A novel task assessing intention and emotion attribution: Italian standardization and normative data of the Story-based Empathy Task. Neurol Sci 2015;36:1907–12.
[102] Bertoux M, Volle E, De Souza L, et al. Neural correlates of the mini-SEA (Social cognition and Emotional Assessment) in behavioral variant frontotemporal dementia. Brain Imaging Behav 2014b;8:1–6.
[103] Palmer K, Berger A, Monastero R, et al. Predictors of progression from mild cognitive impairment to Alzheimer disease. Neurology 2007;68:1596–602.
[104] Ruthirakuhan M, Herrmann N, Vieira D, et al. The roles of apathy and depression in predicting alzheimer disease: a longitudinal analysis in older adults with mild cognitive impairment. Am J Geriatr Psychiatry 2019;27:873–82.
[105] Gulpers B, Ramakers I, Hamel R, et al. Anxiety as a predictor for cognitive decline and dementia: a systematic review and meta-analysis. Am J Geriatr Psychiatry 2016;24:823–42.
[106] Zheng F, Zhong B, Song X, et al. Persistent depressive symptoms and cognitive decline in older adults. Br J Psychiatry 2018;213:638–44.
[107] Ismail Z, Smith EE, Geda Y, et al. Neuropsychiatric symptoms as early manifestations of emergent dementia: provisional diagnostic criteria for mild behavioral impairment. Alzheimer's Dement 2016;12:195–202.
[108] Cummings JL, Mega M, Gray K, et al. The Neuropsychiatric Inventory: comprehensive assessment of psychopathology in dementia. Neurology 1994;44:2308–14.
[109] Kertesz A, Davidson W, Fox H. Frontal behavioral inventory: diagnostic criteria for frontal lobe dementia. Can J Neurol Sci 1997;24:29–36.

[110] Husain M. Alzheimer's disease: time to focus on the brain, not just molecules. Brain 2017;140(2):251−3.
[111] Hamilton M. The assessment of anxiety states by rating. Br J Med Psychiatry 1959;32:50−5.
[112] Hamilton M. A rating scale for depression. J Neurol Neurosurg Psychiatry 1978;133:429−35.
[113] Yesavage JA, Brink TL, Rose TL, et al. Development and validation of a geriatric depression screening scale: a preliminary report. J Psychiat Res 1982;17:37−49.
[114] Dubois B, Feldman HH, Jacova C, Hampel H, Molinuevo JL, Blennow K, DeKosky ST, Gauthier S, Selkoe D, Bateman R, Cappa S, Crutch S, Engelborghs S, Frisoni GB, Fox NC, Galasko D, Habert M-O, Jicha GA, Nordberg A, Cummings JL. Advancing research diagnostic criteria for Alzheimer's disease: the IWG-2 criteria. Lancet Neurol 2014;13(6):614−29. https://doi.org/10.1016/S1474-4422(14)70090-0.
[115] Spinelli EG, Mandelli ML, Miller ZA, Santos-Santos MA, Wilson SM, Agosta F, et al. Typical and atypical pathology in primary progressive aphasia variants. Annal Neurol 2017;81:430−43.
[116] Ciafone J, Little B, Thomas AJ, et al. The neuropsychological profile of Mild Cognitive Impairment in Lewy body dementias. J Int Neuropsychol Soc 2020;26:210−25.
[117] Tang-Wai DF, Josephs KA, Boeve BF, Dickson DW, Parisi JE, Petersen RC. Pathologically confirmed corticobasal degeneration presenting with visuospatial dysfunction. Neurology 2003;61(8):1134−5.
[118] McKeith IG, Ferman TJ, Thomas AJ, et al. Research criteria for the diagnosis of prodromal dementia with Lewy bodies. Neurology 2020;94:743−55.
[119] Pressman PS, Miller BL. Diagnosis and management of behavioral variant frontotemporal dementia. Biol Psychiatry 2014;75:574−81.
[120] Kamminga J, Kumfor F, Burrell JR, et al. Differentiating between right-lateralised semantic dementia and behavioural-variant frontotemporal dementia: an examination of clinical characteristics and emotion processing. J Neurol Neurosurg Psychiatry 2015;86:1082−8.
[121] Ossenkoppele R, Pijnenburg YA, Perry DC, Cohn-Sheehy BI, Scheltens NM, Vogel JW, et al. The behavioural/dysexecutive variant of Alzheimer's disease: clinical, neuroimaging and pathological features. Brain 2015;138:2732−49.
[122] Wong S, Strudwick J, Devenney E, Hodges JR, Piguet O, Kumfor F. Frontal variant of Alzheimer's disease masquerading as behavioural-variant frontotemporal dementia: a case study comparison. Neurocase 2019;25:48−58.
[123] Sikkes SA, Knol DL, Pijnenburg YA, De Lange-de Klerk ES, Uitdehaag BM, Scheltens P. Validation of the Amsterdam IADL Questionnaire©, a new tool to measure instrumental activities of daily living in dementia. Neuroepidemiology 2013;41:35−41.
[124] Hansson O, Batrla R, Brix B, Carrillo MC, Corradini V, Edelmayer RM, Esquivel RN, Hall C, Lawson J, Bastard NL, Molinuevo JL, Nisenbaum LK, Rutz S, Salamone SJ, Teunissen CE, Traynham C, Umek RM, Vanderstichele H, Vandijck M, Blennow K. The Alzheimer's Association international guidelines for handling of cerebrospinal fluid for routine clinical measurements of amyloid β and tau. Alzheimers Dementia 2021;17(9):1575−82. https://doi.org/10.1002/alz.12316.
[125] Bocchetta M, Galluzzi S, Kehoe PG, Aguera E, Bernabei R, Bullock R, Ceccaldi M, Dartigues J, Mendonça A, Didic M, Eriksdotter M, Félician O, Frölich L, Gertz H, Hallikainen M, Hasselbalch SG, Hausner L, Heuser I, Jessen F, Frisoni GB. The use of

biomarkers for the etiologic diagnosis of MCI in Europe: an EADC survey. Alzheimers Dementia 2015;11(2):195. https://doi.org/10.1016/j.jalz.2014.06.006.

[126] Windon C, Iaccarino L, Mundada N, Allen I, Boxer AL, Byrd D, ADNI. Comparison of plasma and CSF biomarkers across ethnoracial groups in the ADNI. Alzheimers Dement Diagnosis Assessment Dis Monitoring 2022;14(1):e12315.

[127] Leuzy A, Mattsson-Carlgren N, Cullen NC, Stomrud E, Palmqvist S, La Joie R, Iaccarino L, Zetterberg H, Rabinovici G, Blennow K, Janelidze S, Hansson O. Robustness of CSF Aβ42/40 and Aβ42/P-tau181 measured using fully automated immunoassays to detect AD-related outcomes. Alzheimers Dement 2023:12897. https://doi.org/10.1002/alz.12897.

[128] Buchhave P. Cerebrospinal fluid levels ofβ-amyloid 1-42, but not of tau, are fully changed already 5 to 10 Years before the onset of Alzheimer dementia. Arch Gen Psychiatr 2012;69(1):98. https://doi.org/10.1001/archgenpsychiatry.2011.155.

[129] Mattsson N, Lönneborg A, Boccardi M, Blennow K, Hansson O. Clinical validity of cerebrospinal fluid Aβ42, tau, and phospho-tau as biomarkers for Alzheimer's disease in the context of a structured 5-phase development framework. Neurobiol Aging 2017;52:196−213. https://doi.org/10.1016/j.neurobiolaging.2016.02.034.

[130] Hansson O, Blennow K, Zetterberg H, Dage J. Blood biomarkers for Alzheimer's disease in clinical practice and trials. Nat Aging 2023;3(5):506−19. https://doi.org/10.1038/s43587-023-00403-3.

[131] Struyfs H, Niemantsverdriet E, Goossens J, Fransen E, Martin J-J, De Deyn PP, Engelborghs S. Cerebrospinal fluid P-Tau181P: biomarker for improved differential dementia diagnosis. Front Neurol 2015;6. https://doi.org/10.3389/fneur.2015.00138.

[132] Tang W, Huang Q, Yao Y-Y, Wang Y, Wu Y-L, Wang Z-Y. Does CSF p-tau181 help to discriminate Alzheimer's disease from other dementias and mild cognitive impairment? A meta-analysis of the literature. J Neural Transm 2014;121(12):1541−53. https://doi.org/10.1007/s00702-014-1226-y.

[133] Frölich L, Peters O, Lewczuk P, Gruber O, Teipel SJ, Gertz HJ, Jahn H, Jessen F, Kurz A, Luckhaus C, Hüll M, Pantel J, Reischies FM, Schröder J, Wagner M, Rienhoff O, Wolf S, Bauer C, Schuchhardt J, Kornhuber J. Incremental value of biomarker combinations to predict progression of mild cognitive impairment to Alzheimer's dementia. Alzheimers Res Ther 2017;9(1):84. https://doi.org/10.1186/s13195-017-0301-7.

[134] Constantinides VC, Paraskevas GP, Boufidou F, Bourbouli M, Pyrgelis E-S, Stefanis L, Kapaki E. CSF Aβ42 and Aβ42/Aβ40 ratio in Alzheimer's disease and frontotemporal dementias. Diagnostics 2023;13(4):783. https://doi.org/10.3390/diagnostics13040783.

[135] Lei D, Mao C, Li J, Huang X, Sha L, Liu C, Dong L, Xu Q, Gao J. CSF biomarkers for early-onset Alzheimer's disease in Chinese population from PUMCH dementia cohort. Front Neurol 2023;13:1030019. https://doi.org/10.3389/fneur.2022.1030019.

[136] Paterson RW, Toombs J, Slattery CF, Nicholas JM, Andreasson U, Magdalinou NK, Blennow K, Warren JD, Mummery CJ, Rossor MN, Lunn MP, Crutch SJ, Fox NC, Zetterberg H, Schott JM. Dissecting IWG-2 typical and atypical Alzheimer's disease: insights from cerebrospinal fluid analysis. J Neurol 2015;262(12):2722−30. https://doi.org/10.1007/s00415-015-7904-3.

[137] Forgrave LM, Ma M, Best JR, DeMarco ML. The diagnostic performance of neurofilament light chain in CSF and blood for Alzheimer's disease, frontotemporal dementia, and amyotrophic lateral sclerosis: a systematic review and meta-analysis. Alzheimers Dement Diagnosis Assessment Dis Monitoring 2019;11(1):730−43. https://doi.org/10.1016/j.dadm.2019.08.009.

[138] Lombardi G, Crescioli G, Cavedo E, Lucenteforte E, Casazza G, Bellatorre A-G, Lista C, Costantino G, Frisoni G, Virgili G, Filippini G. Structural magnetic resonance imaging for the early diagnosis of dementia due to Alzheimer's disease in people with mild cognitive impairment. Cochrane Database Syst Rev 2020. https://doi.org/10.1002/14651858.CD009628.pub2.

[139] Frisoni GB, Fox NC, Jack CR, Scheltens P, Thompson PM. The clinical use of structural MRI in Alzheimer disease. Nat Rev Neurol 2010;6(2):67−77. https://doi.org/10.1038/nrneurol.2009.215.

[140] Ten Kate M, Ingala S, Schwarz AJ, Fox NC, Chételat G, Van Berckel BNM, Ewers M, Foley C, Gispert JD, Hill D, Irizarry MC, Lammertsma AA, Molinuevo JL, Ritchie C, Scheltens P, Schmidt ME, Visser PJ, Waldman A, Wardlaw J, Barkhof F. Secondary prevention of Alzheimer's dementia: neuroimaging contributions. Alzheimers Res Ther 2018;10(1):112. https://doi.org/10.1186/s13195-018-0438-z.

[141] Frisoni GB, Bocchetta M, Chetelat G, Rabinovici GD, De Leon MJ, Kaye J, Reiman EM, Scheltens P, Barkhof F, Black SE, Brooks DJ, Carrillo MC, Fox NC, Herholz K, Nordberg A, Jack CR, Jagust WJ, Johnson KA, Rowe CC, For ISTAART's NeuroImaging Professional Interest Area. Imaging markers for Alzheimer disease: which vs how. Neurology 2013;81(5):487−500. https://doi.org/10.1212/WNL.0b013e31829d86e8.

[142] Crutch SJ, Lehmann M, Schott JM, Rabinovici GD, Rossor MN, Fox NC. Posterior cortical atrophy. Lancet Neurol 2012;11(2):170−8. https://doi.org/10.1016/S1474-4422(11)70289-7.

[143] Cerami C, Crespi C, Della Rosa PA, Dodich A, Marcone A, Magnani G, Coppi E, Falini A, Cappa SF, Perani D. Brain changes within the visuo-spatial attentional network in posterior cortical atrophy. J Alzheimer Dis 2014;43(2):385−95. https://doi.org/10.3233/JAD-141275.

[144] Lehmann M, Barnes J, Ridgway GR, Ryan NS, Warrington EK, Crutch SJ, Fox NC. Global gray matter changes in posterior cortical atrophy: a serial imaging study. Alzheimers Dementia 2012;8(6):502−12. https://doi.org/10.1016/j.jalz.2011.09.225.

[145] Whitwell JL, Jack CR, Kantarci K, Weigand SD, Boeve BF, Knopman DS, Drubach DA, Tang-Wai DF, Petersen RC, Josephs KA. Imaging correlates of posterior cortical atrophy. Neurobiol Aging 2007;28(7):1051−61. https://doi.org/10.1016/j.neurobiolaging.2006.05.026.

[146] Migliaccio R, Agosta F, Rascovsky K, Karydas A, Bonasera S, Rabinovici GD, Miller BL, Gorno-Tempini ML. Clinical syndromes associated with posterior atrophy: early age at onset AD spectrum. Neurology 2009;73(19):1571−8. https://doi.org/10.1212/WNL.0b013e3181c0d427.

[147] Whitwell JL, Jack CR, Boeve BF, Parisi JE, Ahlskog JE, Drubach DA, Senjem ML, Knopman DS, Petersen RC, Dickson DW, Josephs KA. Imaging correlates of pathology in corticobasal syndrome. Neurology 2010;75(21):1879−87. https://doi.org/10.1212/WNL.0b013e3181feb2e8.

[148] Cerami C, Dodich A, Iannaccone S, Magnani G, Marcone A, Guglielmo P, Vanoli G, Cappa SF, Perani D. Individual brain metabolic signatures in corticobasal syndrome. J Alzheimers Dis 2020;76(2):517−28. https://doi.org/10.3233/JAD-200153.

[149] Möller C, Hafkemeijer A, Pijnenburg YAL, Rombouts SARB, Van Der Grond J, Dopper E, Van Swieten J, Versteeg A, Pouwels PJW, Barkhof F, Scheltens P, Vrenken H, Van Der Flier WM. Joint assessment of white matter integrity, cortical and subcortical atrophy to distinguish AD from behavioral variant FTD: a two-center study. Neuroimage Clin 2015;9:418−29. https://doi.org/10.1016/j.nicl.2015.08.022.

[150] Chételat G, Arbizu J, Barthel H, Garibotto V, Law I, Morbelli S, Van De Giessen E, Agosta F, Barkhof F, Brooks DJ, Carrillo MC, Dubois B, Fjell AM, Frisoni GB, Hansson O, Herholz K, Hutton BF, Jack CR, Lammertsma AA, Drzezga A. Amyloid-PET and 18F-FDG-PET in the diagnostic investigation of Alzheimer's disease and other dementias. Lancet Neurol 2020;19(11):951–62. https://doi.org/10.1016/S1474-4422(20)30314-8.

[151] Villemagne VL. Amyloid imaging: past, present and future perspectives. Ageing Res Rev 2016;30:95–106. https://doi.org/10.1016/j.arr.2016.01.005.

[152] Therriault J, Zimmer ER, Benedet AL, Pascoal TA, Gauthier S, Rosa-Neto P. Staging of Alzheimer's disease: past, present, and future perspectives. Trends Mol Med 2022;28(9):726–41. https://doi.org/10.1016/j.molmed.2022.05.008.

[153] Grothe MJ, Barthel H, Sepulcre J, Dyrba M, Sabri O, Teipel SJ, For the Alzheimer's Disease Neuroimaging Initiative. In vivo staging of regional amyloid deposition. Neurology 2017;89(20):2031–8. https://doi.org/10.1212/WNL.0000000000004643.

[154] Zhong Y, Karlawish J, Johnson MK, Neumann PJ, Cohen JT. The potential value of β-amyloid imaging for the diagnosis and management of dementia: a survey of clinicians. Alzheimer Dis Assoc Disord 2017;31(1):27–33. https://doi.org/10.1097/WAD.0000000000000168.

[155] Mormino EC, Betensky RA, Hedden T, Schultz AP, Amariglio RE, Rentz DM, Johnson KA, Sperling RA. Synergistic effect of β-amyloid and neurodegeneration on cognitive decline in clinically normal individuals. JAMA Neurol 2014;71(11):1379. https://doi.org/10.1001/jamaneurol.2014.2031.

[156] Wirth M, Villeneuve S, Haase CM, Madison CM, Oh H, Landau SM, Rabinovici GD, Jagust WJ. Associations between Alzheimer disease biomarkers, neurodegeneration, and cognition in cognitively normal older people. JAMA Neurol 2013. https://doi.org/10.1001/jamaneurol.2013.4013.

[157] Roberts BR, Lind M, Wagen AZ, Rembach A, Frugier T, Li Q-X, Ryan TM, McLean CA, Doecke JD, Rowe CC, Villemagne VL, Masters CL. Biochemically-defined pools of amyloid-β in sporadic Alzheimer's disease: correlation with amyloid PET. Brain 2017;140(5):1486–98. https://doi.org/10.1093/brain/awx057.

[158] Mattsson N, Insel PS, Donohue M, Landau S, Jagust WJ, Shaw LM, Trojanowski JQ, Zetterberg H, Blennow K, Weiner MW. Independent information from cerebrospinal fluid amyloid-β and florbetapir imaging in Alzheimer's disease. Brain 2015;138(3):772–83. https://doi.org/10.1093/brain/awu367.

[159] Palmqvist S, Mattsson N, Hansson O, for the Alzheimer's Disease Neuroimaging Initiative. Cerebrospinal fluid analysis detects cerebral amyloid-β accumulation earlier than positron emission tomography. Brain 2016;139(4):1226–36. https://doi.org/10.1093/brain/aww015.

[160] Schindler SE, Gray JD, Gordon BA, Xiong C, Batrla-Utermann R, Quan M, Wahl S, Benzinger TLS, Holtzman DM, Morris JC, Fagan AM. Cerebrospinal fluid biomarkers measured by Elecsys assays compared to amyloid imaging. Alzheimers Dement 2018;14(11):1460–9. https://doi.org/10.1016/j.jalz.2018.01.013.

[161] Johnson KA, Minoshima S, Bohnen NI, Donohoe KJ, Foster NL, Herscovitch P, Karlawish JH, Rowe CC, Carrillo MC, Hartley DM, Hedrick S, Pappas V, Thies WH. Appropriate use criteria for amyloid PET: a report of the amyloid imaging task force, the society of nuclear medicine and molecular imaging, and the Alzheimer's association. Alzheimers Dement 2013;9(1). https://doi.org/10.1016/j.jalz.2013.01.002.

[162] Rabinovici GD, Jagust WJ, Furst AJ, Ogar JM, Racine CA, Mormino EC, O'Neil JP, Lal RA, Dronkers NF, Miller BL, Gorno-Tempini ML. Aβ amyloid and glucose

metabolism in three variants of primary progressive aphasia. Ann Neurol 2008;64(4):388−401. https://doi.org/10.1002/ana.21451.

[163] Kato T, Inui Y, Nakamura A, Ito K. Brain fluorodeoxyglucose (FDG) PET in dementia. Ageing Res Rev 2016;30:73−84. https://doi.org/10.1016/j.arr.2016.02.003.

[164] Cerami C, Della Rosa PA, Magnani G, Santangelo R, Marcone A, Cappa SF, Perani D. Brain metabolic maps in Mild Cognitive Impairment predict heterogeneity of progression to dementia. Neuroimage Clin 2015;7:187−94. https://doi.org/10.1016/j.nicl.2014.12.004.

[165] Caminiti SP, Ballarini T, Sala A, Cerami C, Presotto L, Santangelo R, Fallanca F, Vanoli EG, Gianolli L, Iannaccone S, Magnani G, Perani D, Parnetti L, Eusebi P, Frisoni G, Nobili F, Picco A, Scarpini E. FDG-PET and CSF biomarker accuracy in prediction of conversion to different dementias in a large multicentre MCI cohort. Neuroimage Clin 2018;18:167−77. https://doi.org/10.1016/j.nicl.2018.01.019.

[166] Perani D, Cerami C, Caminiti SP, Santangelo R, Coppi E, Ferrari L, Pinto P, Passerini G, Falini A, Iannaccone S, Cappa SF, Comi G, Gianolli L, Magnani G. Cross-validation of biomarkers for the early differential diagnosis and prognosis of dementia in a clinical setting. Eur J Nucl Med Mol Imag 2016;43(3):499−508. https://doi.org/10.1007/s00259-015-3170-y.

[167] Sala A, Caprioglio C, Santangelo R, Vanoli EG, Iannaccone S, Magnani G, Perani D. Brain metabolic signatures across the Alzheimer's disease spectrum. Eur J Nucl Med Mol Imag 2020;47(2):256−69. https://doi.org/10.1007/s00259-019-04559-2.

[168] Paquin V, Therriault J, Pascoal TA, Rosa-Neto P, Gauthier S. Frontal variant of Alzheimer disease differentiated from frontotemporal dementia using in vivo amyloid and tau imaging. Cognit Behav Neurol 2020;33(4):288−93. https://doi.org/10.1097/WNN.0000000000000251.

[169] Wolters EE, Dodich A, Boccardi M, Corre J, Drzezga A, Hansson O, Nordberg A, Frisoni GB, Garibotto V, Ossenkoppele R. Clinical validity of increased cortical uptake of [18F]flortaucipir on PET as a biomarker for Alzheimer's disease in the context of a structured 5-phase biomarker development framework. Eur J Nucl Med Mol Imag 2021;48(7):2097−109. https://doi.org/10.1007/s00259-020-05118-w.

[170] Singleton E, Hansson O, Pijnenburg YAL, La Joie R, Mantyh WG, Tideman P, Stomrud E, Leuzy A, Johansson M, Strandberg O, Smith R, Berendrecht E, Miller BL, Iaccarino L, Edwards L, Strom A, Wolters EE, Coomans E, Visser D, Ossenkoppele R. Heterogeneous distribution of tau pathology in the behavioural variant of Alzheimer's disease. J Neurol Neurosurg Psychiatr 2021;92(8):872−80. https://doi.org/10.1136/jnnp-2020-325497.

[171] Groot C, Smith R, Collij LE, Mastenbroek SE, Stomrud E, Binette AP, Hansson O. Tau positron emission tomography for predicting dementia in individuals with mild cognitive impairment. JAMA Neurol 2024. https://doi.org/10.1001/jamaneurol.2024.1612.

[172] Jessen F, Wolfsgruber S, Kleineindam L, Spottke A, Altenstein S, Bartels C, et al. Subjective cognitive decline and stage 2 of Alzheimer disease in patients from memory centers. Alzheimer's Dement 2023;19(2):487−97.

[173] Dubois B, Feldman HH, Jacova C, Cummings JL, Dekosky ST, Barberger-Gateau P, et al. Revising the definition of Alzheimer's disease: a new lexicon. Lancet Neurol 2010;9:1118−27.

[174] De Renzi E, Motti F, Nichelli P. Imitating gestures: a quantitative approach to ideomotor apraxia. Arch Neurol 1980;37:6−10.

[175] Smits LL, Flapper M, Sistermans N, et al. Apraxia in mild cognitive impairment and Alzheimer's disease: validity and reliability of the Van Heugten test for apraxia. Dement Geriatr Cogn Disord 2014;38:55−64.

Chapter 6

Novel approaches to diagnosis

Chapter 6a

Opportunities arising from the tech revolution

Giedrė Čepukaitytė and Dennis Chan
Institute of Cognitive Neuroscience, University College London, London, United Kingdom

Introduction

Digital devices, such as smartphones, smartwatches and home assistants, are used by the overwhelming majority of the population in high income countries, and a growing number of individuals in low and middle income countries [1]. Facilitated by wireless communications between personal devices and remote servers, these sensor-packed gadgets are becoming ever more embedded into everyday life, being used to organize day-to-day activities and track physical and physiological outputs alongside their entertainment and communication value. The resulting stream of multidimensional big data is transferred to large data centers and, through machine learning, provide the parent technology companies with insights into the behaviors of their technology users.

Given their capabilities and the worldwide extent of their usage, unsurprisingly there is growing interest in the use of these digital technologies for medical purposes and - in the context of Alzheimer's disease (AD) - to detect early pathology, track progression and response to interventions [2]. Such digital biomarkers, with their potential to measure multiple aspects of human behavior, can provide functional readouts of disease to complement biomarker-based detection of AD molecular pathology, in order to aid diagnostic evaluation. Crucially in terms of overcoming the limitations of current tests used in routine clinical practice, digital tests could provide greater diagnostic sensitivity and specificity than legacy pen-and-paper cognitive tests without the penalty of language, educational and cultural confounds that compromise the interpretation of these legacy tests. Furthermore, in probing everyday behaviors, digital measures deliver ecologically valid outcomes with relevance to real life activities.

The comparatively low price, wide commercial availability and user-oriented design of digital devices make them scalable and easy to use. The high-quality internal sensors present in these devices—such as the accelerometer, microphone and gyroscope—can be utilized to collect multidimensional datasets, potentially as continuous time series spanning hours or days [3], to which machine learning algorithms can be applied in order to extract additional features of disease that may be invisible to traditional analytic approaches applied to unidimensional datasets. Such measures can then be used to track behavioral changes associated with AD over time, either in isolation or in combination with other health big data, such as blood biomarkers [4], genetics, and information in the medical health records, with one ultimate aim being to uncover a digital "fingerprint" of disease to aid provision of personalized intervention plans [5].

The aim of this chapter is to outline how digital technologies, big data and machine learning could aid early detection of AD at population scale. We will start by introducing the two complementary approaches to digital assessment of behavior, active testing and passive sensing, and will outline how these approaches can be combined to enable digital phenotyping [5]. This will be followed by a discussion on the benefits and risks of big data and machine learning approaches. The chapter concludes with a discussion on the promise of personalized diagnostics and considerations on how current healthcare systems could be adapted to maximize the benefits and reduce the potential downsides of using digital technology in clinical practice.

Digital testing

Although it is typically decline in episodic memory that leads affected individuals and their carers to seek medical advice, many other functions are impaired as a result of early AD pathology. For example, increasing evidence indicates that spatial navigation is one of the first abilities impaired in AD, predating memory decline [6], consistent with the entorhinal cortex and hippocampus being cortical sites initially affected in AD [7] and the role of these regions in navigation [8]. Early deficits are also apparent in speech and language [9], attention [10], executive function [11] and processing speed [12]. Beyond these cognitive functions, anxiety, mood disturbance and changes in social behavior are observed from the earliest stages of AD [13,14]. Finally, preclinical and prodromal stages of AD are associated with disturbances in sleep [15] circadian rhythms [16], alterations in gait [17], pupillary responses [18] and eye movement patterns [19]. Despite this, these latter functions are not routinely evaluated within a clinical diagnostic assessment for early AD, and if tested, assessments will typically involve qualitative (clinical interview, physical examination) rather than quantitative evaluation, with associated issues of subjectivity and bias.

App-based active testing

Active testing probes functions of interest by asking participants to complete tasks following written or audio instructions provided by the app. The abundance of well-validated pen-and-paper and computer-based tests that can be converted to an app format is one reason behind the popularity of this approach, while another reason is the degree of experimental control permitted by the use of active testing, with its predetermined test design and outcomes. Examples of apps that focus primarily on active testing include the neotiv-App [20], Mezurio [21,22], Boston Remote Assessment for Neurocognitive Health BRANCH; [23], Altoida DNS [24,25], Sea Hero Quest [26,27] and Roche mobile app used in research on Parkinson's disease [28]. While some apps contain a single gamified task [26,27], others combine a variety of different tests [20–22], many of which have been shown to be sensitive to early pathological changes preceding neurodegeneration. In doing so, they may be of greater value for early AD diagnosis that digitized versions of legacy pen and paper tests, such as the Mini-Mental State Examination (MMSE) (for an overview, see Ref. [29]), which were historically designed as screening tests for dementia and have limited sensitivity and specificity for the earlier, pre-dementia, stages of AD [30].

In part to address the limitations of these legacy cognitive tests, many of which were developed before the era of cognitive neuroscience and the associated understanding of brain-behavior relationships, new approaches to active digital testing aim to leverage current knowledge about disease pathophysiology and the neural underpinnings of human behavior. For example, the application of spatial behavioral tests is informed by the knowledge that AD neurodegeneration is observed initially in medial temporal lobe regions such as the entorhinal cortex (EC) and hippocampus and that neurons within these regions, such as EC grid cells and hippocampal place cells, exhibit spatially-modulated firing activity [31,32]. Computer- and VR-based tests of allocentric spatial navigation and memory, designed to reflect EC and hippocampal function, have been found to have high sensitivity and specificity for prodromal AD [33,34] with spatial test performance also impaired in people at risk of AD prior to symptom onset [35,36]. Sea Hero Quest, a gamified smartphone-based navigation task, has been successfully deployed at a large scale across multiple nations and has been shown to distinguish carriers of apolipoprotein E (APOE) ε4 allele, the biggest genetic risk factor for sporadic AD [37], from non-carriers [27].

Another advantage of the active approach is the ability to increase the frequency of test application. Tests used in clinic require trained staff for their application and test frequency is limited by the (in)frequency of clinic appointments, typically with months between appointments. This external constraint on cognitive assessment limits evaluation of gradually progressive, non-discretized, diseases such as AD. It is likely that the intermittent nature of clinical assessments contributes to the variable performance of legacy pen-and-paper tests, such as MMSE, when predicting conversion to AD [38]. While infrequent testing is problematic, higher frequency application of legacy tests carries its own difficulties, since the practice effect associated with frequent testing results in artificially higher scores on repeat testing, leading to an underestimation of any progression in cognitive decline over time [39]. While this can be offset to some extent by the use of alternate versions of each test, these are not available for some of the commonest tests used in clinical practice, such as the MMSE. Careful design of active digital tasks, including a sufficient number of practice trials, automated changes in task versions and large stimuli databases, minimizes practice effects, enabling frequent longitudinal follow-ups [20–22]. As a result, active tests are much better at capturing changes occurring over time, with benefits for longitudinal monitoring of disease progression and treatment outcomes.

Passive sensing

In contrast to active testing, passive sensing involves capturing data arising from naturally occurring everyday activities or human-device interactions without asking participants to engage with the device in any specific way. Provided that the device is fit for purpose (i.e., it includes appropriate sensors, can be used in an intended way and is suitable for running the required app [40]); this approach is inherently low in burden for participants, a crucial consideration for acceptability, compliance and adherence. Apart from downloading and setting up the app, the user is simply required to carry the device around and perform everyday activities as normal. As a result, this approach enables high-frequency or even continuous collection of data over prolonged periods of time.

Consumer-grade digital devices in principle lend themselves well to the capture of such information via passive sensing of behaviors [2]. The global positioning system (GPS) and magnetometers, used in smartphones and smartwatches as an aid for planning routes or measuring distance travelled, can be employed to track real-world navigation and social participation [41,42]. The microphone, which enables voice calls and hands-free control of digital devices, is suitable for capturing changes in speech and language [9]. Tapping, typing, scrolling and zooming patterns, recorded using touch-sensitive screens on smartphones and tablets, may not only reveal deficits in fine motor control but also, by proxy, cognition [43,44]. Other sensors, such as accelerometers and gyroscopes, often present in smartphones and smartwatches to quantify mobility as a function of daily step count, can be used to derive other fine-grained measures of gait and balance [28,45]. Additional uses for activity data derived from accelerometers include proxy measures for sleep [15] and circadian disturbances [16]. Front and back facing cameras on digital devices can not only capture photographs and videos of increasingly higher quality, but may also be used to record eye movements and changes in pupil size [46]. Finally, the increasingly capacious internal storage on digital devices enables safekeeping of data until their transfer through network, Wi-Fi or Bluetooth connectivity to other digital devices and remote servers, where information can be stored and subsequent application of machine learning algorithms may extract additional clinically relevant signal.

Cross-cultural utility and ecological validity represent additional advantages of passive sensing. Unlike legacy cognitive tests, performance on which is heavily confounded by language, educational, and other factors [47,48], data resulting from passive recording of functions common across humanity such as spatial navigation, eye movements, gait and sleep, are largely free of such confounds and therefore less biased. Although undoubtedly there are variations in behavior depending on individual's age (e.g. Ref. [49]), sex and gender [50,51] level of expertise (e.g. Ref. [52]), environment (e.g. Refs. [53,54]), or cultural factors (e.g. Refs. [55—57]), passive assessment enables quantification of an individual's real-world functional decline from their own baseline, making these naturally-occurring variations less detrimental when it comes to data interpretation.

Passively captured data on real-world behaviors are generally resistant to practice effects and performance biases. By definition, everyday behaviors are over-practised, therefore the performance is not likely to improve over the measurement window. Passive measurement also offsets the Hawthorne effect, when the knowledge of being observed leads to changes in performance [58]. This is because volitional alterations in behavior, especially in the context of decline, are energetically costly and difficult to maintain over prolonged periods of time in naturally challenging real-world environments.

Despite these theoretical advantages, passive sensing is only beginning to emerge as a option for early detection of AD, especially using off-the-shelf technologies. Some successful examples include passive recording of geolocation, app usage and actigraphy data. Geolocation data have been passively recorded using GPS data loggers fitted into cars to assess navigation ability [59—61]. Measures derived from these data, such as the number of unique destinations visited and days traveled, could distinguish older adults with preclinical AD from healthy controls both cross-sectionally and longitudinally [59,61], and predict conversion to MCI or AD dementia at 45 months with a high area under the curve [60], showing promise for early detection. The Behapp smartphone app has been used to capture similar measures [42], suggesting that passive sensing of navigation is possible using accessible devices independently of the mode of transport. This latter study also captured communication app usage and found differences among individuals at different stages of cognitive impairment [42], providing a novel measure of social behavior for future validation.

Various aspects of sleep architecture, reduced sleep efficiency, reduced time in rapid eye movement sleep and alterations in circadian rhythms are associated with early AD [15,16,62]. Although formal sleep staging requires polysomnography, sleep and circadian disturbances can be detected using accelerometers, easily attached in a form of a wrist band which is cheaper, more convenient and more scalable. Changes in daily physical activity levels derived from actigraphy data were associated with AD biomarker presence in otherwise healthy middle-aged to older adults [16,63]. However, to assess fitness for purpose, further studies will be required to evaluate the diagnostic and predictive ability of these measures and to validate the performance of accelerometers against gold standard polysomnography measures.

Ultimately, a combination of both active and passive approaches, with their complementary strengths and weaknesses, is likely to have the greatest clinical value. Early disease detection models in particular would benefit from rich digital phenotyping collected using a variety of different devices and apps [5] though these advantages will have to be considered in terms of costs, such as user burden, device and data requirements and expense.

The increased burden on users may decrease compliance and result in discontinuation and loss of valuable longitudinal data required for disease detection and prediction models. A number of different types of user burden relevant for app-based testing have been identified, including access, emotional, mental, time and privacy [40], with active testing particularly problematic for all of these bar the privacy burden. Although careful design, such as the use of gamified tasks to aid user engagement and thus compliance and adherence to testing [64], respect for the cultural, educational and linguistic background of intended participants [40] and appropriate task difficulty, may minimize negative effects, there is a limit on how much time an individual can dedicate to active testing each day. While passive sensing is not associated with the same penalties such as the time burden, arguably the predominant drawback of this approach lies in the issue of privacy loss. This problem, alongside options for privacy preservation, is covered at length in a dedicated chapter in this book but is mentioned briefly here for illustrative purposes. The passively recording of everyday behaviors, including communication (be that face-to-face speech or remote digital communication via an electronic device) or geolocation, raises the risk of data breaches revealing potentially sensitive information, such as individual identifiers. While this may be offset by lowering sampling rates, data minimization may also inadvertently lead to less adaptive machine learning algorithms.

Beyond issues of user burden, passive sensing may also be associated with data-related complexities. Depending on the sampling rate and data type, high-frequency or continuous recording of data may generate large files that require copious amounts of storage space, both on devices and data servers. This may be both financially and, due to the high energy requirements of data servers, environmentally costly [65]. In addition, large datasets take time to transfer. Ideally, the downloading of data from devices to data servers should occur in the background to enable uninterrupted device use and through wireless rather than network connectivity, avoiding unexpected financial costs to participants [40] associated with data upload.

The pre-processing and analysis of passively recorded data are also challenging. Automated signal extraction tools are required that can cope with missing data and noise from dynamic real-world environments or unexpected patterns of use. For example, gait signal may depend on the walking surface, elevation and wind. Furthermore, some studies suggest that the location of the device in relation to the body also affects signal quality [66]. While detailed instructions and diaries may help to enhance the quality of data or account for confounds in the analysis, these will increase user burden [40]. To avoid this, it may be possible to derive the required information automatically from other sensors, with a downside of increasing data dimensionality. Finally, it should be noted that observed effects may differ in the real-world as compared to lab-generated data [54]. This means that a thorough validation of passive real-world measures against disease biomarkers and longitudinal development of AD is required, irrespective of whether lab-based studies have been carried out.

Regardless of the approach, as with any new tools, all measures derived from digital testing need to be validated against disease biomarkers and longitudinal clinical outcomes. Validation against amyloid and tau biomarkers helps ensure that digital measures correlate with AD molecular pathology and can be considered a readout of the effect of disease on brain function, while demonstration of ability to predict the future development of dementia delivers the outcome that is arguably the most relevant in clinical practice, augmenting the predictive ability of biomarkers in isolation.

For adoption into clinical practice, more work is required to demonstrate that digital measures are superior to currently used pen-and-paper tests in terms of diagnostic sensitivity and specificity as well as predictive ability. While the number of commercially available digital tests boasting early detection potential is rapidly increasing, few validation studies have compared the performance of digital and legacy tests, and these have only shown minor advantages (e.g. Ref. [67]), in turn indicating limited value both for healthcare systems and patients. Notably, this reflects limitations in the design of the tests evaluated and not digital tests in general. Given the emergence of anti-amyloid therapies [68,69], use of which is currently restricted in part by limited access to tests with high diagnostic sensitivity and specificity for early AD, even a modest advantage in accurately detecting the earliest disease stages - which in time can be enhanced through targeted design improvements - may bring significant benefits to those affected by AD.

Big data

The term big data refers to datasets that are too large and rich to be analyzed using traditional statistical approaches. They are increasingly collected and utilized in different areas, including logistics to optimize route planning and minimize fuel costs [70,71], in farming to predict and increase yields [72], and by big tech to personalize services and advertising. In medicine, preliminary evidence suggests that big data can help predict metabolic diseases before the onset of symptoms and enable mapping of pathophysiology to molecular mechanisms [73]. In the field of cancer research, where big data from a number of omics have flourished, there has been an explosion in novel research that may ultimately lead to scientific breakthroughs [74].

Given that sporadic AD has multiple etiologies, with disease outcomes driven by a complex interplay between genetics and the environment, there is a growing interest in the application of big data approaches to the detection of AD and prediction of disease progression [75,76], with different datasets potentially contributing in additive fashion. Health records, for example, include qualitative data from visits, medication use, hospitalizations, blood test results, health history, occupation, postcodes lived and lifestyle factors [77], and have added value in identifying emerging environmental risks, such as air pollution [78]. Digitization of medical records into accurate, searchable and secure electronic health records (EHRs) that follow a uniform standard will enable their sharing for research purposes. By comparison, data on fluid and imaging biomarkers, genetics (in particular, the APOE ε4 status) and neurophysiological test results [79,80], if not available through EHRs, are routinely collected as part of research studies. Research study data are becoming increasingly more accessible through large repositories, such as the Dementias Platform UK Data Portal [81], Image and Data Archive [82] and others. However, their linkage to other data, such as EHRs, is not always straightforward since, for identity protection, linkage files that connect data to individuals are deleted soon after the study completion date. Therefore, timely data sharing agreements are required to access these data. Finally, there is the emerging potential of digital datasets that may supplement clinical and research study assessments, sparsely sampled as a consequence of the frequency of clinic or research visits, with high-frequency or even continuous longitudinal data. This combination of EHRs, research and digital big data would enable implementation of machine learning algorithms to extract additional disease-related features that may not be uncovered from analyses of more limited datasets.

The first hurdle toward employing big data approaches is obtaining the data. Truly big data may require engaging with many sources, from EHRs, data repositories, research institutions and technology providers to individuals and their many digital devices to obtain both retrospective and prospective longitudinal data. This may involve lengthy applications and complex data sharing agreements between institutions to ensure data will be stored and processed in accordance to regulations, such as the General Data Protection Regulation (GDPR) in the EU and the UK [83], and with any pre-existing agreements between institutions and individuals (e.g., informed consent), as well as to define a priori the intellectual property rights. The GDPR, for instance, has enabled residents of the EU and UK to take ownership of their own data [83], which may enable data acquisition through donation by individuals [84]. Although this may be an excellent way to obtain data, particularly those collected using digital devices, it may raise unforeseen issues to the data donors down the line, such as the inability to control how their data are used.

Accessing digital data directly from the devices of consenting individuals is also not straightforward. Devices differ in hardware, operating systems and their release dates, and their application programming interfaces (APIs) [85]. Infrastructure that enables integration of data from multiple sensors across different devices and applications at the same time is required to make the process less laborious. One example of such infrastructure is RADAR-base, which also provides a dashboard to track the progress on data collection [86].

The data should also be prepared for downstream processing and analysis. This involves ensuring that data from multiple sources are consistently labelled and scored. For this, well-maintained data dictionaries and meta information are required. However, even then, significant time investment from data wranglers will be needed to standardize data across datasets. This would be made easier by employing standardized data structures for their storage, such as those already globally used for neuroimaging data (e.g., Brain Imaging Data Structure [BIDS]; [87]), in addition to creating standards for labeling common variables found in clinical research and employing them as the data are being collected. However, given that the movement toward shared standards is often voluntary, an incentive system is required to encourage this, potentially from bodies that fund research.

To achieve meaningful insights, individuals' data derived from multiple sources must be linked. For the purposes of data protection, this cannot be done using identifiable information, such as names, surnames or initials. Normally, research studies assign each participant with a unique code that enables identifier-free linkage of records. If participants agree to their data being openly shared, further steps are taken to anonymize and even suppress data to reduce the risk of identities being reverse engineered [88]. However, compiling retrospective big data from multiple sources requires linkage, which creates both logistical and privacy issues. In the future, this process could be made easier and safer through the creation of a global research ID [89], an identification code that enables access to data for health research purposes. This ID could be added to EHRs by default. If patients consented to their data being linked, it could also be included in research study and digital data, enabling smoother linkage in the future. The legislation governing the use of the global research ID would ensure that linked files are used appropriately.

Another consideration in relation to big data is their storage. Data servers are energy-demanding, with an increasing share of global energy consumption dedicated to their manufacturing, running, and cooling [65]. In order to detect

AD in preclinical stages, at which lifestyle and treatment interventions are likely to be most effective, data on approximately two billion of Earth's inhabitants over the age of 40 may have to be collected [90]. While currently not feasible, such population-wide approaches in the future would greatly increase the energy demands of big data storage, with associated negative implications for climate change, though these may be mitigated in part by strategies for increasing energy efficiency and data minimization [91,92].

Machine learning

The main purpose of machine learning algorithms is to automate laborious and time-consuming procedures, such as signal extraction, and find patterns in complex data. For example, with the help of machine learning, characteristics of passively recorded digital language, gait and eye movement data can be separated from noise for further processing. Machine learning algorithms also replace traditional statistical approaches in the analysis of big multidimensional datasets. This reduces the possibility of biases as well as type I error, whereby null hypothesis is rejected even though it is true [93]. A range of machine learning approaches have been developed that can be harnessed for early detection and prediction of AD.

Supervised machine learning involves training algorithms on human-labeled data. For example, if retrospective digital data associated with a definite AD diagnosis are used for training, the algorithm will learn the relationship between patterns in data and subsequent outcome of AD. Supervised learning includes regression (e.g., linear and logistic) and classification algorithms (e.g., support vector machine, decision tree, k-nearest neighbor and random forest) [94]. Random forest algorithms have been successfully utilized for speech and GPS data analysis, distinguishing preclinical cases of AD [59] and MCI patients [95] from healthy controls. Unsupervised learning, on the other hand, involves supplying the algorithm with unlabeled data and allowing it to group datapoints based on naturally occurring patterns. Examples of unsupervised learning are clustering (e.g., k-means and hierarchical) and dimensionality reduction algorithms (e.g., principle component analysis (PCA) and t-distributed stochastic neighbor embedding) [94]. Unsupervized algorithms provide an excellent means for data-driven analysis. In addition, without prior knowledge, they can detect relationships between predictors, enabling their grouping for model simplification.

Deep learning algorithms, such as convolutional neural networks (CNNs), are becoming increasingly popular for image, text and speech analysis. Deep learning can be both supervised and unsupervised and mimics human learning by using neural network architecture in addition to large datasets, achieving very high performance. So far, this powerful technique has been implemented successfully to identify prodromal AD from MRI scans, outperforming traditional CSF biomarkers [96]. However, the complexity of deep learning algorithms makes it difficult to identify exactly which information contributed to the observed outcomes. This can lead to a lack of trust of medical professionals in this approach, with deep learning algorithms being perceived as a "black box" whose workings and outputs are inaccessible and opaque [97] which may limit the future use of deep learning in healthcare settings.

The quality of machine learning outputs depends on how rigorously models are implemented. The first stage before model fitting is data pre-processing. While the exact procedures depend on data (e.g., sensor, imaging or neuropsychology test results) and the research question, pre-processing often involves feature extraction, normalization and dimensionality reduction steps. Decisions made at this stage affect model performance [98]. Particular care should be taken not to introduce biases, for example, by outlier removal [94]. Machine learning models are then constructed and trained. Overfitting, when the sample size is too low for the level of dimensionality of the dataset, may emerge as the main issue at this stage, resulting in models that do not generalize to test data [99]. Models trained on digital data, which are highly dimensional, may be at a particular risk of overfitting, making dimensionality reduction an important initial step. Additionally, to estimate model performance accurately, training data must be representative of those used for model testing [100]. Random splitting of data may not always be the best option, as real-world data are often non-random and vary systematically (e.g., over time; [100]), suggesting that careful matching of training and test data may yield better results. For real world scenarios, this translates to matching the sample used for model training to the general population in terms of demographic characteristics and disease severity. Otherwise, any models applied at population scale, especially in the context of AD detection, may risk increasing pre-existing health inequalities [101]. Notably, many of the cohorts with retrospective data that are used to train machine learning algorithms (e.g., Alzheimer's Disease Neuroimaging Initiative [102]) are composed of participants that are mostly white, highly educated and come from higher socio-economic status than the general population, which may pose problems to the generalizability of these algorithms to other ethnic and demographic groups. Finally, model performance, validity and generalizability are evaluated, with the best models selected using cross-validation. As machine learning algorithms are not error proof, careful checks must be carried out at this stage to see whether the algorithm has produced expected outcomes.

Ethical issues to consider if machine learning algorithms are to be used in the prediction and detection of AD. Although machine learning models have the capacity to outperform humans, in particular at early disease stages, factors such as the quality of data supplied for training and pre-processing steps taken, may affect their performance [94,98,100]. False prediction based on inadequate algorithms may have negative consequences on individuals' lives, including the aforementioned discrimination and stigma in case of a false AD diagnosis [103], as well as inadequate care and life arrangements in case of a false 'all clear' verdict. Steps should be taken to obtain training data that are representative of the population as a whole, which may mean reaching out and working closely with communities that are already underrepresented in the current healthcare system. Otherwise, machine learning approaches may reduce, rather than increase, access to early detection of AD [101] (Fig. 6.1).

Personalized diagnostics

An often-quoted objective of app-based testing, big data and machine learning in combination is the enablement of personalized diagnostics of AD. Digital phenotypes, with their theoretical ability to characterize healthy cognition and behavior, may permit detection of within-individual changes indicative of disease onset, while powerful well-validated machine learning algorithms trained on representative high-quality data could potentially extract features of disease that are specific to the individual. These findings could then be used by healthcare professionals to assign personalized interventions, such as individual-specific lifestyle changes or drug treatments, to maximize clinical benefit. Furthermore, machine learning approaches can be employed to track the effectiveness of interventions, finetuning the provision of personalized treatments over time.

To date, most projects aiming to derive big data on brain health using digital technologies have been limited to relatively small samples, in the order of hundreds (e.g., Refs. [3,104]). This is understandable given that the volume of data produced by these studies is already extremely large, requiring substantial investment in terms of time and money to organize data collection, storage, processing and mining. For future implementation in healthcare, these studies will need replication in populations with sample sizes that are larger by several orders of magnitude.

While health big data approaches are far from being routinely implemented, many governments view utilization of big data for personalized diagnostics as a strategic priority [105–107]. This is encouraging, as it will ultimately lead to funding and implementation of secure data environments, and integration of digital technologies, omics and machine learning, though without a shared incentive, this may be harder to achieve in countries with private healthcare systems [89].

Further work is also needed to tackle health inequalities, in order for personalized diagnostics to benefit the entire population. The generalizability of machine learning outputs relies on representative training data [89,100]. Health inequalities are widespread, even in G7 countries with universal healthcare, such as the UK where individuals from ethnic minorities and deprived areas have poorer health outcomes (data for England; [108]). Furthermore, in Europe alone, a scoping review commissioned by the World Health Organization has identified many segments of the population that may be left behind by digital medicine because of lower technology use, which

Opportunities arising from the tech revolution Chapter | 6a | 111

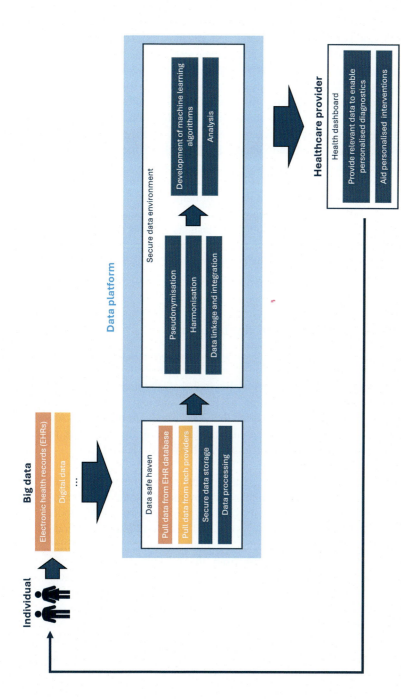

FIGURE 6.1 **Data flow from individuals to healthcare providers.** Big data, such as electronic health records (EHRs) and digital device data, will have to pass through a data platform in order to be useful for personalized diagnostics and benefit individuals through interventions. Such data platforms should enable pulling of data from various sources, including EHR databases and technology providers, and storing these data in a secure fashion. Furthermore, secure data environments should have the capability of linking and harmonizing pseudonymized data from different sources, preparing it for machine learning algorithms and analysis. Insights derived from the analysis tools can then be delivered to healthcare providers though easily interpretable dashboards to help inform and personalize interventions.

include people that are older, live in rural areas or come from lower socio-economic backgrounds [109]. Therefore, it may be that certain parts of the society would not benefit from personalized diagnostics, because disease detection models would not work as well for them due to their exclusion from training data and because they would not be able to take part in health assessments due to their lower technology use, especially if personal devices are required [2]. This is far from being the intention, and frameworks such as CLEARS (C—culture, L—limiting conditions, E—education, A—age, R—residence, S—socioeconomic status), which aim to identify and mitigate digital exclusion in underserved individuals [110], should be incorporated into future healthcare strategies alongside other measures to tackle health inequalities in order for personalized diagnostics to serve the entire population.

Finally, there are also concerns that instead of helping, big data approaches will increase the burden on the already overloaded healthcare systems by increasing the number of unneeded follow-up tests as a result of type I error [111]. The only solution to this problem is to design the system in a way that it would allow to estimate with great objectivity the need for follow-up assessments, and an objective "all clear for now" maybe reduce rather than increase the demand on clinicians' time.

Risks associated with usage of digital tools

The earlier section has already touched on ethical risks associated with digital diagnostics and the critical importance of addressing potential inequities in access to technology-based clinical tools. Beyond that, there are also ethical considerations related to the acquisition by passive sensing of real life behaviours such as sleep and social interaction. The implications for privacy preservation, and computer science-based strategies for mitigating risks to privacy, are covered in depth in the chapter by Ghosh and Mascolo.

The relentless churn of advances in hardware and software creates an entirely different, but similarly grave, risk in terms of obsolescence. Digital diagnostics based around devices manufactured by multinational tech corporations are vulnerable to upgrades that may render previously operational tools redundant or unusable. By comparison, approaches based on hardware or code from smaller scale providers, such as small/medium sized commercial entities or academic spinouts, will not benefit from the infrastructure for device maintenance or customer support that would be needed for large scale clinical deployment. To some extent such risks can be mitigated by basing any digital approaches around invariants such as the behaviours to be measured, and not the devices used for the measurement.

One final issue is the risk that a next generation digital approach is not effectively implemented into clinical services that at present are based around analogue legacy practices such as pen-and-paper cognitive testing and qualitative neurological examinations. To ensure that future digital diagnostics are fit for purpose, consideration needs to be given *ab initio* to their implementation within a framework such as that based on systems engineering principles that designs the new clinical service around the technologies and their clinical use cases as well as the differing needs of the various stakeholders from patients and clinicians through to service commissioners. Implementation using systems principles helps ensure that future technologies are deployed to maximal benefit for those concerned while the incorporation of risk management strategies and the capacity for iterative improvement—for instance arising from tech advances or breakthroughs in understanding of disease—helps ensure digital approaches retain their fitness for purpose into the future.

Conclusions

App-based testing, big data and machine learning have considerable potential to improve early AD detection and management, and used together may help achieve the highly desirable goal of personalized diagnosis and treatment, for maximum clinical benefit. However, for any future digitally-based service to be deployed to maximum advantage, major challenges will need to be addressed in terms of ensuring data quality and infrastructure, redesign of health systems and ethical considerations including privacy preservation and avoidance of health inequality.

References

[1] World Bank. Digital progress and trends report 2023. The World Bank; 2024. https://doi.org/10.1596/978-1-4648-2049-6.

[2] Kourtis LC, Regele OB, Wright JM, Jones GB. Digital biomarkers for Alzheimer's disease: the mobile/wearable devices opportunity. Npj Digital Med 2019;2:1–9. https://doi.org/10.1038/s41746-019-0084-2.

[3] Muurling M, de Boer C, Kozak R, Religa D, Koychev I, Verheij H, Nies VJM, Duyndam A, Sood M, Froehlich H, Hannesdottir K, Erdemli G, Lucivero F, Lancaster C, Hinds C, Stravopoulos TG, Nikolopoulos S, Kompatsiaris I, Manyakov NV, Owens AP, Narayan VA, Aarsland D, Visser PJ. Remote monitoring technologies in Alzheimer's disease: design of the RADAR-AD study. Alzheimer's Res Ther 2021;13:89. https://doi.org/10.1186/s13195-021-00825-4.

[4] Planche V, Bouteloup V, Pellegrin I, Mangin J-F, Dubois B, Ousset P-J, Pasquier F, Blanc F, Paquet C, Hanon O, Bennys K, Ceccaldi M, Annweiler C, Krolak-Salmon P, Godefroy O, Wallon D, Sauvee M, Boutoleau-Bretonnière C, Bourdel-Marchasson I, Jalenques I, Chene G, Dufouil C, the MEMENTO Study Group. Validity and performance of blood biomarkers for Alzheimer disease to predict dementia risk in a large clinic-based cohort. Neurology 2023;100:e473–84. https://doi.org/10.1212/WNL.0000000000201479.

[5] Insel TR. Digital phenotyping: technology for a new science of behavior. JAMA 2017;318:1215. https://doi.org/10.1001/jama.2017.11295.

[6] Coughlan G, Laczó J, Hort J, Minihane A-M, Hornberger M. Spatial navigation deficits — overlooked cognitive marker for preclinical Alzheimer disease? Nat Rev Neurol 2018;14:496−506. https://doi.org/10.1038/s41582-018-0031-x.

[7] Braak H, Del Tredici K. The preclinical phase of the pathological process underlying sporadic Alzheimer's disease. Brain 2015;138:2814−33. https://doi.org/10.1093/brain/awv236.

[8] Moser EI, Kropff E, Moser M-B. Place cells, grid cells, and the brain's spatial representation system. Annu Rev Neurosci 2008;31:69−89. https://doi.org/10.1146/annurev.neuro.31.061307.090723.

[9] Verfaillie SCJ, Witteman J, Slot RER, Pruis IJ, Vermaat LEW, Prins ND, Schiller NO, Van De Wiel M, Scheltens P, Van Berckel BNM, Van Der Flier WM, Sikkes SAM. High amyloid burden is associated with fewer specific words during spontaneous speech in individuals with subjective cognitive decline. Neuropsychologia 2019;131:184−92. https://doi.org/10.1016/j.neuropsychologia.2019.05.006.

[10] Sorg C, Myers N, Redel P, Bublak P, Riedl V, Manoliu A, Perneczky R, Grimmer T, Kurz A, Förstl H, Drzezga A, Müller HJ, Wohlschläger AM, Finke K. Asymmetric loss of parietal activity causes spatial bias in prodromal and mild Alzheimer's disease. Biol Psychiatr 2012;71:798−804. https://doi.org/10.1016/j.biopsych.2011.09.027.

[11] Ho JK, Nation DA, for the Alzheimers Disease Neuroimaging Initiative. Neuropsychological profiles and trajectories in preclinical Alzheimer's disease. J Int Neuropsychol Soc 2018;24:693−702. https://doi.org/10.1017/S135561771800022X.

[12] Duke Han S, Nguyen CP, Stricker NH, Nation DA. Detectable neuropsychological differences in early preclinical Alzheimer's disease: a meta-analysis. Neuropsychol Rev 2017;27:305−25. https://doi.org/10.1007/s11065-017-9345-5.

[13] Dafsari FS, Jessen F. Depression—an underrecognized target for prevention of dementia in Alzheimer's disease. Transl Psychiatry 2020;10:160. https://doi.org/10.1038/s41398-020-0839-1.

[14] Sommerlad A, Sabia S, Singh-Manoux A, Lewis G, Livingston G. Association of social contact with dementia and cognition: 28-year follow-up of the Whitehall II cohort study. PLoS Med 2019;16:e1002862. https://doi.org/10.1371/journal.pmed.1002862.

[15] Peter-Derex L, Yammine P, Bastuji H, Croisile B. Sleep and Alzheimer's disease. Sleep Med Rev 2015;19:29−38. https://doi.org/10.1016/j.smrv.2014.03.007.

[16] Spira AP, Zipunnikov V, Raman R, Choi J, Di J, Bai J, Carlsson CM, Mintzer JE, Marshall GA, Porsteinsson AP, Yaari R, Wanigatunga SK, Kim J, Wu MN, Aisen PS, Sperling RA, Rosenberg PB. Brain amyloid burden, sleep, and 24-hour rest/activity rhythms: screening findings from the anti-amyloid treatment in asymptomatic Alzheimer's and longitudinal evaluation of amyloid risk and neurodegeneration studies. Sleep Adv 2021;2:zpab015. https://doi.org/10.1093/sleepadvances/zpab015.

[17] Tian Q, Bair W-N, Resnick SM, Bilgel M, Wong DF, Studenski SA. beta-amyloid deposition is associated with gait variability in usual aging. Gait Posture 2018;61:346−52. https://doi.org/10.1016/j.gaitpost.2018.02.002.

[18] Frost S, Kanagasingam Y, Sohrabi H, Taddei K, Bateman R, Morris J, Benzinger T, Goate A, Masters C, Martins R. Pupil response biomarkers distinguish amyloid precursor protein mutation carriers from non-carriers. CAR 2013;10:790−6. https://doi.org/10.2174/15672050113109990154.

[19] Holden JG, Cosnard A, Laurens B, Asselineau J, Biotti D, Cubizolle S, Dupouy S, Formaglio M, Koric L, Seassau M, Tilikete C, Vighetto A, Tison F. Prodromal Alzheimer's

disease demonstrates increased errors at a simple and automated anti-saccade task. JAD 2018;65:1209−23. https://doi.org/10.3233/JAD-180082.

[20] Öhman F, Berron D, Papp KV, Kern S, Skoog J, Hadarsson Bodin T, Zettergren A, Skoog I, Schöll M. Unsupervised mobile app-based cognitive testing in a population-based study of older adults born 1944. Front Digit Health 2022;4:933265. https://doi.org/10.3389/fdgth.2022.933265.

[21] Lancaster C, Koychev I, Blane J, Chinner A, Wolters L, Hinds C. Evaluating the feasibility of frequent cognitive assessment using the Mezurio smartphone app: observational and interview study in adults with elevated dementia risk. JMIR Mhealth Uhealth 2020b;8:e16142. https://doi.org/10.2196/16142.

[22] Lancaster C, Koychev I, Blane J, Chinner A, Chatham C, Taylor K, Hinds C. Gallery Game: smartphone-based assessment of long-term memory in adults at risk of Alzheimer's disease. J Clin Exp Neuropsychol 2020a;42:329−43. https://doi.org/10.1080/13803395.2020.1714551.

[23] Papp KV, Samaroo A, Chou H, Buckley R, Schneider OR, Hsieh S, Soberanes D, Quiroz Y, Properzi M, Schultz A, García-Magariño I, Marshall GA, Burke JG, Kumar R, Snyder N, Johnson K, Rentz DM, Sperling RA, Amariglio RE. Unsupervised mobile cognitive testing for use in preclinical Alzheimer's disease. Alzheimers Dement 2021;13. https://doi.org/10.1002/dad2.12243.

[24] Buegler M, Harms RL, Balasa M, Meier IB, Exarchos T, Rai L, Boyle R, Tort A, Kozori M, Lazarou E, Rampini M, Cavaliere C, Vlamos P, Tsolaki M, Babiloni C, Soricelli A, Frisoni G, Sanchez-Valle R, Whelan R, Merlo-Pich E, Tarnanas I. Digital biomarker-based individualized prognosis for people at risk of dementia. Alzheimer's Dement 2020;12. https://doi.org/10.1002/dad2.12073.

[25] Seixas AA, Rajabli F, Pericak-Vance MA, Jean-Louis G, Harms RL, Tarnanas I. Associations of digital neuro-signatures with molecular and neuroimaging measures of brain resilience: the altoida large cohort study. Front Psychiatr 2022;13:899080. https://doi.org/10.3389/fpsyt.2022.899080.

[26] Coughlan G, Puthusseryppady V, Lowry E, Gillings R, Spiers H, Minihane A-M, Hornberger M. Test-retest reliability of spatial navigation in adults at-risk of Alzheimer's disease. PLoS One 2020;15:e0239077. https://doi.org/10.1371/journal.pone.0239077.

[27] Coughlan G, Coutrot A, Khondoker M, Minihane AM, Spiers H, Hornberger M. Toward personalized cognitive diagnostics of at-genetic-risk Alzheimer's disease. Proc Natl Acad Sci USA 2019;116:9285−92. https://doi.org/10.1073/pnas.1901600116.

[28] Lipsmeier F, Taylor KI, Postuma RB, Volkova-Volkmar E, Kilchenmann T, Mollenhauer B, Bamdadian A, Popp WL, Cheng W-Y, Zhang Y-P, Wolf D, Schjodt-Eriksen J, Boulay A, Svoboda H, Zago W, Pagano G, Lindemann M. Reliability and validity of the Roche PD Mobile Application for remote monitoring of early Parkinson's disease. Sci Rep 2022;12:12081. https://doi.org/10.1038/s41598-022-15874-4.

[29] Thabtah F, Peebles D, Retzler J, Hathurusingha C. Dementia medical screening using mobile applications: a systematic review with a new mapping model. J Biomed Inf 2020;111:103573. https://doi.org/10.1016/j.jbi.2020.103573.

[30] Arevalo-Rodriguez I, Smailagic N, Roqué-Figuls M, Ciapponi A, Sanchez-Perez E, Giannakou A, Pedraza OL, Bonfill Cosp X, Cullum S. Mini-Mental State Examination (MMSE) for the early detection of dementia in people with mild cognitive impairment (MCI). Cochrane Database Syst Rev 2021;2021. https://doi.org/10.1002/14651858.CD010783.pub3.

[31] Hafting T, Fyhn M, Molden S, Moser M-B, Moser EI. Microstructure of a spatial map in the entorhinal cortex. Nature 2005;436:801−6. https://doi.org/10.1038/nature03721.

[32] O'Keefe J, Dostrovsky J. The hippocampus as a spatial map. Preliminary evidence from unit activity in the freely-moving rat. Brain Res 1971;34:171−5. https://doi.org/10.1016/0006-8993(71)90358-1.

[33] Howett D, Castegnaro A, Krzywicka K, Hagman J, Marchment D, Henson R, Rio M, King JA, Burgess N, Chan D. Differentiation of mild cognitive impairment using an entorhinal cortex-based test of virtual reality navigation. Brain 2019;142:1751−66. https://doi.org/10.1093/brain/awz116.

[34] Moodley K, Minati L, Contarino V, Prioni S, Wood R, Cooper R, D'Incerti L, Tagliavini F, Chan D. Diagnostic differentiation of mild cognitive impairment due to Alzheimer's disease using a hippocampus-dependent test of spatial memory: diagnostic Differentiation of Mild Cognitive Impairment. Hippocampus 2015;25:939−51. https://doi.org/10.1002/hipo.22417.

[35] Newton C, Pope M, Rua C, Henson R, Ji Z, Burgess N, Rodgers CT, Stangl M, Dounavi M, Castegnaro A, Koychev I, Malhotra P, Wolbers T, Ritchie K, Ritchie CW, O'Brien J, Su L, Chan D, for the PREVENT Dementia Research Programme. Entorhinal-based path integration selectively predicts midlife risk of Alzheimer's disease. Alzheimer's Dement 2024;13733. https://doi.org/10.1002/alz.13733.

[36] Ritchie K, Carrière I, Howett D, Su L, Hornberger M, O'Brien JT, Ritchie CW, Chan D. Allocentric and egocentric spatial processing in middle-aged adults at high risk of late-onset Alzheimer's disease: the PREVENT dementia study. JAD 2018;65:885−96. https://doi.org/10.3233/JAD-180432.

[37] Farrer LA. Effects of age, sex, and ethnicity on the association between apolipoprotein E genotype and Alzheimer disease. JAMA 1997;278:1349. https://doi.org/10.1001/jama.1997.03550160069041.

[38] Chen Y, Qian X, Zhang Y, Su W, Huang Y, Wang X, Chen X, Zhao E, Han L, Ma Y. Prediction models for conversion from mild cognitive impairment to Alzheimer's disease: a systematic review and meta-analysis. Front Aging Neurosci 2022;14:840386. https://doi.org/10.3389/fnagi.2022.840386.

[39] Goldberg TE, Harvey PD, Wesnes KA, Snyder PJ, Schneider LS. Practice effects due to serial cognitive assessment: implications for preclinical Alzheimer's disease randomized controlled trials. Alzheimer's Dement 2015;1:103−11. https://doi.org/10.1016/j.dadm.2014.11.003.

[40] Suh H, Shahriaree N, Hekler EB, Kientz JA. Developing and validating the user burden scale: a tool for assessing user burden in computing systems. In: Proceedings of the 2016 CHI Conference on Human Factors in Computing Systems. Presented at the CHI'16: CHI Conference on Human Factors in Computing Systems. ACM, San Jose California USA; 2016. p. 3988−99. https://doi.org/10.1145/2858036.2858448.

[41] Ghosh A, Puthusseryppady V, Chan D, Mascolo C, Hornberger M. Machine learning detects altered spatial navigation features in outdoor behaviour of Alzheimer's disease patients. Sci Rep 2022;12:3160. https://doi.org/10.1038/s41598-022-06899-w.

[42] Muurling M, Reus LM, de Boer C, Wessels SC, Jagesar RR, Vorstman JAS, Kas MJH, Visser PJ. Assessment of social behavior using a passive monitoring app in cognitively normal and cognitively impaired older adults: observational study. JMIR Aging 2022;5:e33856. https://doi.org/10.2196/33856.

[43] Lang C, Gries C, Lindenberg KS, Lewerenz J, Uhl S, Olsson C, Samzelius J, Landwehrmeyer GB. Monitoring the motor phenotype in Huntington's disease by analysis

of keyboard typing during real life computer use. JHD 2021;10:259—68. https://doi.org/10.3233/JHD-200451.

[44] Meulemans C, Leijten M, Van Waes L, Engelborghs S, De Maeyer S. Cognitive writing process characteristics in Alzheimer's disease. Front Psychol 2022;13:872280. https://doi.org/10.3389/fpsyg.2022.872280.

[45] Arora S, Baig F, Lo C, Barber TR, Lawton MA, Zhan A, Rolinski M, Ruffmann C, Klein JC, Rumbold J, Louvel A, Zaiwalla Z, Lennox G, Quinnell T, Dennis G, Wade-Martins R, Ben-Shlomo Y, Little MA, Hu MT. Smartphone motor testing to distinguish idiopathic REM sleep behavior disorder, controls, and PD. Neurology 2018;91:1528—38. https://doi.org/10.1212/WNL.0000000000006366.

[46] Valliappan N, Dai N, Steinberg E, He J, Rogers K, Ramachandran V, Xu P, Shojaeizadeh M, Guo L, Kohlhoff K, Navalpakkam V. Accelerating eye movement research via accurate and affordable smartphone eye tracking. Nat Commun 2020;11:4553. https://doi.org/10.1038/s41467-020-18360-5.

[47] Rosselli M, Uribe IV, Ahne E, Shihadeh L. Culture, ethnicity, and level of education in Alzheimer's disease. Neurotherapeutics 2022;19:26—54. https://doi.org/10.1007/s13311-022-01193-z.

[48] Sbordone RJ, Purisch AD. Hazards of blind analysis of neuropsychological test data in assessing cognitive disability: the role of confounding factors. NRE 1996;7:15—26. https://doi.org/10.3233/NRE-1996-7103.

[49] Zadik S, Benady A, Gutwillig S, Florentine MM, Solymani RE, Plotnik M. Age related changes in gait variability, asymmetry, and bilateral coordination — when does deterioration starts? Gait Posture 2022;96:87—92. https://doi.org/10.1016/j.gaitpost.2022.05.009.

[50] Ko S, Tolea MI, Hausdorff JM, Ferrucci L. Sex-specific differences in gait patterns of healthy older adults: results from the Baltimore Longitudinal Study of Aging. J Biomech 2011;44:1974—9. https://doi.org/10.1016/j.jbiomech.2011.05.005.

[51] Nazareth A, Huang X, Voyer D, Newcombe N. A meta-analysis of sex differences in human navigation skills. Psychon Bull Rev 2019;26:1503—28. https://doi.org/10.3758/s13423-019-01633-6.

[52] Maguire EA, Gadian DG, Johnsrude IS, Good CD, Ashburner J, Frackowiak RSJ, Frith CD. Navigation-related structural change in the hippocampi of taxi drivers. Proc Natl Acad Sci USA 2000;97:4398—403. https://doi.org/10.1073/pnas.070039597.

[53] Coutrot A, Manley E, Goodroe S, Gahnstrom C, Filomena G, Yesiltepe D, Dalton RC, Wiener JM, Hölscher C, Hornberger M, Spiers HJ. Entropy of city street networks linked to future spatial navigation ability. Nature 2022;604:104—10. https://doi.org/10.1038/s41586-022-04486-7.

[54] Mc Ardle R, Del Din S, Donaghy P, Galna B, Thomas AJ, Rochester L. The impact of environment on gait assessment: considerations from real-world gait analysis in dementia subtypes. Sensors 2021;21:813. https://doi.org/10.3390/s21030813.

[55] Barone TL. Is the siesta an adaptation to disease?: a cross-cultural examination. Hum Nat 2000;11:233—58. https://doi.org/10.1007/s12110-000-1012-4.

[56] Ekirch AR. Segmented sleep in preindustrial societies. Sleep 2016;39:715—6. https://doi.org/10.5665/sleep.5558.

[57] Worthman CM. Developmental cultural ecology of sleep. In: Sleep and development: familial and socio-cultural considerations. Oxford University Press; 2011.

[58] Ardestani MM, Hornby TG. Effect of investigator observation on gait parameters in individuals with stroke. J Biomech 2020;100:109602. https://doi.org/10.1016/j.jbiomech.2020.109602.

[59] Bayat S, Babulal GM, Schindler SE, Fagan AM, Morris JC, Mihailidis A, Roe CM. GPS driving: a digital biomarker for preclinical Alzheimer disease. Alz Res Therapy 2021;13:115. https://doi.org/10.1186/s13195-021-00852-1.

[60] Di X, Shi R, DiGuiseppi C, Eby DW, Hill LL, Mielenz TJ, Molnar LJ, Strogatz D, Andrews HF, Goldberg TE, Lang BH, Kim M, Li G. Using naturalistic driving data to predict mild cognitive impairment and dementia: preliminary findings from the longitudinal research on aging drivers (LongROAD) study. Geriatrics 2021;6:45. https://doi.org/10.3390/geriatrics6020045.

[61] Roe CM, Stout SH, Rajasekar G, Ances BM, Jones JM, Head D, Benzinger TLS, Williams MM, Davis JD, Ott BR, Warren DK, Babulal GM. A 2.5-year longitudinal assessment of naturalistic driving in preclinical Alzheimer's disease. JAD 2019;68:1625–33. https://doi.org/10.3233/JAD-181242.

[62] Lucey BP, Wisch J, Boerwinkle AH, Landsness EC, Toedebusch CD, McLeland JS, Butt OH, Hassenstab J, Morris JC, Ances BM, Holtzman DM. Sleep and longitudinal cognitive performance in preclinical and early symptomatic Alzheimer's disease. Brain 2021;144:2852–62. https://doi.org/10.1093/brain/awab272.

[63] Gao L, Li P, Gaba A, Musiek E, Ju YS, Hu K. Fractal motor activity regulation and sex differences in preclinical Alzheimer's disease pathology. Alzheimers Dement 2021;13. https://doi.org/10.1002/dad2.12211.

[64] Lumsden J, Edwards EA, Lawrence NS, Coyle D, Munafò MR. Gamification of cognitive assessment and cognitive training: a systematic review of applications and efficacy. JMIR Serious Games 2016;4:e11. https://doi.org/10.2196/games.5888.

[65] IEA, 2017. Digitalization and energy. Paris.

[66] Kuntapun J, Silsupadol P, Kamnardsiri T, Lugade V. Smartphone monitoring of gait and balance during irregular surface walking and obstacle crossing. Front Sports Act Living 2020;2:560577. https://doi.org/10.3389/fspor.2020.560577.

[67] Muurling M, De Boer C, Vairavan S, Harms RL, Chadha AS, Tarnanas I, Luis EV, Religa D, Gjestsen MT, Galluzzi S, Ibarria Sala M, Koychev I, Hausner L, Gkioka M, Aarsland D, Visser PJ, Brem A-K. Augmented reality versus standard tests to assess cognition and function in early Alzheimer's disease. Npj Digit Med 2023;6:234. https://doi.org/10.1038/s41746-023-00978-6.

[68] Sims JR, Zimmer JA, Evans CD, Lu M, Ardayfio P, Sparks J, Wessels AM, Shcherbinin S, Wang H, Monkul Nery ES, Collins EC, Solomon P, Salloway S, Apostolova LG, Hansson O, Ritchie C, Brooks DA, Mintun M, Skovronsky DM, TRAILBLAZER-ALZ 2 Investigators, Abreu R, Agarwal P, Aggarwal P, Agronin M, Allen A, Altamirano D, Alva G, Andersen J, Anderson A, Anderson D, Arnold J, Asada T, Aso Y, Atit V, Ayala R, Badruddoja M, Badzio-jagiello H, Bajacek M, Barton D, Bear D, Benjamin S, Bergeron R, Bhatia P, Black S, Block A, Bolouri M, Bond W, Bouthillier J, Brangman S, Brew B, Brisbin S, Brisken T, Brodtmann A, Brody M, Brosch J, Brown C, Brownstone P, Bukowczan S, Burns J, Cabrera A, Capote H, Carrasco A, Cevallos Yepez J, Chavez E, Chertkow H, Chyrchel-paszkiewicz U, Ciabarra A, Clemmons E, Cohen D, Cohen R, Cohen I, Concha M, Costell B, Crimmins D, Cruz-pagan Y, Cueli A, Cupelo R, Czarnecki M, Darby D, Dautzenberg Pl j, De Deyn P, De La Gandara J, Deck K, Dibenedetto D, Dibuono M, Dinnerstein E, Dirican A, Dixit S, Dobryniewski J, Drake R, Drysdale P, Duara R, Duffy J, Ellenbogen A, Faradji V, Feinberg M, Feldman R, Fishman S, Flitman S, Forchetti C, Fraga I, Frank A, Frishberg B, Fujigasaki H, Fukase H, Fumero I, Furihata K, Galloway C, Gandhi R, George K, Germain M, Gitelman D, Goetsch N, Goldfarb D, Goldstein M, Goldstick L, Gonzalez Rojas Y, Goodman I,

Greeley D, Griffin C, Grigsby E, Grosz D, Hafner K, Hart D, Henein S, Herskowitz B, Higashi S, Higashi Y, Ho G, Hodgson J, Hohenberg M, Hollenbeck L, Holub R, Hori T, Hort J, Ilkowski J, Ingram KJ, Isaac M, Ishikawa M, Janu L, Johnston M, Julio W, Justiz W, Kaga T, Kakigi T, Kalafer M, Kamijo-M, Kaplan J, Karathanos M, Katayama S, Kaul S, Keegan A, Kerwin D, Khan U, Khan A, Kimura N, Kirk G, Klodowska G, Kowa H, Kutz C, Kwentus J, Lai R, Lall A, Lawrence M, Lee E, Leon R, Linker G, Lisewski P, Liss J, Liu C, Losk S, Lukaszyk E, Lynch J, Macfarlane S, Macsweeney J, Mannering N, Markovic O, Marks D, Masdeu J, Matsui Y, Matsuishi K, Mcallister P, Mcconnehey B, Mcelveen A, Mcgill L, Mecca A, Mega M, Mensah J, Mickielewicz A, Minaeian A, Mocherla B, Murphy C, Murphy P, Nagashima H, Nair A, Nair M, Nardandrea J, Nash M, Nasreddine Z, Nishida Y, Norton J, Nunez L, Ochiai J, Ohkubo T, Okamura Y, Okorie E, Olivera E, O'mahony J, Omidvar O, Ortiz-Cruz D, Osowa A, Papka M, Parker A, Patel P, Patel A, Patel M, Patry C, Peckham E, Pfeffer M, Pietras A, Plopper M, Porsteinsson A, Poulin Robitaille R, Prins N, Puente O, Ratajczak M, Rhee M, Ritter A, Rodriguez R, Rodriguez Ables L, Rojas J, Ross J, Royer P, Rubin J, Russell D, Rutgers SM, Rutrick S, Sadowski M, Safirstein B, Sagisaka T, Scharre D, Schneider L, Schreiber C, Schrift M, Schulz P, Schwartz H, Schwartzbard J, Scott J, Selem L, Sethi P, Sha S, Sharlin K, Sharma S, Shiovitz T, Shiwach R, Sladek M, Sloan B, Smith A, Solomon P, Sorial E, Sosa E, Stedman M, Steen S, Stein L, Stolyar A, Stoukides J, Sudoh S, Sutton J, Syed J, Szigeti K, Tachibana H, Takahashi Y, Tateno A, Taylor JD, Taylor K, Tcheremissine O, Thebaud A, Thein S, Thurman L, Toenjes S, Toji H, Toma M, Tran D, Trueba P, Tsujimoto M, Turner R, Uchiyama A, Ussorowska D, Vaishnavi S, Valor E, Vandersluis J, Vasquez A, Velez J, Verghese C, Vodickova-borzova K, Watson D, Weidman D, Weisman D, White A, Willingham K, Winkel I, Winner P, Winston J, Wolff A, Yagi H, Yamamoto H, Yathiraj S, Yoshiyama Y, Zboch M. Donanemab in early symptomatic Alzheimer disease: the TRAILBLAZER-ALZ 2 randomized clinical trial. JAMA 2023;330:512. https://doi.org/10.1001/jama.2023.13239.

[69] Van Dyck CH, Swanson CJ, Aisen P, Bateman RJ, Chen C, Gee M, Kanekiyo M, Li D, Reyderman L, Cohen S, Froelich L, Katayama S, Sabbagh M, Vellas B, Watson D, Dhadda S, Irizarry M, Kramer LD, Iwatsubo T. Lecanemab in early Alzheimer's disease. N Engl J Med 2023;388:9−21. https://doi.org/10.1056/NEJMoa2212948.

[70] Boyd B. Big data in maritime transport. Mewburn Ellis; 2023. URL, https://www.mewburn.com/news-insights/big-data-in-maritime-transport.

[71] Smith J, Hall S, Coombs G, Byrne J, Thorne M, Brearley A, Long D, Meredith M, Fox M. Autonomous passage planning for a polar vessel (preprint). Artificial Intell Robotics 2022. https://doi.org/10.31223/X5KP90.

[72] Osinga SA, Paudel D, Mouzakitis SA, Athanasiadis IN. Big data in agriculture: between opportunity and solution. Agric Syst 2022;195:103298. https://doi.org/10.1016/j.agsy.2021.103298.

[73] Schüssler-Fiorenza Rose SM, Contrepois K, Moneghetti KJ, Zhou W, Mishra T, Mataraso S, Dagan-Rosenfeld O, Ganz AB, Dunn J, Hornburg D, Rego S, Perelman D, Ahadi S, Sailani MR, Zhou Y, Leopold SR, Chen J, Ashland M, Christle JW, Avina M, Limcaoco P, Ruiz C, Tan M, Butte AJ, Weinstock GM, Slavich GM, Sodergren E, McLaughlin TL, Haddad F, Snyder MP. A longitudinal big data approach for precision health. Nat Med 2019;25:792−804. https://doi.org/10.1038/s41591-019-0414-6.

[74] Jiang P, Sinha S, Aldape K, Hannenhalli S, Sahinalp C, Ruppin E. Big data in basic and translational cancer research. Nat Rev Cancer 2022;22:625−39. https://doi.org/10.1038/s41568-022-00502-0.

[75] Ienca M, Vayena E, Blasimme A. Big data and dementia: charting the route ahead for research, ethics, and policy. Front Med 2018;5:13. https://doi.org/10.3389/fmed.2018.00013.

[76] Perakslis E, Riordan H, Friedhoff L, Nabulsi A, Pich EM. A call for a global 'bigger' data approach to Alzheimer disease. Nat Rev Drug Discov 2019;18:319—20. https://doi.org/10.1038/nrd.2018.86.

[77] NHS, 2021. Your health records. Url https://www.nhs.uk/using-the-nhs/about-the-nhs/your-health-records/.

[78] Carey IM, Anderson HR, Atkinson RW, Beevers SD, Cook DG, Strachan DP, Dajnak D, Gulliver J, Kelly FJ. Are noise and air pollution related to the incidence of dementia? A cohort study in London, England. BMJ Open 2018;8:e022404. https://doi.org/10.1136/bmjopen-2018-022404.

[79] Albert MS, DeKosky ST, Dickson D, Dubois B, Feldman HH, Fox NC, Gamst A, Holtzman DM, Jagust WJ, Petersen RC, Snyder PJ, Carrillo MC, Thies B, Phelps CH. The diagnosis of mild cognitive impairment due to Alzheimer's disease: recommendations from the National Institute on Aging-Alzheimer's Association workgroups on diagnostic guidelines for Alzheimer's disease. Alzheimer's Dement 2011;7:270—9. https://doi.org/10.1016/j.jalz.2011.03.008.

[80] Dubois B, Hampel H, Feldman HH, Scheltens P, Aisen P, Andrieu S, Bakardjian H, Benali H, Bertram L, Blennow K, Broich K, Cavedo E, Crutch S, Dartigues J, Duyckaerts C, Epelbaum S, Frisoni GB, Gauthier S, Genthon R, Gouw AA, Habert M, Holtzman DM, Kivipelto M, Lista S, Molinuevo J, O'Bryant SE, Rabinovici GD, Rowe C, Salloway S, Schneider LS, Sperling R, Teichmann M, Carrillo MC, Cummings J, Jack CR. Proceedings of the Meeting of the International Working Group (IWG) and the American Alzheimer's Association on "The Preclinical State of AD"; July 23, 2015; Washington DC, USA, 2016. Preclinical Alzheimer's disease: Definition, natural history, and diagnostic criteria. Alzheimer's Dement 2016;12:292—323. https://doi.org/10.1016/j.jalz.2016.02.002.

[81] Bauermeister S, Orton C, Thompson S, Barker RA, Bauermeister JR, Ben-Shlomo Y, Brayne C, Burn D, Campbell A, Calvin C, Chandran S, Chaturvedi N, Chêne G, Chessell IP, Corbett A, Davis DHJ, Denis M, Dufouil C, Elliott P, Fox N, Hill D, Hofer SM, Hu MT, Jindra C, Kee F, Kim C-H, Kim C, Kivimaki M, Koychev I, Lawson RA, Linden GJ, Lyons RA, Mackay C, Matthews PM, McGuiness B, Middleton L, Moody C, Moore K, Na DL, O'Brien JT, Ourselin S, Paranjothy S, Park K-S, Porteous DJ, Richards M, Ritchie CW, Rohrer JD, Rossor MN, Rowe JB, Scahill R, Schnier C, Schott JM, Seo SW, South M, Steptoe M, Tabrizi SJ, Tales A, Tillin T, Timpson NJ, Toga AW, Visser P-J, Wade-Martins R, Wilkinson T, Williams J, Wong A, Gallacher JEJ. The dementias platform UK (DPUK) data portal. Eur J Epidemiol 2020;35:601—11. https://doi.org/10.1007/s10654-020-00633-4.

[82] Crawford KL, Neu SC, Toga AW. The image and data archive at the laboratory of neuro imaging. Neuroimage 2016;124:1080—3. https://doi.org/10.1016/j.neuroimage.2015.04.067.

[83] The European Parliament and the Council of the European Union. REGULATION (EU) 2016/679 OF THE EUROPEAN PARLIAMENT AND OF THE COUNCIL of 27 April 2016 on the protection of natural persons with regard to the processing of personal data and on the free movement of such data, and repealing Directive 95/46/EC (General Data Protection Regulation). 2016.

[84] Skatova A, Goulding J. Psychology of personal data donation. PLoS One 2019;14:e0224240. https://doi.org/10.1371/journal.pone.0224240.
[85] Karthan M, Martin R, Holl F, Swoboda W, Kestler HA, Pryss R, Schobel J. Enhancing mHealth data collection applications with sensing capabilities. Front Public Health 2022;10:926234. https://doi.org/10.3389/fpubh.2022.926234.
[86] Ranjan Y, Rashid Z, Stewart C, Conde P, Begale M, Verbeeck D, Boettcher S, The H, Dobson R, Folarin A, The RADAR-CNS Consortium. RADAR-base: open source mobile health platform for collecting, monitoring, and analyzing data using sensors, wearables, and mobile devices. JMIR Mhealth Uhealth 2019;7:e11734. https://doi.org/10.2196/11734.
[87] Gorgolewski KJ, Auer T, Calhoun VD, Craddock RC, Das S, Duff EP, Flandin G, Ghosh SS, Glatard T, Halchenko YO, Handwerker DA, Hanke M, Keator D, Li X, Michael Z, Maumet C, Nichols BN, Nichols TE, Pellman J, Poline J-B, Rokem A, Schaefer G, Sochat V, Triplett W, Turner JA, Varoquaux G, Poldrack RA. The brain imaging data structure, a format for organizing and describing outputs of neuroimaging experiments. Sci Data 2016;3:160044. https://doi.org/10.1038/sdata.2016.44.
[88] Ohm P. Broken promises of privacy: responding to the surprising failure of anonymization. UCLA Law Rev 2009;57:1701.
[89] Agrawal R, Prabakaran S. Big data in digital healthcare: lessons learnt and recommendations for general practice. Heredity 2020;124:525—34. https://doi.org/10.1038/s41437-020-0303-2.
[90] United Nations Department of Economic and Social Affairs Population Division. World population prospects 2022. 2022. Online Edition.
[91] Jones N. How to stop data centres from gobbling up the world's electricity. Nature 2018;561:163—6. https://doi.org/10.1038/d41586-018-06610-y.
[92] Masanet E, Shehabi A, Lei N, Smith S, Koomey J. Recalibrating global data center energy-use estimates. Science 2020;367:984—6. https://doi.org/10.1126/science.aba3758.
[93] Cobb AN, Benjamin AJ, Huang ES, Kuo PC. Big data: more than big data sets. Surgery 2018;164:640—2. https://doi.org/10.1016/j.surg.2018.06.022.
[94] Silva-Spínola A, Baldeiras I, Arrais JP, Santana I. The road to personalized medicine in Alzheimer's disease: the use of artificial intelligence. Biomedicines 2022;10:315. https://doi.org/10.3390/biomedicines10020315.
[95] Liang X, Batsis JA, Zhu Y, Driesse TM, Roth RM, Kotz D, MacWhinney B. Evaluating voice-assistant commands for dementia detection. Comput Speech Lang 2022;72:101297. https://doi.org/10.1016/j.csl.2021.101297.
[96] Feng X, Provenzano FA, Small SA, for the Alzheimers Disease Neuroimaging Initiative. A deep learning MRI approach outperforms other biomarkers of prodromal Alzheimer's disease. Alz Res Therapy 2022;14:45. https://doi.org/10.1186/s13195-022-00985-x.
[97] Fogel AL, Kvedar JC. Artificial intelligence powers digital medicine. Npj Digital Med 2018;1:5. https://doi.org/10.1038/s41746-017-0012-2.
[98] Burdack J, Horst F, Giesselbach S, Hassan I, Daffner S, Schöllhorn WI. Systematic comparison of the influence of different data preprocessing methods on the performance of gait classifications using machine learning. Front Bioeng Biotechnol 2020;8:260. https://doi.org/10.3389/fbioe.2020.00260.
[99] Berisha V, Krantsevich C, Hahn PR, Hahn S, Dasarathy G, Turaga P, Liss J. Digital medicine and the curse of dimensionality. Npj Digit Med 2021;4:153. https://doi.org/10.1038/s41746-021-00521-5.
[100] Riley P. Three pitfalls to avoid in machine learning. Nature 2019;572:27—9. https://doi.org/10.1038/d41586-019-02307-y.

[101] Ford E, Milne R, Curlewis K. Ethical issues when using digital biomarkers and artificial intelligence for the early detection of dementia. WIREs Data Min Knowl 2023;13:e1492. https://doi.org/10.1002/widm.1492.

[102] Liu Y, Mazumdar S, Bath PA. An unsupervised learning approach to diagnosing Alzheimer's disease using brain magnetic resonance imaging scans. Int J Med Inf 2023;173:105027. https://doi.org/10.1016/j.ijmedinf.2023.105027.

[103] Price WN, Cohen IG. Privacy in the age of medical big data. Nat Med 2019;25:37−43. https://doi.org/10.1038/s41591-018-0272-7.

[104] Koychev I, Lawson J, Chessell T, Mackay C, Gunn R, Sahakian B, Rowe JB, Thomas AJ, Rochester L, Chan D, Tom B, Malhotra P, Ballard C, Chessell I, Ritchie CW, Raymont V, Leroi I, Lengyel I, Murray M, Thomas DL, Gallacher J, Lovestone S. Deep and Frequent Phenotyping study protocol: an observational study in prodromal Alzheimer's disease. BMJ Open 2019;9:e024498. https://doi.org/10.1136/bmjopen-2018-024498.

[105] Cuggia M, Combes S. The French health data hub and the German medical informatics initiatives: two national projects to promote data sharing in healthcare. Yearb Med Inform 2019;28:195−202. https://doi.org/10.1055/s-0039-1677917.

[106] Ministry of Health, Ministry of Finance, Danish Regions and Local Government Denmark. Digital health strategy 2018−2022, A coherent and trustworthy health network for all. 2018.

[107] NHS. The NHS long term plan. 2019.

[108] Williams E, Buck D, Babalola G, Maguire D. What are health inequalities?. 2022.

[109] World Health Organization. Equity within digital health technology within the WHO European Region: a scoping review (No. WHO/EURO:2022-6810-46576-67595). Copenhagen: WHO Regional Office for Europe; 2022.

[110] Wilson S, Tolley C, McArdle R, Slight R, Slight S. Who is most at risk of digital exclusion within healthcare? Int J Pharm Pract 2024;32:i3−4. https://doi.org/10.1093/ijpp/riae013.004.

[111] Snyder M, Zhou W. Big data and health. Lancet Digital Health 2019;1:e252−4. https://doi.org/10.1016/S2589-7500(19)30109-8.

Chapter 6b

Advances in imaging

Emrah Düzel[a,b,c] and Jose Bernal[a,b,d]
[a]Institute of Cognitive Neurology and Dementia Research, Otto-von-Guericke University Magdeburg, Magdeburg, Germany; [b]German Center for Neurodegenerative Diseases, Magdeburg, Germany; [c]Institute of Cognitive Neuroscience, University College London, London, United Kingdom; [d]Centre for Clinical Brain Sciences, University of Edinburgh, Edinburgh, United Kingdom

Neurodegeneration and MRI

MRI is a fundamental readout for neurodegeneration, according to the ATN (amyloid/tau/neurodegeneration) biological diagnosis of Alzheimer's disease (AD) [1], which enables the cross-sectional measurement of brain volumes and the longitudinal tracking of volume loss or atrophy. The pivotal role of MRI in quantifying brain atrophy has been instrumental in driving significant advancements in both the acquisition and analysis technologies in recent years, and these advancements have proven fruitful. Volumetric MRI measurements can explain a considerable amount of the variability in cognitive performance and cognitive change over time and even predict the clinical and biomarker status of an individual [2].

Bayesian modeling of structural MRI data with a cortical voxel-based morphometry approach showed [3] that cross-sectional volumetric patterns in large cohorts indeed provide evidence for a monotonic brain volume decline compatible with an ATN-like progression for amyloid-conversion first, tau-conversion second, neurodegeneration-conversion (hippocampal atrophy) last. However, in addition, many cortical regions showed a non-monotonic volume relationship with AD stages, which is compatible with an early amyloid-related tissue expansion or sampling effects, e.g., due to brain reserve. Moreover, brain regions affected earlier by tau tangle deposition (Braak stage I-IV, medial temporal lobe (MTL), limbic system) showed stronger evidence for volume loss than those involved in late Braak stages (V/VI, cortical regions) [3].

Data driven methods of MRI morphometric analysis have also been able to capture biological heterogeneity in AD using statistical clustering approaches. These studies have identified distinct atrophy subtypes based on cross-sectional structural MRI for instance using SuStaIn which is publicly available (https://github.com/ucl-pond/pySuStaIn) [4]. It is now clear that even in individuals who present as prodromal AD with a predominantly memory-based clinical impairment and no clinically evident posterior cortical atrophy, a limbic-predominant and a hippocampal-sparing subtype of atrophy can be identified [5]. Longitudinal studies will be important to understand how these atrophy subtypes are related to clinical progression over time.

While the empirical evidence supporting the predictive value of volumetric MRI for cognitive performance is compelling, studies with disease-modifying treatments (DMTs) do not show a strong correspondence between changes in cognitive decline rates and changes in MRI atrophy rates. That is, the rate of change in MRI atrophy does not parallel the rate of change in cognition in patients treated with DMTs [6]. If it did, there would be a clearer relationship between slowing of cognitive decline, clearance of amyloid-plaques (seen using amyloid positron emission tomography (PET)) with DMTs and reduced rate of MRI atrophy progression. To some extent, this has prevented volumetric MRI from being a primary outcome measure in clinical trials and in the management of DMTs in healthcare. Many experts would nonetheless agree that if a therapy could effectively decelerate MRI atrophy, it would represent a major breakthrough with significant clinically relevance.

Three complementary factors may help explain why MRI is better related to cognition and pathology cross-sectionally than longitudinally. First, MRI volumetric measures may be tightly related to brain reserve and/or cognitive reserve [7] and therefore explain the off-set of trajectories and their slope in relation to individual dispositions rather than to interventions. Second, there might exist a more dynamic aspect of brain physiology tied to cognition, e.g., synaptic function, cerebrovascular function, and brain waste clearance, which conventional volumetric measures do not capture. Third, vascular supply and clearance could impact cognition, with relevant imaging indicators beyond conventional volumetric measures and only accessible through advanced technology, particularly in terms of medical image analysis.

Within this chapter, we discuss advanced MR readouts for reserve, clearance, synaptic function and vascular pathology, with the goal of providing a perspective on complementary factors that can influence cognitive performance.

Reserve and resilience

According to the new NIH Working Group Reserve & Resilience criteria [7], resilience is defined as an overarching umbrella term for maintaining cognition and withstanding pathology. Maintenance, brain reserve and cognitive reserve are therefore subsumed under resilience.

Brain reserve

Brain reserve refers to the "capital" of neural resources (e.g., numbers of neurons and synapses) with which individuals start off when challenged by a pathology or perturbation, but—unlike cognitive reserve—it does not involve an active adaptation of functional cognitive processes. Defining brain reserve as an intercept also implies that while the volume of a brain region may enable better-than-expected cognitive performance at a certain stage of brain pathology, it might not translate into slower decline over time.

The notion of brain reserve can explain why cross-sectional MRI volumetry shows a strong correlation with cognitive performance but does not demonstrate the same level of correlation with cognitive decline [2]. A recent example of brain reserve compatible with the new NIH Working Group criteria was reported in Ref. [8]. Cognitively unimpaired elderly participants with no memory complaints but amyloid positivity, according to cerebrospinal fluid (CSF) readings, can have larger hippocampal CA1 subfields compared to those who are amyloid negative. Conversely, those who have memory complaints (subjective cognitive decline, [9]) and amyloid positivity tend to have smaller CA1 subfields than those with no amyloid pathology. For these two statements to be compatible, brain reserve—a higher "capital" of neural resources in the hippocampus—must enable some individuals to remain without memory complaints despite having amyloid pathology [8].

An interesting question is whether observations on brain reserve also point toward potential intervention targets. Although the NIH Working Group criteria see brain reserve as a life span accumulation of neural resources, there is of course also the possibility that increasing the volume of CA1 (and its related network partners) would provide a similar benefit. One example for such an intervention is exercise [10–12].

Cognitive reserve

Cognitive reserve represents the brain's adaptive ability to maintain cognitive performance in the context of life-course-related brain changes, injuries, or diseases [7]. It relates to flexibility and efficiency of cognitive processes, allowing individuals to sustain cognitive function longer by compensating through alternative networks or active cognitive strategies. In brief, cognitive reserve moderates the relationship between pathology (insults, disease, or risk factors) and cognitive decline over time. Investigating the neural underpinning of cognitive reserve requires measures of cognitive function in conjunction with a demonstration that a particular brain activity pattern moderates decline over time [7]. fMRI measures of task-related brain activity combining tasks tapping into different cognitive domains indicates that anterior cingulate cortex activity can provide task-invariant cognitive reserve against decline in old age [13]. In the preclinical and prodromal AD spectrum, encoding-related brain activity has also been associated with cognitive reserve [14]. A multivariate moderation analysis of task-related fMRI identified a cognitive reserve-related activity pattern underlying successful memory encoding that moderated the detrimental effect of Alzheimer's pathological load on cognitive performance. Cognitive reserve was primarily represented by a more pronounced expression of the task-active (novelty encoding) network encompassing the default mode network, anterior cingulate cortex and inferior temporal regions including the fusiform gyrus, leading to the extraction of personalized fMRI-based cognitive reserve scores that moderated the impact of Alzheimer's pathology (a score combining hippocampal atrophy, CSF p-tau levels and CSF Aβ42/40 ratio) on cognitive performance and were positively associated with years of education [14]. Related research has also been performed with resting-state fMRI [13]. Cognitive reserve (or resilience) was associated with better preservation of segregation of functional resting state networks in the brain despite increasing levels of tau-pathology [15]. Higher hub connectivity of the inferior frontal junction was associated with higher cognitive reserve against the impact of brain pathology including tau [16]. These results were consistently observed in sporadic and familial AD, and were reproduced in several cohorts of AD [17].

These data indicate that certain individuals can maintain brain activity in the default mode network and a network encompassing parts of the visual processing stream for encoding despite having Alzheimer's pathology and this activity maintenance fulfills the definition of cognitive reserve. The mechanisms underpinning this cognitive reserve are unknown, highlighting the nascent stage of research in this area. However, such assessments suggest that key modifiers of cognitive decline and potentially relevant entry points for targeted intervention can be identified with advanced MRI techniques.

Structural traits

A third aspect, not outlined in the NIH Working Group criteria but potentially essential alongside brain reserve and cognitive reserve, is the presence of stable (non-modifiable) structural traits in the brain architecture. Stable structural traits include cortical folding patterns, which are likely to influence how well individuals adapt to new challenges [18]. Emerging evidence suggests that stable structural traits, presumably unresponsive to training, could significantly influence plasticity [19,20]. Although technology to quantify folding is developing [18,21], this aspect of the brain morphology has not been intensely studied in the context of AD.

Another "trait" like aspect that is accessible with advanced ultrahigh-field 7T time-of-flight imaging (TOF) MR imaging is individual vascular supply patterns in the brain, in particular in the hippocampus [22–25]. This will be discussed in more detail below in the section on cerebrovascular imaging.

Synaptic dysfunction

The hallmarks of AD pathology are extracellular amyloid aggregates and intracellular accumulation of hyperphosphorylated tau [26,27]. Tau accumulation in medial temporal and lateral temporal regions correlates with memory performance [28–30] and CSF levels of total-tau correlate with memory performance and with hippocampal novelty responses to novel images [31]. An important question is how tau and amyloid pathology interact in causing synaptic dysfunction (for a review see Ref. [32]). If amyloid pathology is permissive for tau-related hippocampal dysfunction, this would indicate that removing amyloid pathology could improve brain function even if there is no corresponding volumetric change in the MRI.

A recent study in the DZNE's Longitudinal Cognitive Impairment and Dementia Study (DELCODE) cohort showed that CSF total-tau and phospho-tau levels were negatively associated with memory performance and with novelty responses in the hippocampus and amygdala, in interaction with CSF Aβ42 levels [31]. Irrespective of diagnosis (non-complaining individuals, subjective complainers and mild cognitive impairment patients), age, education, sex, Aβ42 levels determined the relationship between tau and hippocampal novelty responses and this relationship could not be explained by higher phospho-tau pathology in the amyloid positive group. This shows that the presence of amyloid-pathology is permissive for tau-related hippocampal synaptic dysfunction and, in turn, indicates that in the predementia stage of AD, removing amyloid-pathology could improve memory and hippocampal function.

A study conducted in the BioFinder cohort in Sweden [28,33] indicated that synaptic function measured through functional connectivity is present in amyloid-beta-positive cognitively unimpaired individuals, mainly as decreased functional connectivity between the medial temporal lobe and regions in the anterior-temporal system, most prominently between left medial prefrontal cortex. In contrast, amyloid-beta-positive individuals with mild cognitive impairment exhibited reduced connectivity between the medial temporal lobe and posterior-medial regions, predominantly between the anterior hippocampus and posterior cingulate cortex. Importantly, reduced medial temporal-cortical functional connectivity networks were associated with more rapid longitudinal memory decline as shown by linear mixed-effects regression analysis.

To summarize, understanding the relationship between synaptic dysfunction, AD pathology and neurodegeneration is still in its infancy. But it is already clear that synaptic function captures variance in cognition and decline that is not captured with volumetric measures and this is very well compatible with the known pathophysiology from animal studies that amyloid and tau pathology, in interaction, cause synaptic dysfunction.

Cerebrovascular imaging

Cerebral blood vessels are active and dynamic organs that transport blood throughout the brain, delivering oxygen and nutrients to brain cells and collecting cellular waste [34]. These roles inherently make them essential for ensuring the optimal functioning and maintenance of the brain [25,34,35]. It is therefore unsurprizing that their malfunction is a contributing factor to cognitive decline and dementia [36—42]. Interruptions in blood supply to the brain, even as brief as 1 min, can be extremely harmful and result in the loss of approximately one million neurons, 14 billion synapses, and 12 km of myelinated fibers [43].

Advances in medical imaging techniques enable the visualization of cerebral blood vessels from both structural and functional perspectives. These advancements encompass imaging the structure of major blood vessels, measuring the flow of blood within them, and assessing microcirculation aspects, including cerebral perfusion [43,44].

Time of flight (TOF) imaging

TOF MR angiography (MRA) provides high-contrast images of blood vessels, facilitating detailed visualization of vascular structures and blood flow dynamics without the need for contrast agents [44]. This makes it particularly valuable for assessing vascular supply as a structural trait—and its contributions to the capacity of the brain to withstand cerebrovascular and neurodegenerative pathologies [22—25].

High-resolution 7T TOF MRA, for instance, enables visualization of the cerebral microvasculature at the level of the hippocampus and allows for the assessment of blood supply to the hippocampus across different individuals (Fig. 1) [45]. Around 60% of individuals have a dual (augmented) hippocampal vascular supply originating from both the anterior choroidal artery and the posterior cerebral artery [25]. Those individuals have higher anterior hippocampal volumes and better memory performance in old age in the presence of cerebral small vessel disease (CSVD) [25]. It is conceivable that dual hippocampal vascularization as visible on 7T imaging, for example, counteracts age-related microvascular alterations, such as cerebral blood flow (CBF) or cerebrovascular reactivity (CVR) reduction, and thus preserves neurovascular plasticity. Furthermore, hippocampal vascularization patterns might modulate MTL vascular and network plasticity, which is highly variable in older adults. Given the small size of hippocampal vessels, particularly those originating in the anterior choroidal artery, the assessment of supply pattern can only be done at 7T (Fig. 1) [25]. Therefore, data on how supply patterns as stable "traits" influence disease progression in AD is challenging to investigate.

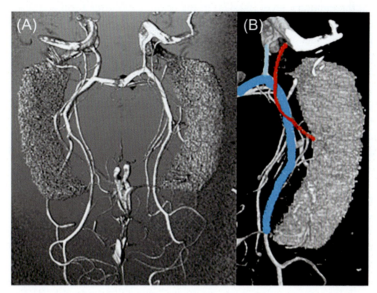

FIG. 1 TOF MRA image acquired with isotropic voxel dimensions of 0.3 mm using 7 T scanner. (A) TOF MRA reconstruction of the hippocampus. (B) Delineation of main vessels involved in hippocampal vascular supply (light blue, red, and cyan for posterior cerebral artery, anterior choroidal artery, and posterior communicating artery, respectively).

In addition to providing insights into the type of vascular supply (e.g., augmented or single), advanced image processing known as vessel distance mapping (VDM) at 7T [22] can yield quantitative measurements of the proximity of hippocampal voxels to the nearest blood vessels. With VDM, researchers can understand how the spatial arrangement and density of blood vessels in the hippocampal region contribute to maintaining cognitive functions, particularly in the context of aging or vascular pathology. VDM studies have demonstrated that reduced hippocampal vascular supply is associated with poorer cognitive performance in subjects with vascular pathology, but not in healthy controls, suggesting a significant role of vascularization in cognitive resilience [22]. A recent post-mortem study using 7T imaging showed that entorhinal vessel density correlated with tau pathology [46]. Further utilization of VDM could offer insights into the degree to which these observations translate to other pathologies, including AD, and to which vascular supply conditions training-induced plasticity, especially in light of physical and cognitive interventions.

Perfusion and permeability imaging

Cerebral perfusion serves as the mechanism for transporting energy into the brain, facilitating the delivery of oxygen and nutrients to tissues through blood flow at the capillary level [34,47]. In the brain, capillaries form a singular and tightly packed layer of endothelial cells, collectively known as the endothelium, which acts as a selective barrier between the blood and the brain, referred to as the blood-brain barrier (BBB). This specialized microvascular arrangement regulates the movement of substances into and out of brain tissues, thereby shielding the brain from harmful agents and pathogens circulating in the bloodstream [48]. Perfusion and permeability imaging deal with evaluating physiological aspects, such as blood flow, volume, and oxygenation, capillary density, and BBB permeability [49,50].

Cerebral perfusion and permeability are typically assessed through two main methods, which differ in whether or not they utilize an exogenous contrast agent [43,49]. In contrast-enhanced imaging, for instance, a series of MRI scans are taken before and after injecting an exogenous, often Gadolinium-based, contrast agent intravenously [50,51]. The degree to which susceptibility and relativity change across various tissues due to the arrival or presence of a contrast agent indicates abnormal microvasculature. In dynamic contrast-enhanced MRI (DCE-MRI), the contrast agent causes relaxation time of water molecules to decrease in T1-w and, hence, its accumulation in the extracellular extravascular space due to an abnormally permeable BBB would result in signal enhancement [51]. Research conducted in humans using this sophisticated imaging technology has demonstrated that the impairment of the BBB in the hippocampus, cortex, and throughout the entire brain develops with aging and correlates with cognitive performance as well as with CSVD and AD [50,52–61].

An alternative stream that has also been gaining terrain is the usage of endogenous tracers, such as in arterial spin labeling (ASL) MRI [49]. ASL differs from contrast bolus techniques in that it is non-invasive and thus mitigates concerns related to contrast agent safety and potential adverse reactions. The primary physiological parameter assessed with ASL is CBF, which governs the rate of oxygen and nutrient delivery to the capillary bed [43,49,62,63]. However, the time taken by the labeled blood to travel from the labeling region to the imaging region, referred to arterial transit time, has gained importance in the context of brain function and has also gained significance in the realm of brain function [63]. This is in part because its necessity for accurate CBF quantification but also because it mirrors brain's oxygen and nutrient demands. ASL-derived parameters have so far shown that hypoperfusion is characteristic of CSVD but also AD. While promising, it is important to emphasize that low signal-to-noise ratio continues to pose a significant challenge, along with susceptibility to imaging artifacts, particularly motion artifacts. Ongoing efforts to improve imaging protocols, optimize hardware, and software capabilities, and develop advanced motion correction techniques may nonetheless lead to better measurement of perfusion in the coming years [43].

Glymphatic imaging

The evolution of neuroimaging technologies has opened the door to observing intricate yet minuscule brain structures that were once overlooked, not due to their lesser importance in comparison to larger, more prominent structures, but rather because routine MRI lacked the necessary sensitivity to detect them. The glymphatic system is a prime example of this phenomenon.

Perivascular spaces

The lymphatic system transports soluble substances, proteins, and fluid from the interstitial space back into the circulatory system. Its network spans the entirety of the body, with the exception of the central nervous system, which, despite its high metabolic demands, does not possess lymphatic vessels. Instead, emerging evidence points toward its dependence on a complex network of perivascular compartments, termed the glymphatic system [64,65]. These perivascular spaces (PVS), also known as Virchow-Robin spaces, surround small (perforating) blood vessels within the brain [66,67] and serve as pathways for fluid transport and movement, thereby aiding in the removal of metabolic waste products and maintenance of brain homeostasis.

Despite their microscopic size, PVS possess the capacity to undergo enlargement, ultimately becoming salient on MRI scans [39,67]. When visible, PVS appear as rounded, ovoid, or tubular structures, displaying an intensity profile akin to that of CSF (Fig. 2), e.g., hyperintensity in T2-weighted imaging and hypointensity in T1-weighed and FLAIR imaging [39,67]. These spaces become visible predominantly across two specific cerebral regions: the basal ganglia, where they appear in the lentiform nuclei and internal and external capsules, and centrum semiovale, where they appear as linear formations that extend inwardly, perpendicular to the brain's surface, from juxtacortical white matter toward the lateral ventricles, following the course of perforating brain vessels [67]. PVS can, nevertheless, be found enlarged across the hippocampus, midbrain, pons, and cerebellar white matter.

The expansion of PVS has intrigued researchers, yet the precise mechanisms underlying this phenomenon remain elusive, being a significant area of ongoing investigation [67,69]. Cross-sectional lifespan as well as longitudinal studies of PVS have already provided compelling evidence of their dynamic nature and propensity for change over time [79–83] (Fig. 2A) as well as of their potential intertwinement with other markers of cerebrovascular dysfunction, such as white matter hyperintensities (WMH) [74,77,78] (Fig. 2B). Studies have also found PVS linked to aging [80–82], cardiovascular risk factors [68,79,84,85], CSVD [39,68,69,82], and inflammation [70–73]. Mounting evidence also suggests an association between PVS alterations and elevated levels of Aβ and tau [82,86,87], and vascular Aβ

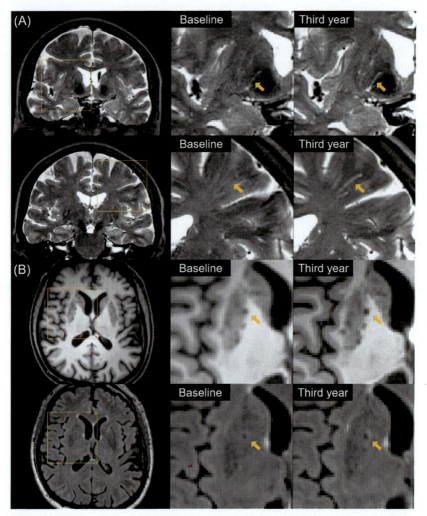

FIG. 2 PVS can grow in size over time. (A) Longitudinal T2-weighted imaging shows PVS enlarging over a 3-year period (*yellow arrows*), emphasizing aging contributes to PVS changes. Other conditions, such as CSVD [39,68,69], inflammation [70–73] and vascular Aβ deposition [74–76], may also contribute to PVS enlargement. (B) Longitudinal T1-weighted (*top*) and FLAIR (*bottom*) imaging shows WMH can appear in the vicinity of PVS (*yellow arrow*). This co-occurrence hints to a potential interplay between these two phenomena and raises intriguing questions about their common underlying pathophysiological mechanisms [74,77,78], e.g., whether extravasation of fluid from PVS into the brain parenchyma ultimately results in the development of certain WMH or not.

deposition in cerebral amyloid angiopathy [74–76], in alignment with the proposed role of PVS in brain waste clearance [88–90]. Taken together, this all suggests that PVS enlargement is reflective of a state of dysregulated brain homeostasis [91] and could potentially serve as an early biomarker of dysfunction.

The development, validation, and deployment of computational methodologies for quantifying PVS have significantly eased the exploration of PVS, especially in large-scale studies where traditional qualitative assessments of PVS would otherwise demand substantial time and labor resources [92–95]. By leveraging computational techniques, researchers can nowadays analyze vast quantities of imaging data, extract quantitative metrics related to PVS burden and morphology (total PVS counts and volumes as well as individual PVS lengths, widths, and volumes) [67,96–99]. While structural imaging has long been the cornerstone of PVS research, there exists a noticeable gap in the exploration of dynamic imaging techniques within this field, as well as the integration of multiple imaging modalities. Dynamic imaging approaches present an exciting opportunity to delve deeper into the vascular mechanisms underlying PVS-related alterations. For instance, techniques such as DCE-MRI, blood-oxygenation-level dependent (BOLD) MRI, and ASL MRI offer avenues to assess parameters like permeability, cerebrovascular reactivity, and perfusion, which are crucial for understanding vascular integrity, cerebral blood flow, and cerebral circulation and their impact on PVS dynamics over time [39,42,67,83,95].

Future research could prioritize refining imaging protocols, optimizing image acquisition parameters, and developing advanced analytical methods. International collaborations have also been established to investigate and meta-analyze perivascular spaces and brain waste clearance. These collaborations, exemplified by initiatives such as those led by the Small Vessel Disease Consortium (https://www.small-vessel-disease.org/), the Translational Approach toward Understanding Brain Waste Clearance in Cerebral Amyloid Angiopathy (https://www.fondationleducq.org/network/a-translational-approach-towards-understanding-brain-waste-clearance-in-cerebral-amyloid-angiopathy/), and the International Perivascular Spaces Meta-analysis Collaboration (https://clinical-brain-sciences.ed.ac.uk/row-fogo-centre-research-ageing-and-brain/our-research/row-fogo-research-projects/international), serve to streamline the research process and broaden the scope of current investigations. By pooling resources and expertize across multiple centers and cohorts, these collaborations seek to enable more comprehensive examinations of PVS across diverse populations and clinical cohorts, ultimately advancing our understanding of PVS in the context of brain health in the coming years.

Inferior frontal sulcal hyperintensities

CSF located in areas immediately superior to the cribriform plate, particularly within the inferior frontal sulci, can exhibit abnormally elevated signal intensities when observed through T2-weighted FLAIR imaging [100] (Fig. 3). These anomalies, referred to as inferior frontal sulcal hyperintensities (IFSH),

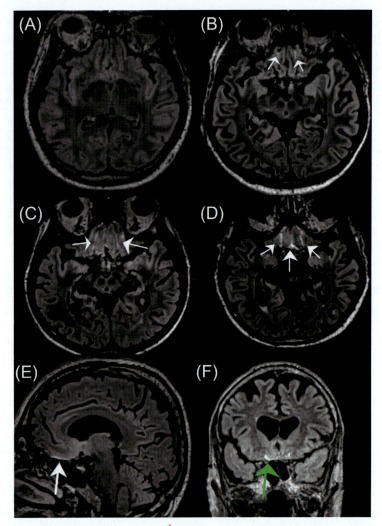

FIG. 3 IFSH as seen on fluid-attenuated inversion recovery (FLAIR) sequences. (A–D) FLAIR imaging for four distinct participants with IFSH scores of 0, 2, 5, and 8, respectively. (E–D) IFSH on both sagittal and coronal planes.

are hypothesized to reflect compromised CSF clearance and glymphatic dysfunction [101]. This stem from the notion that CSF may traverse into meningeal lymphatics in close proximity to the cribriform plate, with stagnation of protein, blood, and cellular debris in this vicinity potentially leading to the emergence of these hyperintensities [67,100–102].

Despite being in its nascent stages, a handful of studies have embarked on investigations concerning nasal turbinates, the cribriform plate, and inferior frontal sulcal hyperintensities (IFSH) in human subjects. Among the pioneering endeavors in this area, one study [102] utilized dynamic PET imaging with 18F-THK5117, a tracer devised to detect tau pathology, to scrutinize the involvement of human nasal turbinates in CSF clearance—a thoroughly studied region in animal models—and whether this process was somewhat compromised in AD patients. The findings from the study suggested that human nasal turbinates do play a role in the CSF clearance system and exhibited abnormalities in AD participants compared to controls [101] introduced a novel visual rating system to evaluate inferior frontal sulcal hyperintensities (IFSH), uncovering a heightened prevalence of these hyperintensities among elderly individuals and those with increased burdens of PVS. A separate investigation involving non-demented participants [103] aimed to discern whether IFSH could serve as indicators of CSVD and the accumulation of amyloid and tau in brain tissue. Through the application of amyloid and tau PET, the study revealed a correlation between IFSH and CSVD, though no association was found with PET-derived amyloid or tau concentrations. Future investigation is nonetheless warranted to comprehensively understand the origins and exact nature of IFSH.

White matter hyperintensities

The movement of fluid within PVS appears to be contingent upon vascular pulsations [67,104]. It is therefore expected that a dysfunction of the cerebral vascular system leads to a deterioration of the glymphatic system and vice versa [67].

WMH stand out as a defining feature of CSVD [105]. WMH are dynamic and diffuse alterations occurring within the periventricular and deep white matter regions of the brain, which appear hyperintense on T2-weighted and FLAIR MRI and hypodense on computed tomography and, upon histopathological examination, show demyelination, axon loss, and gliosis [106,107]. WMH can be found in most (>90%) sexagenarian adults [108] and, although historically regarded as "silent", they are associated with a broad spectrum of clinical manifestations, including apathy, fatigue, delirium, depression, disruptions in physical function, progressive cognitive decline, and increased risk of stroke and dementia [37,105].

WMH are not only influenced by genetic and lifestyle factors [39,107,109−113] but also by the onset and progression of neurodegenerative and cerebrovascular diseases [39,114−119]. For years, CSVD has been considered the primary factor contributing to the emergence and evolution of WMH [85,120−122]—reason why they are often labeled "WMH of presumed vascular origins".

Recent investigations, however, have broadened our comprehension of WMH, particularly in the context of AD. These investigations have revealed that individuals across the AD spectrum can exhibit increased global and posterior volumes of WMH, despite minimal exposure to cardiovascular risk factors [106,120,123–126], and that the spatial distribution of WMH within the brain is somewhat distinct for CSVD and AD [108,120,123,124,127–135]. Collectively, these findings undermine the conventional belief that vascular dysfunction is the exclusive catalyst behind these cerebral abnormalities and, instead, emphasize the multifaceted origins of WMH formation. The involvement of neurodegenerative and vascular processes in the formation and progression of WMH offers a compelling explanation for the persistent correlations observed between WMH and cortical neurodegeneration, even after adjustments for demographic variables and traditional cardiovascular risk factors [110,125,136–139].

The temporal interplay between WMH and cortical atrophy is under examination from two complementary perspectives: the cerebrovascular hypothesis and the neurodegenerative hypothesis [108,117]. The cerebrovascular hypothesis suggests that ischemic and hypoxic insults, in the form of WMH [85,107,140,141], may initially lead to a perilesional reduction in oxygen, nutrient, and trophic support [85,142]. These insults may progressively impair the function and metabolic requirements of compromised white matter tracts and their associated cortical regions, ultimately resulting in cortical atrophy [116,117,138,140,143] (Fig. 4). Conversely, the neurodegenerative hypothesis suggests that cortical neurodegeneration may trigger demyelination and axonal loss, contributing to the formation of WMH [108,117,126,133,144], particularly when coupled with amyloid or tau pathologies [108,141,145]. Although longitudinal evidence supporting these hypotheses is still limited [146], ongoing research endeavors are actively investigating these intricate dynamics [119].

FIG. 4 WMH are associated with cortical neurodegeneration. It has been proposed and to some extent shown that ischemic and hypoxic damage, in the form of lacunes and WMH [85,107,140,141], may initially lead to a perilesional reduction in oxygen, nutrient, and trophic support [85,142]. These insults may progressively impair the function and metabolic requirements of compromised white matter tracts and their associated cortical regions, ultimately resulting in cortical atrophy [116,117,138,140,143]. The opposite direction, wherein cortical neurodegeneration triggers demyelination and axonal loss and contribute to the formation of WMH [108,117,126,133,144] is also discussed in the literature.

Summary

It has become increasingly clear that MRI atrophy over time is just one piece of the puzzle that, while essential, do not fully account for the overall variability in cognitive performance and its changes over time in the AD spectrum. Alongside potentially non-modifiable structural traits, reserve and synaptic function currently offers a better explanation for the disparities between MRI-based measurements of atrophy and cognitive trajectories. Through the lenses of resilience and reserve, preserving cognition requires not only the accumulation of neural resources throughout the lifespan but also the brain's adaptive capacity to cope with or compensate for pathology sustained throughout life.

Comprehensive and quantitative multimodal imaging of structural, functional and vascular brain integrity and physiology is evolving into an integral component of an advanced personalized medicine approach for managing AD. The most recent advancements in both structural and functional MRI scanner technology, along with sophisticated data analysis techniques, are now enabling us to investigate brain fluids, categorize distinct disease subtypes, and compute personalized cognitive reserve scores on the basis of brain function. A comprehensive grasp of these interrelated components offers potential devising precise interventions to safeguard cognitive function.

References

[1] Jack CR, Bennett DA, Blennow K, Carrillo MC, Dunn B, Haeberlein SB, Holtzman DM, Jagust W, Jessen F, Karlawish J, Liu E, Molinuevo JL, Montine T, Phelps C, Rankin KP, Rowe CC, Scheltens P, Siemers E, Snyder HM, Sperling R, Elliott C, Masliah E, Ryan L, Silverberg N. NIA-AA Research Framework: toward a biological definition of Alzheimer's disease. Alzheimers Dement 2018;14:535−62. https://doi.org/10.1016/j.jalz.2018.02.018.

[2] Nemali A, Vockert N, Berron D, Maas A, Bernal J, Yakupov R, et al. Gaussian Process-based prediction of memory performance and biomarker status in ageing and Alzheimer's disease—a systematic model evaluation. Med Image Anal 2023;90:102913. https://doi.org/10.1016/j.media.2023.102913.

[3] Heinzinger N, Maass A, Berron D, Yakupov R, Peters O, Fiebach J, et al. Exploring the ATN classification system using brain morphology. Alzheimers Res Ther 2023;15:50. https://doi.org/10.1186/s13195-023-01185-x.

[4] Dong A, Toledo JB, Honnorat N, Doshi J, Varol E, Sotiras A, Wolk D, Trojanowski JQ, Davatzikos C. Heterogeneity of neuroanatomical patterns in prodromal Alzheimer's disease: links to cognition, progression and biomarkers. Brain 2017;140:735−47. https://doi.org/10.1093/aww335.

[5] Baumeister H, Vogel JW, Insel PS, Kleineidam L, Wolfsgruber S, Stark M, et al. A generalizable data-driven model of atrophy heterogeneity and progression in memory clinic settings. Brain 2024;147(7):2400−13. https://doi.org/10.1093/brain/awae118.

[6] Mintun MA, Lo AC, Duggan Evans C, Wessels AM, Ardayfio PA, Andersen SW, et al. Donanemab in early Alzheimer's disease. N Engl J Med 2021;384:1691−704. https://doi.org/10.1056/NEJMoa2100708.

[7] Stern Y, Albert M, Barnes CA, Cabeza R, Pascual-Leone A, Rapp PR. A framework for concepts of reserve and resilience in aging. Neurobiol Aging 2023;124:100−3. https://doi.org/10.1016/j.neurobiolaging.2022.10.015.

[8] Yildirim Z, Delen F, Berron D, Baumeister H, Ziegler G, Schütze H, et al. Brain reserve contributes to distinguishing preclinical Alzheimer's stages 1 and 2. Alzheimers Res Ther 2023;15:43. https://doi.org/10.1186/s13195-023-01187-9.

[9] Jessen F, Wolfsgruber S, Kleineindam L, Spottke A, Altenstein S, Bartels C, Berger M, Brosseron F, Daamen M, Dichgans M, Dobisch L, Ewers M, Fenski F, Fliessbach K, Freiesleben SD, Glanz W, Görß D, Gürsel S, Janowitz D, Kilimann I, Kobeleva X, Lohse A, Maier F, Metzger C, Munk M, Preis L, Sanzenbacher C, Spruth E, Rauchmann B, Vukovich R, Yakupov R, Weyrauch AS, Ziegler G, Schmid M, Laske C, Perneczky R, Schneider A, Wiltfang J, Teipel S, Bürger K, Priller J, Peters O, Ramirez A, Boecker H, Heneka MT, Wagner M, Düzel E. Subjective cognitive decline and stage 2 of Alzheimer disease in patients from memory centers. Alzheimers Dement 2023a;9:487−97. https://doi.org/10.1002/alz.12674.

[10] Duzel E, Van Praag H, Sendtner M. Can physical exercise in old age improve memory and hippocampal function? Brain 2016;139:662−73. https://doi.org/10.1093/brain/awv407.

[11] Maass A, Düzel S, Brigadski T, Goerke M, Becke A, Sobieray U, Neumann K, Lövdén M, Lindenberger U, Bäckman L, Braun-Dullaeus R, Ahrens D, Heinze HJ, Müller NG, Lessmann V, Sendtner M, Düzel E. Relationships of peripheral IGF-1, VEGF and BDNF levels to exercise-related changes in memory, hippocampal perfusion and volumes in older adults. Neuroimage 2016;131:142−54. https://doi.org/10.1016/j.neuroimage.2015.10.084.

[12] Maass A, Düzel S, Goerke M, Becke A, Sobieray U, Neumann K, Lövden M, Lindenberger U, Bäckman L, Braun-Dullaeus R, Ahrens D, Heinze HJ, Müller NG, Düzel E. Vascular hippocampal plasticity after aerobic exercise in older adults. Mol Psychiatry 2015;20:585−93. https://doi.org/10.1038/mp.2014.114.

[13] Stern Y, Gazes Y, Razlighi Q, Steffener J, Habeck C. A task-invariant cognitive reserve network. Neuroimage 2018;178:36−45. https://doi.org/10.1016/j.neuroimage.2018.05.033.

[14] Vockert N, Machts J, Kleineidam L, Nemali A, Incesoy E, Bernal J, et al. Cognitive reserve against Alzheimer's pathology is linked to brain activity during memory formation. Nat Commun 2023;15:9815. https://doi.org/10.1038/s41467-024-53360-9.

[15] Ewers M, Luan Y, Frontzkowski L, Neitzel J, Rubinski A, Dichgans M, et al. Segregation of functional networks is associated with cognitive resilience in Alzheimer's disease. Brain 2021;144(7):2176−85. https://doi.org/10.1093/brain/awab112.

[16] Franzmeier N, Duzel E, Jessen F, Buerger K, Levin J, Duering M, et al. Left frontal hub connectivity delays cognitive impairment in autosomal-dominant and sporadic Alzheimer's disease. Brain 2018;141(4):1186−200. https://doi.org/10.1093/brain/awy008.

[17] Neitzel J, Franzmeier N, Rubinski A, Ewers M. Alzheimer's disease neuroimaging I. Left frontal connectivity attenuates the adverse effect of entorhinal tau pathology on memory. Neurology 2019;93(4):e347−57. https://doi.org/10.1212/WNL.0000000000007822.

[18] Germanaud D, Lefèvre J, Toro R, Fischer C, Dubois J, Hertz-Pannier L, Mangin JF. Larger is twistier: spectral analysis of gyrification (SPANGY) applied to adult brain size polymorphism. Neuroimage 2012;63:1257−72. https://doi.org/10.1016/j.neuroimage.2012.07.053.

[19] Makin TR, Krakauer JW. Against cortical reorganisation. Elife 2023;12:e84716. https://doi.org/10.7554/eLife.84716.

[20] Walhovd KB, Lövden M, Fjell AM. Timing of lifespan influences on brain and cognition. Trends Cognit Sci 2023;27(10):901−15. https://doi.org/10.1016/j.tics.2023.07.001.

[21] Borchert RJ, Azevedo T, Badhwar AP, Bernal J, Betts M, Bruffaerts R, Burkhart MC, Dewachter I, Gellersen HM, Low A, Lourida I, Machado L, Madan CR, Malpetti M, Mejia J, Michopoulou S, Muñoz-Neira C, Pepys J, Peres M, Phillips V, Ramanan S, Tamburin S, Tantiangco HM, Thakur L, Tomassini A, Vipin A, Tang E, Newby D, Ranson JM, Llewellyn DJ, Veldsman M, Rittman T. Artificial intelligence for diagnostic and prognostic neuroimaging in dementia: a systematic review. Alzheimers Dement 2023:1−20. https://doi.org/10.1002/alz.13412.

[22] Garcia-Garcia B, Mattern H, Vockert N, Yakupov R, Schreiber F, Spallazzi M, Perosa V, Haghikia A, Speck O, Düzel E, Maass A, Schreiber S. Vessel distance mapping: a novel methodology for assessing vascular-induced cognitive resilience. Neuroimage 2023;274. https://doi.org/10.1016/j.neuroimage.2023.120094.

[23] Georgakis MK, Fang R, Düring M, Wollenweber FA, Bode FJ, Stösser S, Kindlein C, Hermann P, Liman TG, Nolte CH, Kerti L, Ikenberg B, Bernkopf K, Poppert H, Glanz W, Perosa V, Janowitz D, Wagner M, Neumann K, Speck O, Dobisch L, Düzel E, Gesierich B, Dewenter A, Spottke A, Waegemann K, Görtler M, Wunderlich S, Endres M, Zerr I, Petzold G, Dichgans M. Cerebral small vessel disease burden and cognitive and functional outcomes after stroke: a multicenter prospective cohort study. Alzheimers Dement 2023;19:1152−63. https://doi.org/10.1002/alz.12744.

[24] Perosa V, Rotta J, Yakupov R, Kuijf HJ, Schreiber F, Oltmer JT, et al. Implications of quantitative susceptibility mapping at t Tesla MRI for microbleeds detection in cerebral small vessel disease. Front Neurol 2023;14:1112312. https://doi.org/10.3389/fneur.2023.1112312.

[25] Perosa V, Priester A, Ziegler G, Cardenas-Blanco A, Dobisch L, Spallazzi M, Assmann A, Maass A, Speck O, Oltmer J, Heinze HJ, Schreiber S, Düzel E. Hippocampal vascular reserve associated with cognitive performance and hippocampal volume. Brain 2020;143:622−34. https://doi.org/10.1093/brain/awz383.

[26] Selkoe DJ. The advent of Alzheimer treatments will change the trajectory of human aging. Nat Aging 2024;4:453−63. https://doi.org/10.1038/s43587-024-00611-5.

[27] Selkoe DJ. Alzheimer's disease is a synaptic failure. Science (1979) 2002;298:789−91. https://doi.org/10.1126/science.1074069.

[28] Berron D, Vogel JW, Insel PS, Pereira JB, Xie L, Wisse LEM, Yushkevich PA, Palmqvist S, Mattsson-Carlgren N, Stomrud E, Smith R, Strandberg O, Hansson O. Early stages of tau pathology and its associations with functional connectivity, atrophy and memory. Brain 2021;144:2771−83. https://doi.org/10.1093/brain/awab114.

[29] Maass A, Lockhart SN, Harrison TM, Bell RK, Mellinger T, Swinnerton K, Baker SL, Rabinovici GD, Jagust WJ. Entorhinal tau pathology, episodic memory decline, and neurodegeneration in aging. J Neurosci 2018;38:530−43. https://doi.org/10.1523/JNEUROSCI.2028-17.2017.

[30] Ossenkoppele R, Schonhaut DR, Schöll M, Lockhart SN, Ayakata N, Baker SL, O'Neil JP, Janabi M, Lazaris A, Vossel KA, Kramer JH, Gorno-Tempini ML, Miller BL, Jagust WJ, Rabinovici GD. Tau PET patterns mirror clinical and neuroanatomical variability in Alzheimer's disease. Brain 2016;139:1551−67. https://doi.org/10.1093/brain/aww041.

[31] Düzel E, Berron D, Schütze H, Cardenas-Blanco A, Metzger C, Betts M, Ziegler G, Chen Y, Dobisch L, Bittner D, Glanz W, Reuter M, Spottke A, Rudolph J, Brosseron F, Buerger K, Janowitz D, Fliessbach K, Heneka M, Laske C, Buchmann M, Nestor P, Peters O, Diesing D, Li S, Priller J, Spruth EJ, Altenstein S, Ramirez A, Schneider A, Kofler B, Speck O, Teipel S, Kilimann I, Dyrba M, Wiltfang J, Bartels C, Wolfsgruber S, Wagner M, Jessen F. CSF total tau levels are associated with hippocampal novelty

[32] irrespective of hippocampal volume. Alzheimers Dement 2018;10:782−90. https://doi.org/10.1016/j.dadm.2018.10.003.
[32] Busche MA, Hyman BT. Synergy between amyloid-β and tau in Alzheimer's disease. Nat Neurosci 2020;23:1183−93. https://doi.org/10.1038/s41593-020-0687-6.
[33] Edgar CJ, Vradenburg G, Hassenstab J. The 2018 revised FDA guidance for early Alzheimer's disease: establishing the meaningfulness of treatment effects. J Prev Alzheimers Dis 2019;6:223−7. https://doi.org/10.14283/jpad.2019.30.
[34] Bit A, Suri JS, Ranjani A. Anatomy and physiology of blood vessels. Flow Dynamics and Tissue Engineering of Blood Vessels; 2020. p. 1−16. https://doi.org/10.1088/978-0-7503-2088-7ch1.
[35] Schreiber S, Difrancesco JC. Impaired occipital cerebrovascular reactivity as a biomarker for vascular β-amyloid. Neurology 2020;95(10):415−6. https://doi.org/10.1212/WNL.0000000000010207.
[36] Binnewijzend MAA, Benedictus MR, Kuijer JPA, van der Flier WM, Teunissen CE, Prins ND, Wattjes MP, van Berckel BNM, Scheltens P, Barkhof F. Cerebral perfusion in the predementia stages of Alzheimer's disease. Eur Radiol 2016;26:506−14. https://doi.org/10.1007/s00330-015-3834-9.
[37] Clancy U, Gilmartin D, Jochems ACC, Knox L, Doubal FN, Wardlaw JM. Neuropsychiatric symptoms associated with cerebral small vessel disease: a systematic review and meta-analysis. Lancet Psychiatr 2021;8:225−36. https://doi.org/10.1016/s2215-0366(20)30431-4.
[38] Clancy U, Appleton JP, Arteaga C, Doubal FN, Bath PM, Wardlaw JM. Clinical management of cerebral small vessel disease: a call for a holistic approach. Chin Med J 2020;134:127−42. https://doi.org/10.1097/CM9.0000000000001177.
[39] Duering M, Biessels GJ, Brodtmann A, Chen C, Cordonnier C, de Leeuw F-E, Debette S, Frayne R, Jouvent E, Rost NS, Ter Telgte A, Al-Shahi Salman R, Backes WH, Bae H-J, Brown R, Chabriat H, De Luca A, DeCarli C, Dewenter A, Doubal FN, Ewers M, Field TS, Ganesh A, Greenberg S, Helmer KG, Hilal S, Jochems ACC, Jokinen H, Kuijf H, Lam BYK, Lebenberg J, MacIntosh BJ, Maillard P, Mok VCT, Pantoni L, Rudilosso S, Satizabal CL, Schirmer MD, Schmidt R, Smith C, Staals J, Thrippleton MJ, van Veluw SJ, Vemuri P, Wang Y, Werring D, Zedde M, Akinyemi RO, Del Brutto OH, Markus HS, Zhu Y-C, Smith EE, Dichgans M, Wardlaw JM. Neuroimaging standards for research into small vessel disease-advances since 2013. Lancet Neurol 2023;4422:2−4. https://doi.org/10.1016/S1474-4422(23)00131-X.
[40] Fouda AY, Fagan SC, Ergul A. Brain vasculature and cognition. Arterioscler Thromb Vasc Biol 2019;39(4):593−602. https://doi.org/10.1161/ATVBAHA.118.311906.
[41] Gorelick PB, Furie KL, Iadecola C, Smith EE, Waddy SP, Lloyd-Jones DM, et al. Defining optimal brain health in adults: a presidential advisory from the American Heart Association/American Stroke Association. Stroke 2017;48(10):e284−303. https://doi.org/10.1161/STR.0000000000000148.
[42] Wardlaw JM, Smith EE, Biessels GJ, Cordonnier C, Fazekas F, Frayne R, Lindley RI, O'Brien JT, Barkhof F, Benavente OR, Black SE, Brayne C, Breteler M, Chabriat H, DeCarli C, de Leeuw FE, Doubal F, Duering M, Fox NC, Greenberg S, Hachinski V, Kilimann I, Mok V, Oostenbrugge R van, Pantoni L, Speck O, Stephan BCM, Teipel S, Viswanathan A, Werring D, Chen C, Smith C, van Buchem M, Norrving B, Gorelick PB, Dichgans M. Neuroimaging standards for research into small vessel disease and its contribution to ageing and neurodegeneration. Lancet Neurol 2013;12:822−38. https://doi.org/10.1016/S1474-4422(13)70124-8.

[43] van Osch MJP, Teeuwisse WM, Chen Z, Suzuki Y, Helle M, Schmid S. Advances in arterial spin labelling MRI methods for measuring perfusion and collateral flow. J Cerebr Blood Flow Metabol 2018;38(9):1461−80. https://doi.org/10.1177/0271678X17713434.
[44] Macdonald ME, Frayne R. Cerebrovascular MRI: a review of state-of-the-art approaches, methods and techniques. NMR Biomed 2015;28:767−91. https://doi.org/10.1002/nbm.3322.
[45] Spallazzi M, Dobisch L, Becke A, Berron D, Stucht D, Oeltze-Jafra S, et al. Hippocampal vascularization patterns: a high-resolution 7 Tesla time-of-flight magnetic resonance angiography study. Neuroimage Clin 2019;21:101609. https://doi.org/10.1016/j.nicl.2018.11.019.
[46] Llamas Rodriguez J, van der Kouwe AJW, Oltmer J, Rosenblum E, Mercaldo N, Fischl B, et al. Entorhinal vessel density correlates with phosphorylated tau and TDP-43 pathology. Alzheimers Dement 2024;20(7):4649−62.
[47] Tucker W, Arora Y, Mahajan K. Anatomy, blood vessels. In: StatPearls. StatPearls Publishing, Treasure Island; 2021. p. 1−6.
[48] Daneman R. Prat A. The blood-brain barrier. Cold Spring Harb Perspect Biol 2015;7(1):a020412. https://doi.org/10.1101/cshperspect.a020412.
[49] Essig M, Shiroishi MS, Nguyen TB, Saake M, Provenzale JM, Enterline D, et al. Perfusion MRI: the five most frequently asked technical questions. Am J Roentgenol 2013;200(1):24−34. https://doi.org/10.2214/AJR.12.9543.
[50] Thrippleton MJ, Backes WH, Sourbron S, Ingrisch M, van Osch MJP, Dichgans M, Fazekas F, Ropele S, Frayne R, van Oostenbrugge RJ, Smith EE, Wardlaw JM. Quantifying blood-brain barrier leakage in small vessel disease: review and consensus recommendations. Alzheimers Dement 2019;44:1−19. https://doi.org/10.1016/j.jalz.2019.01.013.
[51] Heye AK, Culling RD, Valdés Hernández MDC, Thrippleton MJ, Wardlaw JM. Assessment of blood-brain barrier disruption using dynamic contrast-enhanced MRI. A systematic review. Neuroimage Clin 2014;6:262−74. https://doi.org/10.1016/j.nicl.2014.09.002.
[52] Chagnot A, Barnes SR, Montagne A. Magnetic resonance imaging of blood−brain barrier permeability in dementia. Neuroscience 2021;474:14−29. https://doi.org/10.1016/j.neuroscience.2021.08.003.
[53] Montagne A, Barnes SR, Sweeney MD, Halliday MR, Sagare AP, Zhao Z, Toga AW, Jacobs RE, Liu CY, Amezcua L, Harrington MG, Chui HC, Law M, Zlokovic BV. Blood-Brain barrier breakdown in the aging human hippocampus. Neuron 2015;85:296−302. https://doi.org/10.1016/j.neuron.2014.12.032.
[54] Raja R, Rosenberg GA, Caprihan A. MRI measurements of Blood-Brain Barrier function in dementia: a review of recent studies. Neuropharmacology 2018;134:259−71. https://doi.org/10.1016/j.neuropharm.2017.10.034.
[55] van de Haar HJ, Burgmans S, Jansen JFA, van Osch MJP, van Buchem MA, Muller M, et al. Blood-brain barrier leakage in patients with early Alzheimer disease. Radiology 2016;281(2):527−35. https://doi.org/10.1148/radiol.2017164043.
[56] Verheggen ICM, de Jong JJ, van Boxtel MPJ, Gronenschild EHBM, Palm WM, Postma AA, et al. Increase in blood − brain barrier leakage in healthy, older adults. Geroscience 2020;42(4):1183−93. https://doi.org/10.1007/s11357-020-00211-2.
[57] Villringer K, Grittner U, Brunecker P, Khalil AA. DCE-MRI blood − brain barrier assessment in acute ischemic stroke. Neurology 2017;88(5):433−40. https://doi.org/10.1212/WNL.0000000000003566.
[58] Wang H, Golob EJ, Su MY. Vascular volume and blood-brain barrier permeability measured by dynamic contrast enhanced MRI in hippocampus and cerebellum of patients

with MCI and normal controls. J Magn Reson Imag 2006;24:695−700. https://doi.org/10.1002/jmri.20669.

[59] Wardlaw JM, Makin SJ, Valdés Hernández MC, Armitage PA, Heye AK, Chappell FM, Muñoz-Maniega S, Sakka E, Shuler K, Dennis MS, Thrippleton MJ. Blood-brain barrier failure as a core mechanism in cerebral small vessel disease and dementia: evidence from a cohort study. Alzheimers Dement 2017;13:634−43. https://doi.org/10.1016/j.jalz.2016.09.006.

[60] Wardlaw JM, Doubal F, Armitage P, Chappell F, Carpenter T, Muñoz Maniega S, Farrall A, Sudlow C, Dennis M, Dhillon B. Lacunar stroke is associated with diffuse Blood-Brain barrier dysfunction. Ann Neurol 2009;65:194−202. https://doi.org/10.1002/ana.21549.

[61] Zhang E, Wong SM, Van De Haar HJ, Staals J, Jansen JFA, Jeukens CRLPN, Hofman PAM, Van Oostenbrugge RJ, Backes WH. Blood-brain barrier leakage is more widespread in patients with cerebral small vessel disease. Neurology 2017;88:426−32. https://doi.org/10.1212/WNL.0000000000003556.

[62] Neumann K, Günther M, Düzel E, Schreiber S. Microvascular impairment in patients with cerebral small vessel disease assessed with arterial spin labeling magnetic resonance imaging: a pilot study. Front Aging Neurosci 2022;14:871612. https://doi.org/10.3389/fnagi.2022.871612.

[63] Neumann K, Schidlowski M, Günther M, Stöcker T, Düzel E. Reliability and reproducibility of Hadamard encoded pseudo-continuous arterial spin labeling in healthy elderly. Front Neurosci 2021;15:711898. https://doi.org/10.3389/fnins.2021.711898.

[64] Hlauschek G, Nicolo JP, Sinclair B, Law M, Yasuda CL, Cendes F, et al. Role of the glymphatic system and perivascular spaces as a potential biomarker for post-stroke epilepsy. Epilepsia Open 2024;9(1):60−76. https://doi.org/10.1002/epi4.12877.

[65] Jessen NA, Munk ASF, Lundgaard I, Nedergaard M. The glymphatic system: a beginner's guide. Neurochem Res 2015;40:2583−99. https://doi.org/10.1007/s11064-015-1581-6.

[66] Troili F, Cipollini V, Moci M, Morena E, Palotai M, Rinaldi V, Romano C, Ristori G, Giubilei F, Salvetti M, Orzi F, Guttmann CRG, Cavallari M. Perivascular unit: this must Be the place. The anatomical crossroad between the immune, vascular and nervous system. Front Neuroanat 2020;14. https://doi.org/10.3389/fnana.2020.00017.

[67] Wardlaw JM, Benveniste H, Nedergaard M, Zlokovic BV, Mestre H, Lee H, Doubal FN, Brown R, Ramirez J, MacIntosh BJ, Tannenbaum A, Ballerini L, Rungta RL, Boido D, Sweeney M, Montagne A, Charpak S, Joutel A, Smith KJ, Black SE. Perivascular spaces in the brain: anatomy, physiology and pathology. Nat Rev Neurol 2020;16:137−53. https://doi.org/10.1038/s41582-020-0312-z.

[68] Francis F, Ballerini L, Wardlaw JM. Perivascular spaces and their associations with risk factors, clinical disorders and neuroimaging features: a systematic review and meta-analysis. Int J Stroke 2019;14:359−71. https://doi.org/10.1177/1747493019830321.

[69] Okar SV, Hu F, Shinohara RT, Beck ES, Reich DS, Ineichen BV. The etiology and evolution of magnetic resonance imaging-visible perivascular spaces: systematic review and meta-analysis. Front Neurosci 2023;17:1−13. https://doi.org/10.3389/fnins.2023.1038011.

[70] Aribisala BS, Wiseman S, Morris Z, Valdés-Hernández MC, Royle NA, Maniega SM, Gow AJ, Corley J, Bastin ME, Starr J, Deary IJ, Wardlaw JM. Circulating inflammatory markers are associated with magnetic resonance imaging-visible perivascular spaces but not directly with white matter hyperintensities. Stroke 2014;45:605−7. https://doi.org/10.1161/STROKEAHA.113.004059.

[71] Ineichen BV, Okar SV, Proulx ST, Engelhardt B, Lassmann H, Reich DS. Perivascular spaces and their role in neuroinflammation. Neuron 2022;110:3566−81. https://doi.org/10.1016/j.neuron.2022.10.024.

[72] Satizabal CL, Zhu Y-C, Dufouil C, Tzourio C. Inflammatory proteins and the severity of dilated Virchow-Robin spaces in the elderly. J Alzheim Dis 2012;33:323−8. https://doi.org/10.3233/JAD-2012-120874.

[73] Zimmerman B, Rypma B, Gratton G, Fabiani M. Age-related changes in cerebrovascular health and their effects on neural function and cognition: a comprehensive review. Psychophysiology 2021;58:1−39. https://doi.org/10.1111/psyp.13796.

[74] Charidimou A, Boulouis G, Pasi M, Auriel E, van Etten ES, Haley K, Ayres A, Schwab KM, Martinez-Ramirez S, Goldstein JN, Rosand J, Viswanathan A, Greenberg SM, Gurol ME. MRI-visible perivascular spaces in cerebral amyloid angiopathy and hypertensive arteriopathy. Neurology 2017;88:1157−64. https://doi.org/10.1212/WNL.0000000000003746.

[75] Perosa V, Oltmer J, Munting LP, Freeze WM, Auger CA, Scherlek AA, van der Kouwe AJ, Iglesias JE, Atzeni A, Bacskai BJ, Viswanathan A, Frosch MP, Greenberg SM, van Veluw SJ. Perivascular space dilation is associated with vascular amyloid-β accumulation in the overlying cortex. Acta Neuropathol 2022;143:331−48. https://doi.org/10.1007/s00401-021-02393-1.

[76] Van Veluw SJ, Biessels GJ, Bouvy WH, Spliet WGM, Zwanenburg JJM, Luijten PR, Macklin EA, Rozemuller AJM, Gurol ME, Greenberg SM, Viswanathan A, Martinez-Ramirez S. Cerebral amyloid angiopathy severity is linked to dilation of juxtacortical perivascular spaces. J Cerebr Blood Flow Metab 2016;36:576−80. https://doi.org/10.1177/0271678X15620434.

[77] Barnes A, Ballerini L, Valdés Hernández M del C, Chappell FM, Muñoz Maniega S, Meijboom R, et al. Topological relationships between perivascular spaces and progression of white matter hyperintensities: a pilot study in a sample of the Lothian Birth Cohort 1936. Front Neurol 2022;13:889884. https://doi.org/10.3389/fneur.2022.889884.

[78] Loos CMJ, Klarenbeek P, van Oostenbrugge RJ, Staals J. Association between perivascular spaces and progression of white matter hyperintensities in lacunar stroke patients. PLoS One 2015;10:e0137323. https://doi.org/10.1371/journal.pone.0137323.

[79] Kern KC, Nasrallah IM, Bryan RN, Reboussin DM, Wright CB. Intensive systolic blood pressure treatment remodels brain perivascular spaces: a secondary analysis of the Systolic Pressure Intervention Trial (SPRINT). Neuroimage Clin 2023;40:103513. https://doi.org/10.1016/j.nicl.2023.103513.

[80] Kim HG, Shin N-Y, Nam Y, Yun E, Yoon U, Lee HS, Ahn KJ. MRI-Visible dilated perivascular space in the brain by age: the human connectome project. Radiology 2023;306:1−9. https://doi.org/10.1148/radiol.213254.

[81] Lynch KM, Sepehrband F, Toga AW, Choupan J. Brain perivascular space imaging across the human lifespan. Neuroimage 2023;271:120009. https://doi.org/10.1016/j.neuroimage.2023.120009.

[82] Menze I, Bernal J, Kaya P, Aki Ç, Pfister M, Geisendörfer J, et al. Perivascular space enlargement accelerates in ageing and Alzheimer's disease pathology: evidence from a three-year longitudinal multicentre study. Alz Res Ther 2024;16:242. https://doi.org/10.1186/s13195-024-01603-8.

[83] Vikner T, Karalija N, Eklund A, Malm J, Lundquist A, Gallewicz N, Dahlin M, Lindenberger U, Riklund K, Bäckman L, Nyberg L, Wåhlin A. 5-Year associations among

[84] Mestre H, Tithof J, Du T, Song W, Peng W, Sweeney AM, Olveda G, Thomas JH, Nedergaard M, Kelley DH. Flow of cerebrospinal fluid is driven by arterial pulsations and is reduced in hypertension. Nat Commun 2018;9:4878. https://doi.org/10.1038/s41467-018-07318-3.

[85] Ungvari Z, Toth P, Tarantini S, Prodan CI, Sorond F, Merkely B, Csiszar A. Hypertension-induced cognitive impairment: from pathophysiology to public health. Nat Rev Nephrol 2021a;17:639−54. https://doi.org/10.1038/s41581-021-00430-6.

[86] Vilor-Tejedor N, Ciampa I, Operto G, Falcón C, Suárez-Calvet M, Crous-Bou M, Shekari M, Arenaza-Urquijo EM, Milà-Alomà M, Grau-Rivera O, Minguillon C, Kollmorgen G, Zetterberg H, Blennow K, Guigo R, Molinuevo JL, Gispert JD, Beteta A, Brugulat A, Cacciaglia R, Cañas A, Deulofeu C, Cumplido I, Dominguez R, Emilio M, Fauria K, Fuentes S, Hernandez L, Huesa G, Huguet J, Marne P, Menchón T, Polo A, Pradas S, Rodriguez-Fernandez B, Sala-Vila A, Sánchez-Benavides G, Salvadó G, Soteras A, Vilanova M. Perivascular spaces are associated with tau pathophysiology and synaptic dysfunction in early Alzheimer's continuum. Alzheimers Res Ther 2021;13:135. https://doi.org/10.1186/s13195-021-00878-5.

[87] Wang M-L, Yu M-M, Wei X-E, Li W-B, Li Y-H. Association of enlarged perivascular spaces with Aβ and tau deposition in cognitively normal older population. Neurobiol Aging 2021;100:32−8. https://doi.org/10.1016/j.neurobiolaging.2020.12.014.

[88] Iliff JJ, Wang M, Liao Y, Plogg BA, Peng W, Gundersen GA, Benveniste H, Vates GE, Deane R, Goldman SA, Nagelhus EA, Nedergaard M. A paravascular pathway facilitates CSF flow through the brain parenchyma and the clearance of interstitial solutes, including amyloid β. Sci Transl Med 2012;4:1−12. https://doi.org/10.1126/scitranslmed.3003748.

[89] Rasmussen MK, Mestre H, Nedergaard M. The glymphatic pathway in neurological disorders. Lancet Neurol 2018;17:1016−24. https://doi.org/10.1016/S1474-4422(18)30318-1.

[90] Tarasoff-Conway JM, Carare RO, Osorio RS, Glodzik L, Butler T, Fieremans E, Axel L, Rusinek H, Nicholson C, Zlokovic BV, Frangione B, Blennow K, Ménard J, Zetterberg H, Wisniewski T, De Leon MJ. Clearance systems in the brain - implications for Alzheimer disease. Nat Rev Neurol 2015;11:457−70. https://doi.org/10.1038/nrneurol.2015.119.

[91] Braun M, Iliff JJ. The impact of neurovascular, blood-brain barrier, and glymphatic dysfunction in neurodegenerative and metabolic diseases. In: Söderbom G, Esterline R, Oscarsson J, Mattson MP, editors. International review of neurobiology. 1st ed.154. Elsevier Inc.; 2020. p. 413−36.

[92] Barisano G, Lynch KM, Sibilia F, Lan H, Shih NC, Sepehrband F, et al. Imaging perivascular space structure and function using brain MRI. Neuroimage 2022;257:119329. https://doi.org/10.1016/j.neuroimage.2022.119329.

[93] Moses J, Sinclair B, Law M, O'Brien TJ, Vivash L. Automated methods for detecting and quantitation of enlarged perivascular spaces on MRI. J Magn Reson Imag 2023;57(1):11−24. https://doi.org/10.1002/jmri.28369.

[94] Pham W, Lynch M, Spitz G, O'Brien T, Vivash L, Sinclair B, et al. A critical guide to the automated quantification of perivascular spaces in magnetic resonance imaging. Front Neurosci 2022;16:1021311. https://doi.org/10.3389/fnins.2022.1021311.

[95] Waymont JMJ, Valdés Hernández MC, Bernal J, Duarte Coello R, Brown R, Chappell FM, et al. A systematic review and meta-analysis of automated methods for quantifying enlarged perivascular spaces in the brain. Neuroimage 2024;297:120685. https://doi.org/10.1016/j.neuroimage.2024.120685.

[96] Ballerini L, Booth T, Valdés Hernández M del C, Wiseman S, Lovreglio R, Muñoz Maniega S, Morris Z, Pattie A, Corley J, Gow A, Bastin ME, Deary IJ, Wardlaw J. Computational quantification of brain perivascular space morphologies: associations with vascular risk factors and white matter hyperintensities. A study in the Lothian Birth Cohort 1936. Neuroimage Clin 2020;25:102120. https://doi.org/10.1016/j.nicl.2019.102120.

[97] Bernal J, Valdés-hernández MC, Escudero J, Duarte R, Ballerini L, Mark E. Assessment of PVS enhancement methods using a three-dimensional computational model. Magn Reson Imaging 2022;93:33–51. https://doi.org/10.1016/j.mri.2022.07.016.

[98] Duarte Coello R, Valdés Hernández M del C, Zwanenburg JJM, van der Velden M, Kuijf HJ, De Luca A, et al. Detectability and accuracy of computational measurements of in-silico and physical representations of enlarged perivascular spaces from magnetic resonance images. J Neurosci Methods 2024;403:110039. https://doi.org/10.1016/j.jneumeth.2023.110039.

[99] Valdés Hernández M del C, Duarte Coello R, Xu W, Bernal J, Cheng Y, Ballerini L, et al. Influence of threshold selection and image sequence in in-vivo segmentation of enlarged perivascular spaces. J Neurosci Methods 2024;403:110037. https://doi.org/10.1016/j.jneumeth.2023.110037.

[100] Wardlaw JM, Liebeskind DS. Not just blood: brain fluid systems and their relevance to cerebrovascular diseases. Stroke 2022;53(4):1399–401. https://doi.org/10.1161/STROKEAHA.122.037448.

[101] Zhang JF, Lim HF, Chappell FM, Clancy U, Wiseman S, Valdés-Hernández MC, Garcia DJ, Bastin ME, Doubal FN, Hewins W, Cox SR, Maniega SM, Thrippleton M, Stringer M, Jardine C, McIntyre D, Barclay G, Hamilton I, Kesseler L, Murphy M, Perri C Di, Wu YC, Wardlaw JM. Relationship between inferior frontal sulcal hyperintensities on brain MRI, ageing and cerebral small vessel disease. Neurobiol Aging 2021;106:130–8. https://doi.org/10.1016/j.neurobiolaging.2021.06.013.

[102] De Leon MJ, Li Y, Okamura N, Tsui WH, Saint-Louis LA, Glodzik L, Osorio RS, Fortea J, Butler T, Pirraglia E, Fossati S, Kim HJ, Carare RO, Nedergaard M, Benveniste H, Rusinek H. Cerebrospinal fluid clearance in Alzheimer disease measured with dynamic PET. J Nucl Med 2017;58:1471–6. https://doi.org/10.2967/jnumed.116.187211.

[103] Xu S, Xie L, Zhang Y, Wu X, Hong H, Zhang R, Zeng Q, Li K, Luo X, Zhang M, Sun J, Huang P. Inferior frontal sulcal hyperintensity on FLAIR is associated with small vessel disease but not Alzheimer's pathology. J Alzheim Dis 2023;92:1357–65. https://doi.org/10.3233/JAD-220843.

[104] Mestre H, Kostrikov S, Mehta RI, Nedergaard M. Perivascular spaces, glymphatic dysfunction, and small vessel disease. Physiol Behav 2017;176:139–48. https://doi.org/10.1042/CS20160381.Perivascular.

[105] Wardlaw JM, Smith C, Dichgans M. Small vessel disease: mechanisms and clinical implications. Lancet Neurol 2019;18:684–96. https://doi.org/10.1016/S1474-4422(19)30079-1.

[106] Alber J, Alladi S, Bae HJ, Barton DA, Beckett LA, Bell JM, Berman SE, Biessels GJ, Black SE, Bos I, Bowman GL, Brai E, Brickman AM, Callahan BL, Corriveau RA, Fossati S, Gottesman RF, Gustafson DR, Hachinski V, Hayden KM, Helman AM, Hughes TM, Isaacs JD, Jefferson AL, Johnson SC, Kapasi A, Kern S, Kwon JC, Kukolja J, Lee A, Lockhart SN, Murray A, Osborn KE, Power MC, Price BR, Rhodius-Meester HFM, Rondeau JA, Rosen AC, Rosene DL, Schneider JA, Scholtzova H, Shaaban CE, Silva NCBS, Snyder HM, Swardfager W, Troen AM, van Veluw SJ, Vemuri P, Wallin A, Wellington C, Wilcock DM, Xie SX, Hainsworth AH. White matter hyperintensities in

vascular contributions to cognitive impairment and dementia (VCID): knowledge gaps and opportunities. Alzheimers Dement 2019;5:107−17. https://doi.org/10.1016/j.trci.2019.02.001.

[107] Wardlaw JM, Valdés Hernández MC, Muñoz-Maniega S. What are white matter hyperintensities made of? Relevance to vascular cognitive impairment. J Am Heart Assoc 2015;4:001140. https://doi.org/10.1161/JAHA.114.001140.

[108] Garnier-crussard A, Krolak-salmon P, Garnier-crussard A, Cotton F, Krolak-salmon P. White matter hyperintensities in Alzheimer's disease : beyond vascular contribution. Alzheimers Dement 2023;19(8):3738−48. https://doi.org/10.1002/alz.13057.

[109] Carmelli D, Swan GE, Reed T, Wolf PA, Miller BL, DeCarli C. Midlife cardiovascular risk factors and brain morphology in identical older male twins. Neurology 1999;52:1119−24. https://doi.org/10.1212/wnl.52.6.1119.

[110] Enzinger C, Fazekas F, Matthews PM, Ropele S, Schmidt H, Smith S, Schmidt R. Risk factors for progression of brain atrophy in aging: six-year follow-up of normal subjects. Neurology 2005;64:1704−11. https://doi.org/10.1212/01.WNL.0000161871.83614.BB.

[111] Jochems ACC, Arteaga C, Chappell F, Ritakari T, Hooley M, Doubal F, Maniega SM, Wardlaw JM. Longitudinal changes of white matter hyperintensities in sporadic small vessel disease: a systematic review and meta-analysis. Neurology 2022;99:E2454−63. https://doi.org/10.1212/WNL.0000000000201205.

[112] Ong M, Foo H, Chander RJ, Wen MC, Au WL, Sitoh YY, Tan L, Kandiah N. Influence of diabetes mellitus on longitudinal atrophy and cognition in Parkinson's disease. J Neurol Sci 2017;377:122−6. https://doi.org/10.1016/j.jns.2017.04.010.

[113] Xu J, Li Y, Lin H, Sinha R, Potenza MN. Body mass index correlates negatively with white matter integrity in the fornix and corpus callosum: a diffusion tensor imaging study. Hum Brain Mapp 2013;34:1044−52. https://doi.org/10.1002/hbm.21491.

[114] Brown WR, Moody DM, Thore CR, Challa VR. Apoptosis in leukoaraiosis. Am J Neuroradiol 2000;21:79−82. https://doi.org/10.1097/00005072-199905000-00093.

[115] Jouvent E, Mangin JF, Duchesnay E, Porcher R, Düring M, Mewald Y, Guichard JP, Hervé D, Reyes S, Zieren N, Dichgans M, Chabriat H. Longitudinal changes of cortical morphology in CADASIL. Neurobiol Aging 2012;33:1002.e29−36. https://doi.org/10.1016/j.neurobiolaging.2011.09.013.

[116] Jouvent E, Viswanathan A, Chabriat H. Cerebral atrophy in cerebrovascular disorders. J Neuroimaging 2010;10:213−8. https://doi.org/10.1111/j.1552-6569.2009.00370.x.

[117] Nasrabady SE, Rizvi B, Goldman JE, Brickman AM. White matter changes in Alzheimer's disease: a focus on myelin and oligodendrocytes. Acta Neuropathol Commun 2018;6:22. https://doi.org/10.1186/s40478-018-0515-3.

[118] Obulesu M, Lakshmi MJ. Apoptosis in Alzheimer's disease: an understanding of the physiology, pathology and therapeutic avenues. Neurochem Res 2014;39:2301−12. https://doi.org/10.1007/s11064-014-1454-4.

[119] Ter Telgte A, Van Leijsen EMC, Wiegertjes K, Klijn CJM, Tuladhar AM, De Leeuw FE. Cerebral small vessel disease: from a focal to a global perspective. Nat Rev Neurol 2018;14:387−98. https://doi.org/10.1038/s41582-018-0014-y.

[120] Garnier-Crussard A, Bougacha S, Wirth M, Dautricourt S, Sherif S, Landeau B, Gonneaud J, De Flores R, de la Sayette V, Vivien D, Krolak-Salmon P, Chételat G. White matter hyperintensity topography in Alzheimer's disease and links to cognition. Alzheimers Dement 2022;18:422−33. https://doi.org/10.1002/alz.12410.

[121] Li C, Zhu Y, Ma Y, Hua R, Zhong B, Xie W. Association of cumulative blood pressure with cognitive decline, dementia, and mortality. J Am Coll Cardiol 2022;79:1321−35. https://doi.org/10.1016/j.jacc.2022.01.045.

[122] Palta P, Albert MS, Gottesman RF. Heart health meets cognitive health: evidence on the role of blood pressure. Lancet Neurol 2021;20:854−67. https://doi.org/10.1016/S1474-4422(21)00248-9.

[123] Desmarais P, Gao AF, Lanctôt K, Rogaeva E, Ramirez J, Herrmann N, Stuss DT, Black SE, Keith J, Masellis M. White matter hyperintensities in autopsy-confirmed frontotemporal lobar degeneration and Alzheimer's disease. Alzheimers Res Ther 2021;13:1−16. https://doi.org/10.1186/s13195-021-00869-6.

[124] Pålhaugen L, Sudre CH, Tecelao S, Nakling A, Almdahl IS, Kalheim LF, Cardoso MJ, Johnsen SH, Rongve A, Aarsland D, Bjørnerud A, Selnes P, Fladby T. Brain amyloid and vascular risk are related to distinct white matter hyperintensity patterns. J Cerebr Blood Flow Metabol 2021;41:1162−74. https://doi.org/10.1177/0271678X20957604.

[125] Rizvi B, Sathishkumar M, Kim S, Márquez F, Granger SJ, Larson MS, et al. Posterior white matter hyperintensities are associated with reduced medial temporal lobe subregional integrity and long-term memory in older adults. Neuroimage Clin 2023;37:103308. https://doi.org/10.1016/j.nicl.2022.103308.

[126] Shirzadi Z, Schultz SA, Yau WYW, Joseph-Mathurin N, Fitzpatrick CD, Levin R, et al. Etiology of white matter hyperintensities in autosomal dominant and sporadic Alzheimer disease. JAMA Neurol 2023;80(10):1353−63. https://doi.org/10.1001/jamaneurol.2023.3618.

[127] Bernal J, Schreiber S, Menze I, Ostendorf A, Pfister M, Geisendörfer J, et al. Arterial hypertension and β-amyloid accumulation have spatially overlapping effects on posterior white matter hyperintensity volume: a cross-sectional study. Alzheimers Res Ther 2023;15(1):97. https://doi.org/10.1186/s13195-023-01243-4.

[128] Botz J, Lohner V, Schirmer MD. Spatial patterns of white matter hyperintensities. Syst Rev 2023:1−13. https://doi.org/10.3389/fnagi.2023.1165324.

[129] Englund E. Neuropathology of white matter changes in Alzheimer's disease and vascular dementia. Dement Geriatr Cogn Disord 1998;9:6−12. https://doi.org/10.1159/000051183.

[130] Gaubert M, Lange C, Garnier-Crussard A, Köbe T, Bougacha S, Gonneaud J, de Flores R, Tomadesso C, Mézenge F, Landeau B, de la Sayette V, Chételat G, Wirth M. Topographic patterns of white matter hyperintensities are associated with multimodal neuroimaging biomarkers of Alzheimer's disease. Alzheimers Res Ther 2021;13:1−11. https://doi.org/10.1186/s13195-020-00759-3.

[131] Habes M, Sotiras A, Erus G, Toledo JB, Janowitz D, Wolk DA, Shou H, Bryan NR, Doshi J, Völzke H, Schminke U, Hoffmann W, Resnick SM, Grabe HJ, Davatzikos C. White matter lesions spatial heterogeneity, links to risk factors, cognition, genetics, and atrophy. Neurology 2018;91:E964−75. https://doi.org/10.1212/WNL.0000000000006116.

[132] Huynh K, Piguet O, Kwok J, Dobson-Stone C, Halliday GM, Hodges JR, Landin-Romero R. Clinical and biological correlates of white matter hyperintensities in patients with behavioral-variant frontotemporal dementia and Alzheimer disease. Neurology 2021;96:e1743−54. https://doi.org/10.1212/WNL.0000000000011638.

[133] McAleese KE, Walker L, Graham S, Moya ELJ, Johnson M, Erskine D, Colloby SJ, Dey M, Martin-Ruiz C, Taylor JP, Thomas AJ, McKeith IG, De Carli C, Attems J. Parietal white matter lesions in Alzheimer's disease are associated with cortical neurodegenerative pathology, but not with small vessel disease. Acta Neuropathol 2017;134:459−73. https://doi.org/10.1007/s00401-017-1738-2.

[134] Phuah C, Chen Y, Strain JF, Yechoor N. Association of data-driven white matter hyperintensity spatial signatures with distinct cerebral small vessel disease etiologies. Neurology 2022;99(23):e2535−47. https://doi.org/10.1212/WNL.0000000000201186.

[135] Weaver NA, Doeven T, Barkhof F, Biesbroek JM, Groeneveld ON, Kuijf HJ, Prins ND, Scheltens P, Teunissen CE, van der Flier WM, Biessels GJ. Cerebral amyloid burden is associated with white matter hyperintensity location in specific posterior white matter regions. Neurobiol Aging 2019;84:225−34. https://doi.org/10.1016/j.neurobiolaging.2019.08.001.

[136] Appelman APA, Exalto LG, Van Der Graaf Y, Biessels GJ, Mali WPTM, Geerlings MI. White matter lesions and brain atrophy: more than shared risk factors? A systematic review. Cerebrovasc Dis 2009;28:227−42. https://doi.org/10.1159/000226774.

[137] Dadar M, Manera AL, Ducharme S, Collins DL. White matter hyperintensities are associated with grey matter atrophy and cognitive decline in Alzheimer's disease and frontotemporal dementia. Neurobiol Aging 2022;111:54−63. https://doi.org/10.1016/j.neurobiolaging.2021.11.007.

[138] Lambert C, Benjamin P, Zeestraten E, Lawrence AJ, Barrick TR, Markus HS. Longitudinal patterns of leukoaraiosis and brain atrophy in symptomatic small vessel disease. Brain 2016;139:1136−51. https://doi.org/10.1093/brain/aww009.

[139] Rizvi B, Lao PJ, Chesebro AG, Dworkin JD, Amarante E, Beato JM, Gutierrez J, Zahodne LB, Schupf N, Manly JJ, Mayeux R, Brickman AM. Association of regional white matter hyperintensities with longitudinal Alzheimer-like pattern of neurodegeneration in older adults. JAMA Netw Open 2021:1−13. https://doi.org/10.1001/jamanetworkopen.2021.25166.

[140] Dalby RB, Eskildsen SF, Videbech P, Frandsen J, Mouridsen K, Sørensen L, et al. Oxygenation differs among white matter hyperintensities, intersected fiber tracts and unaffected white matter. Brain Commun 2019;1(1):fcz033. https://doi.org/10.1093/braincomms/fcz033.

[141] van Veluw SJ, Arfanakis K, Schneider JA. Neuropathology of vascular brain health: insights from ex vivo magnetic resonance imaging-histopathology studies in cerebral small vessel disease. Stroke 2022;53:404−15. https://doi.org/10.1161/STROKEAHA.121.032608.

[142] Behl C. Apoptosis and Alzheimer's disease review. JNT (J Neural Transm) 2000;107(11):1325−44. https://doi.org/10.1007/s007020070021.

[143] Mayer C, Frey BM, Schlemm E, Petersen M, Engelke K, Hanning U, Jagodzinski A, Borof K, Fiehler J, Gerloff C, Thomalla G, Cheng B. Linking cortical atrophy to white matter hyperintensities of presumed vascular origin. J Cerebr Blood Flow Metabol 2021;41:1682−91. https://doi.org/10.1177/0271678X20974170.

[144] McAleese KE, Firbank M, Dey M, Colloby SJ, Walker L, Johnson M, Beverley JR, Taylor JP, Thomas AJ, O'Brien JT, Attems J. Cortical tau load is associated with white matter hyperintensities. Acta Neuropathol Commun 2015;3:60. https://doi.org/10.1186/s40478-015-0240-0.

[145] Salvadores N, Gerónimo-Olvera C, Court FA. Axonal degeneration in AD: the contribution of Aβ and tau. Front Aging Neurosci 2020;12:581767. https://doi.org/10.3389/fnagi.2020.581767.

[146] Bernal J, Menze I, Yakupov R, Peters O, Hellmann-Regen J, Freiesleben SD, et al. Longitudinal evidence for a mutually reinforcing relationship between white matter hyperintensities and cortical thickness in cognitively unimpaired older adults. Alz Res Ther 2024;16:240. https://doi.org/10.1186/s13195-024-01606-5.

Chapter 6c

Advances in fluid-based biomarkers

Henrik Zetterberg[a,b,c,d,e,f]

[a]*Department of Psychiatry and Neurochemistry, Institute of Neuroscience & Physiology, Sahlgrenska Academy at the University of Gothenburg, Mölndal, Sweden;* [b]*Clinical Neurochemistry Laboratory, Sahlgrenska University Hospital, Mölndal, Sweden;* [c]*Department of Neurodegenerative Disease, UCL Institute of Neurology, London, United Kingdom;* [d]*UK Dementia Research Institute at UCL, London, United Kingdom;* [e]*Hong Kong Center for Neurodegenerative Diseases, Clear Water Bay, Hong Kong, China;* [f]*Wisconsin Alzheimer's Disease Research Center, University of Wisconsin School of Medicine and Public Health, University of Wisconsin—Madison, Madison, WI, United States*

Introduction

Alzheimer's disease (AD) is a slowly progressive neurodegenerative disease and the most common cause of dementia. The first detectable pathology of the disease is the accumulation of 42 amino acid-long amyloid β (Aβ) protein in extracellular plaques in the brain, which occurs decades before clinical symptom onset [1]. Biomarker studies suggest that Aβ accumulation is followed by increased phosphorylation and secretion of tau [2], a microtubule-binding axonal protein that is highly expressed in cortical neurons [3]. This dysfunctional tau metabolism is strongly associated with neuronal degeneration with the development of intraneuronal neurofibrillary tangles that are composed of hyperphosphorylated and truncated tau proteins [4]. Neurodegeneration eventually translates into the AD clinical syndrome, with cognitive symptoms that worsen as the disease progresses [5]. Four fluid-based biomarkers have been developed into diagnostic tests for these essential brain changes in the AD process: the ratio of 42 to 40 amino acid-long amyloid β peptides (Aβ42/Aβ40), a marker for plaque pathology; total-tau and phosphorylated tau (T-tau and P-tau, respectively), markers for AD-related changes in tau metabolism, phosphorylation, and secretion; and neurofilament light (NfL), a marker for neurodegeneration [6]. Originally, these biomarkers could only be measured in cerebrospinal fluid (CSF), but technological progress resulting in improved analytical sensitivity makes it possible to measure these biomarkers in standard blood samples as well.

Here, I provide an overview of the biomarkers that reflect the core components of AD pathology, including biomarkers for Aβ and tau pathology and neurodegeneration, in line with the amyloid (A), tau (T) and neurodegeneration (N) classification scheme for AD biomarkers [7]. In addition, I describe the work that led to clinical implementation of the CSF biomarkers, and the development of clinically viable and easy-to-use blood tests for AD.

Fluid biomarkers for Aβ pathology

Extracellular deposition of Aβ into plaques, the key pathological feature of AD, has been proposed as a major pathogenic event in the disease [8]. The development of tools to measure Aβ pathology in vivo and before autopsy via biomarkers in CSF started in the 1990s [9], but it was not until 2020 that CSF Aβ42 measurement was fully standardized, a method that uses certified reference materials and methods [10].

AD CSF is characterized by a 50% reduction in the concentration of the 42 amino acid-long and aggregation-prone form of Aβ (Aβ42) [11]. Aβ42 is a secreted cleavage product of amyloid precursor protein (APP) that normally is mobilized from the brain interstitial fluid into the CSF and blood, likely via the glymphatic system [12]. In AD, Aβ42 aggregates in the brain parenchyma, resulting in reduced CSF levels of the protein [13]. The diagnostic accuracy for Aβ pathology can be increased by dividing the concentration of aggregation-prone Aβ42 by the concentration of soluble 40 amino acid Aβ40 as a normalizer for inter-individual differences in Aβ production. The CSF Aβ42/Aβ40 ratio is close to 100% concordant with neuroimaging evidence of cortical β-amyloid deposition obtained from amyloid positron emission tomography (PET) [14], and discordant subjects who are typically CSF-positive and PET-negative often turn PET-positive within a few years [14–16].

For many years, there was not much hope for a reliable blood test for cerebral Aβ pathology [11], but recent findings suggest that plasma Aβ42 in ratio with Aβ40 (measured by immunoprecipitation mass spectrometry or ultrasensitive enzyme-linked immunosorbent assays) reflects cerebral Aβ pathology with relatively high accuracy against both amyloid PET and CSF Aβ42/Aβ40 ratio [17–20]. A recent validation study using a fully automated immunoassay (Elecsys) to measure plasma Aβ42 and Aβ40 further underscores the promising capability of plasma Aβ in clinical laboratory practice [21]. A previous study has published easy-to-use protocols for pre-analytical sample handling that is compatible across all plasma biomarkers for AD [22].

While the technological developments described above have been important for showing high diagnostic performance of plasma Aβ for identification of AD and brain amyloidosis, a contributing factor is likely that most new studies have used amyloid PET as the reference standard. This reduces the risk associated with the historical use of clinical diagnosis as the reference, given that some patients diagnosed on clinical grounds as having probable AD are found to be biomarker-negative and a proportion of cognitively unimpaired elderly may have pre-symptomatic Aβ pathology. Having a proportion of misdiagnosed cases and controls will markedly reduce the chance to find differences in plasma Aβ, given that the Aβ42/Aβ40 ratio is reduced by only 10%–12% in plasma [17–20], compared with 50% in CSF [11]. A complicating factor for plasma Aβ tests is also that the correlation between plasma and CSF levels is weak, which could be explained by production of Aβ peptides in platelets and other non-cerebral tissues. Nevertheless, in spite of these robustness issues, the concordant research findings using high-precision analytical tools still represent an important research advancement toward clinical implementation, perhaps using staged testing (e.g., an Aβ test in blood favoring sensitivity over specificity, followed by a more specific blood-, CSF- or imaging-based test in memory clinics).

Biomarkers for tau pathology

A key pathological feature of AD is the aggregation of hyperphosphorylated forms of the axonal protein tau in the neuronal soma resulting in neurofibrillary tangles, although tau inclusions in neurons or glial cells are also found in some non-AD neurodegenerative dementias such as progressive supranuclear palsy and some forms of frontotemporal dementia [23]. The cornerstone CSF markers total tau (T-tau) and phosphorylated tau (P-tau) have been proposed together with the Aβ42/Aβ40 ratio as biomarkers that can be used to biologically define AD [24] and are considered diagnostic in the research criteria for AD [25]. Both CSF T-tau and P-tau concentrations reflect AD-related pathophysiology, although they do not reflect tau pathology in non-AD tauopathies [26,27]. The most likely explanation for this is that the increased CSF levels of tau are due to increased phosphorylation and secretion of tau from neurons as a neuronal response to Aβ exposure [28,29]. Therefore, CSF T-tau and P-tau may be regarded as predictive markers of AD-type neurodegeneration and tangle formation but not direct markers of these processes (and not markers of non-AD tauopathies, which require improved biomarkers). However, CSF T-tau also increases in disorders with rapid neurodegeneration without amyloid or tau pathology such as Creutzfeldt-Jakob disease [30] and in acute conditions such as stroke and brain trauma [31,32], suggesting that it also may reflect neuronal injury in these conditions. Fully automated T-tau and P-tau assays for clinical use are available [33,34], and standardization work is ongoing in collaborative efforts between IFCC and the Global Biomarker Standardization Consortium.

The three most studied phospho-forms of tau in biomarker studies are P-tau181, P-tau217, and P-tau231. P-tau181 is the classical AD biomarker, and P-tau231 has been suggested to improve the differentiation of AD from frontotemporal dementia [35], but they seem to perform similarly well in this regard. Recent data suggest that CSF P-tau217 may correlate more strongly with tau pathology determined by PET and increase earlier in response to Aβ pathology than CSF P-tau181 [36], intriguing observations that warrant additional research.

Although ultrasensitive plasma T-tau assays can detect neuronal injury in acute brain disorders such as stroke and traumatic brain injury [37,38] similar to when T-tau is measured in CSF (see above), they work relatively poorly in AD settings [39], and the correlation with CSF is weak [40]. A potential explanation for this is that the assay set up may be vulnerable to proteolytic degradation of tau in the blood (the half-life of tau measured using currently available T-tau assays is 10 h [41] compared with around 20 days in CSF) [29]. Another possibility is that currently available T-tau assays in blood may measure peripheral tau; measuring a phospho-form of tau might make the test more CNS-specific.

Recently, we have seen several real breakthroughs in the plasma tau biomarker field. In 2017, Tatebe et al. found that P-tau181 concentration in AD plasma had increased levels compared with control samples, whose concentrations were below the lower limit of quantification, and that plasma and CSF P-tau levels had good correlation [42]. Mielke et al., using an immunoassay with electrochemiluminescence (ECL), demonstrated a correlation between P-tau181 and amyloid and tau PET, indicating that plasma P-tau181 is a good biomarker for brain AD pathology [43]. Using the same ECL immunoassay, Palmqvist et al. confirmed these findings and demonstrated that plasma P-tau181 associates with amyloid PET positivity and correlates strongly with CSF P-tau181 [2]. Interestingly, the change in plasma P-tau181 became significant before amyloid PET but after CSF and plasma Aβ42—i.e., already at sub-PET threshold Aβ pathology [2]. Thus, plasma P-tau181 might be diagnostically useful for the early detection of Aβ-related tau dysmetabolism as well as for disease staging. Recent large validation studies show very similar results [44–47], corroborating plasma P-tau as a robust blood biomarker for AD pathology. Plasma P-tau should be relatively easy to standardize and implement in clinical laboratory practice.

Most data currently available suggest that P-tau217 is earlier and more strongly associated with AD pathology than plasma P-tau181 [47,48], but more head-to-head comparisons are needed before a conclusion can be reached. For example, although plasma P-tau217 measured by the ECL immunoassay showed an AUC of 0.89 to differentiate neuropathologically-defined AD from non-AD in blood samples taken during life in one cohort [47], the corresponding number for plasma P-tau181 measured using Simoa was 0.97 in another cohort [49]. These findings call for further studies comparing different P-tau biomarkers in the same cohort. Interestingly, high plasma P-tau181 is found in tau PET-negative (Braak stage 0) individuals who have evidence of brain amyloidosis via amyloid PET [46], can predict subsequent AD dementia in cognitively-unimpaired individuals and MCI patients [45], and is significantly increased in pre-symptomatic familial AD mutation carriers 16 years before estimated symptom onset [50]. Plasma P-tau181, -217, and -231 assays are now commercially available and they are now being implemented in clinical laboratory practice globally.

Fluid biomarkers for neurodegeneration

Although CSF T-tau might better reflect Aβ-induced tau secretion in AD rather than general neurodegeneration [39], neurofilament light (NfL) has emerged as a strong biomarker candidate for the latter irrespective of the underlying disease process [51]. The biomarker can be measured in both CSF and plasma (or serum), and the correlation between CSF and blood concentrations is good to excellent (r values of 0.70–0.97) [52]. The highest NfL levels are seen in frontotemporal as well as vascular and HIV-associated dementias [53]. However, the findings in familial AD are also quite clear: mutation carriers show a sudden change in their blood NfL levels 10–15 years before expected clinical onset, which probably marks the onset and intensity of the neurodegenerative process [54,55]. In sporadic AD, there is a clear association of increased plasma NfL concentration with Aβ and tau PET positivity as well as with longitudinal neurodegeneration as determined by magnetic resonance imaging (MRI), but with an overlap across groups larger than in familial AD [56]. This difference might be due to the multitude of neurodegenerative changes that may cause NfL increase in people older than 70 years of age.

From research tools to clinical implementation

Since the early 2000s, European memory clinics have used CSF biomarkers to support diagnostic evaluation of patients with suspected AD, and these biomarkers have now been formally approved or recommended by regulatory authorities to support AD diagnostics in patients with clinical symptoms [57]. Additionally, the Alzheimer's Association has published Appropriate Use Criteria—i.e., specific clinical indications when the CSF tests are warranted in the diagnostic assessment of patients with suspected AD [58]. Standard operating procedures for pre-analytical sample handling have been agreed upon and published for both CSF [59] and plasma [22]. Reference methods and materials for CSF Aβ42 assay standardization [10] as well as high-precision clinical chemistry tests on fully automated instruments are in place [60], developments that bode well for full implementation of these biomarkers in clinical laboratory practice with uniform reference limits around the globe. In many European countries, CSF biomarkers are already used in clinical laboratory practice in accordance with country-specific regulations. Work on the reference measurement procedures for CSF Aβ40, T-tau, and P-tau181 is ongoing under the auspices of the International Federation of

Clinical Chemistry and Laboratory Medicine (IFCC) CSF Proteins working group; the Aβ40 part of this work should be concluded during 2020. Similar work has now been initiated for the blood tests as well; Appropriate Use Recommendations were recently published underscoring that the biomarkers currently should be used in symptomatic patients and that positive test results should be confirmed by CSF or imaging biomarkers before too strong diagnostic conclusions are made (this may change as we gain more knowledge on and confidence in the biomarkers) [60]. Plasma NfL is already an available test in clinical laboratory practice in Sweden, United Kingdom, the Netherlands, and France, and many laboratories are now validating the plasma P-tau tests for use in clinical laboratory practice.

Limitations of fluid-based biomarkers

A drawback of fluid biomarkers is the inability to determine brain region-specific changes, which may limit staging of disease severity and their use as progression markers. For plasma P-tau181 and P-tau217, step-wise increases with disease severity have been reported [46], but this is less clear for the other biomarkers. For example, the CSF Aβ42/Aβ40 ratio appears to be a bimodal marker (normal or abnormal) without a clear relationship between the degree of change and the extent of the pathology [14]. Tau and Aβ PET imaging may be done to provide a more direct assessment of disease stage in select clinical cases and in clinical trials.

Biomarkers for AD—how can they be used in the most effective manner now that we have approved disease-modifying therapies?

To date, two anti-Aβ antibodies have received FDA approval as disease-modifying therapies against AD. The recent biomarker breakthroughs described above makes it relatively easy to envision blood-based testing for AD pathology using plasma Aβ42/Aβ40 ratio and plasma P-tau as early test in primary care (likely together with formal cognitive testing). Positive patients could then be referred to a specialized memory clinic to be more closely examined, undergo amyloid PET imaging where available, and begin treatment with an anti-Aβ antibody therapy if Aβ positivity is verified. Plasma P-tau (representing a neuronal reaction to Aβ) and NfL levels (representing neurodegeneration) could be monitored throughout the treatment (e.g., every third month), followed by yearly amyloid PET scans. For anti-Aβ antibodies, repeat MRIs would be needed, at least initially, to monitor amyloid-related imaging abnormalities (ARIA); however, in the future, it is possible that increases in plasma NfL concentration could substitute for MRI to detect clinically-relevant ARIA (amyloid-related imaging abnormalities), although this potential use needs to be formally examined. The patient could then be treated until amyloid PET is negative and plasma P-tau concentration has normalized. Post-treatment, the patient could be followed with annual plasma P-tau and NfL measurements to gauge the potential need for additional therapy. In my view, future clinical trials should incorporate both imaging and fluid biomarker approaches to assess biological response at the same time as they provide the information needed to develop the most effective biomarker algorithm for treatment selection, dose optimization, and drug monitoring.

Acknowledgments

HZ is a Wallenberg Scholar supported by grants from the Swedish Research Council (#2022-01018 and #2019-02397), the European Union's Horizon Europe research and innovation program under grant agreement No 101053962, Swedish State Support for Clinical Research (#ALFGBG-71320), the Alzheimer Drug Discovery Foundation (ADDF), USA (#201809-2016862), the AD Strategic Fund and the Alzheimer's Association (#ADSF-21-831376-C, #ADSF-21-831381-C, and #ADSF-21-831377-C), the Bluefield Project, the Olav Thon Foundation, the Erling-Persson Family Foundation, Stiftelsen för Gamla Tjänarinnor, Hjärnfonden, Sweden (#FO2022-0270), the European Union's Horizon 2020 research and innovation program under the Marie Skłodowska-Curie grant agreement No 860197 (MIRIADE), the European Union Joint Program—Neurodegenerative Disease Research (JPND2021-00694), the National Institute for Health and Care Research University College London Hospitals Biomedical Research Center, and the UK Dementia Research Institute at UCL (UKDRI-1003).

Conflicts of interest

HZ has served at scientific advisory boards and/or as a consultant for Abbvie, Acumen, Alector, Alzinova, ALZPath, Annexon, Apellis, Artery Therapeutics, AZTherapies, CogRx, Denali, Eisai, Nervgen, Novo Nordisk, Optoceutics, Passage Bio, Pinteon Therapeutics, Prothena, Red Abbey Labs, reMYND, Roche, Samumed, Siemens Healthineers, Triplet Therapeutics, and Wave, has given lectures in symposia sponsored by Cellectricon, Fujirebio, Alzecure, Biogen, and Roche, and is a co-founder of Brain Biomarker Solutions in Gothenburg AB (BBS), which is a part of the GU Ventures Incubator Program (outside submitted work).

References

[1] DeTure MA, Dickson DW. The neuropathological diagnosis of Alzheimer's disease. Mol Neurodegener 2019;14:32.
[2] Palmqvist S, Insel PS, Stomrud E, Janelidze S, Zetterberg H, Brix B, et al. Cerebrospinal fluid and plasma biomarker trajectories with increasing amyloid deposition in Alzheimer's disease. EMBO Mol Med 2019;11:e11170.
[3] Kent SA, Spires-Jones TL, Durrant CS. The physiological roles of tau and Abeta: implications for Alzheimer's disease pathology and therapeutics. Acta Neuropathol 2020;140(4):417—47.

[4] Jellinger KA. Neuropathological assessment of the Alzheimer spectrum. J Neural Transm 2020;127:1229−56.
[5] Jack Jr CR, Holtzman DM. Biomarker modeling of Alzheimer's disease. Neuron 2013;80:1347−58.
[6] Blennow K, Hampel H, Weiner M, Zetterberg H. Cerebrospinal fluid and plasma biomarkers in Alzheimer disease. Nat Rev Neurol 2010;6:131−44.
[7] Jack Jr CR, Bennett DA, Blennow K, Carrillo MC, Feldman HH, Frisoni GB, et al. A/T/N: an unbiased descriptive classification scheme for Alzheimer disease biomarkers. Neurology 2016;87:539−47.
[8] Selkoe DJ, Hardy J. The amyloid hypothesis of Alzheimer's disease at 25 years. EMBO Mol Med 2016;8:595−608.
[9] Ashton NJ, Scholl M, Heurling K, Gkanatsiou E, Portelius E, Hoglund K, et al. Update on biomarkers for amyloid pathology in Alzheimer's disease. Biomark Med 2018;12:799−812.
[10] Boulo S, Kuhlmann J, Andreasson U, Brix B, Venkataraman I, Herbst V, et al. First amyloid beta1-42 certified reference material for re-calibrating commercial immunoassays. Alzheimers Dement 2020;16(11):1493−503.
[11] Olsson B, Lautner R, Andreasson U, Ohrfelt A, Portelius E, Bjerke M, et al. CSF and blood biomarkers for the diagnosis of Alzheimer's disease: a systematic review and meta-analysis. Lancet Neurol 2016;15:673−84.
[12] Rasmussen MK, Mestre H, Nedergaard M. The glymphatic pathway in neurological disorders. Lancet Neurol 2018;17:1016−24.
[13] Strozyk D, Blennow K, White LR, Launer LJ. CSF Abeta 42 levels correlate with amyloid-neuropathology in a population-based autopsy study. Neurology 2003;60:652−6.
[14] Hansson O, Lehmann S, Otto M, Zetterberg H, Lewczuk P. Advantages and disadvantages of the use of the CSF Amyloid beta (Abeta) 42/40 ratio in the diagnosis of Alzheimer's Disease. Alzheimer's Res Ther 2019;11:34.
[15] Lewczuk P, Matzen A, Blennow K, Parnetti L, Molinuevo JL, Eusebi P, et al. Cerebrospinal fluid Abeta 42/40 corresponds better than Abeta 42 to amyloid PET in Alzheimer's disease. J Alzheimers Dis 2017;55:813−22.
[16] Mattsson N, Palmqvist S, Stomrud E, Vogel J, Hansson O. Staging beta-amyloid pathology with amyloid positron emission tomography. JAMA Neurol 2019;76(11):1319−29.
[17] Janelidze S, Stomrud E, Palmqvist S, Zetterberg H, van Westen D, Jeromin A, et al. Plasma beta-amyloid in Alzheimer's disease and vascular disease. Sci Rep 2016;6:26801.
[18] Nakamura A, Kaneko N, Villemagne VL, Kato T, Doecke J, Dore V, et al. High performance plasma amyloid-beta biomarkers for Alzheimer's disease. Nature 2018;554:249−54.
[19] Ovod V, Ramsey KN, Mawuenyega KG, Bollinger JG, Hicks T, Schneider T, et al. Amyloid beta concentrations and stable isotope labeling kinetics of human plasma specific to central nervous system amyloidosis. Alzheimers Dement 2017;13:841−9.
[20] Schindler SE, Bollinger JG, Ovod V, Mawuenyega KG, Li Y, Gordon BA, et al. High-precision plasma beta-amyloid 42/40 predicts current and future brain amyloidosis. Neurology 2019;93:e1647−59.
[21] Palmqvist S, Janelidze S, Stomrud E, Zetterberg H, Karl J, Zink K, et al. Performance of fully automated plasma assays as screening tests for Alzheimer disease-related beta-amyloid status. JAMA Neurol 2019;76(9):1060−9.
[22] Rozga M, Bittner T, Batrla R, Karl J. Preanalytical sample handling recommendations for Alzheimer's disease plasma biomarkers. Alzheimers Dement (Amst) 2019;11:291−300.
[23] Irwin DJ. Tauopathies as clinicopathological entities. Parkinsonism Relat Disorders 2016;22(Suppl. 1):S29−33.

[24] Jack Jr CR, Bennett DA, Blennow K, Carrillo MC, Dunn B, Haeberlein SB, et al. NIA-AA Research Framework: toward a biological definition of Alzheimer's disease. Alzheimers Dement 2018;14:535–62.
[25] Dubois B, Feldman HH, Jacova C, Hampel H, Molinuevo JL, Blennow K, et al. Advancing research diagnostic criteria for Alzheimer's disease: the IWG-2 criteria. Lancet Neurol 2014;13:614–29.
[26] Itoh N, Arai H, Urakami K, Ishiguro K, Ohno H, Hampel H, et al. Large-scale, multicenter study of cerebrospinal fluid tau protein phosphorylated at serine 199 for the antemortem diagnosis of Alzheimer's disease. Ann Neurol 2001;50:150–6.
[27] Skillback T, Farahmand BY, Rosen C, Mattsson N, Nagga K, Kilander L, et al. Cerebrospinal fluid tau and amyloid-beta(1-42) in patients with dementia. Brain 2015;138:2716–31.
[28] Maia LF, Kaeser SA, Reichwald J, Hruscha M, Martus P, Staufenbiel M, et al. Changes in amyloid-beta and tau in the cerebrospinal fluid of transgenic mice overexpressing amyloid precursor protein. Sci Transl Med 2013;5:194re2.
[29] Sato C, Barthelemy NR, Mawuenyega KG, Patterson BW, Gordon BA, Jockel-Balsarotti J, et al. Tau kinetics in neurons and the human central nervous system. Neuron 2018;98:861–4.
[30] Skillback T, Rosen C, Asztely F, Mattsson N, Blennow K, Zetterberg H. Diagnostic performance of cerebrospinal fluid total tau and phosphorylated tau in Creutzfeldt-Jakob disease: results from the Swedish Mortality Registry. JAMA Neurol 2014;71:476–83.
[31] Hesse C, Rosengren L, Andreasen N, Davidsson P, Vanderstichele H, Vanmechelen E, et al. Transient increase in total tau but not phospho-tau in human cerebrospinal fluid after acute stroke. Neurosci Lett 2001;297:187–90.
[32] Ost M, Nylen K, Csajbok L, Ohrfelt AO, Tullberg M, Wikkelso C, et al. Initial CSF total tau correlates with 1-year outcome in patients with traumatic brain injury. Neurology 2006;67:1600–4.
[33] Leitao MJ, Silva-Spinola A, Santana I, Olmedo V, Nadal A, Le Bastard N, et al. Clinical validation of the Lumipulse G cerebrospinal fluid assays for routine diagnosis of Alzheimer's disease. Alzheimer's Res Ther 2019;11:91.
[34] Blennow K, Shaw LM, Stomrud E, Mattsson N, Toledo JB, Buck K, et al. Predicting clinical decline and conversion to Alzheimer's disease or dementia using novel Elecsys Abeta(1-42), pTau and tTau CSF immunoassays. Sci Rep 2019;9:19024.
[35] Hampel H, Buerger K, Zinkowski R, Teipel SJ, Goernitz A, Andreasen N, et al. Measurement of phosphorylated tau epitopes in the differential diagnosis of Alzheimer disease: a comparative cerebrospinal fluid study. Arch Gen Psychiatr 2004;61:95–102.
[36] Janelidze S, Stomrud E, Smith R, Palmqvist S, Mattsson N, Airey DC, et al. Cerebrospinal fluid p-tau217 performs better than p-tau181 as a biomarker of Alzheimer's disease. Nat Commun 2020;11:1683.
[37] De Vos A, Bjerke M, Brouns R, De Roeck N, Jacobs D, Van den Abbeele L, et al. Neurogranin and tau in cerebrospinal fluid and plasma of patients with acute ischemic stroke. BMC Neurol 2017;17:170.
[38] Bogoslovsky T, Diaz-Arrastia R. Dissecting temporal profiles of neuronal and axonal damage after mild traumatic brain injury. JAMA Neurol 2016;73:506–7.
[39] Zetterberg H. Review: tau in biofluids - relation to pathology, imaging and clinical features. Neuropathol Appl Neurobiol 2017;43:194–9.
[40] Pereira JB, Westman E, Hansson O, Alzheimer's Disease Neuroimaging I. Association between cerebrospinal fluid and plasma neurodegeneration biomarkers with brain atrophy in Alzheimer's disease. Neurobiol Aging 2017;58:14–29.

[41] Randall J, Mortberg E, Provuncher GK, Fournier DR, Duffy DC, Rubertsson S, et al. Tau proteins in serum predict neurological outcome after hypoxic brain injury from cardiac arrest: results of a pilot study. Resuscitation 2013;84:351−6.

[42] Tatebe H, Kasai T, Ohmichi T, Kishi Y, Kakeya T, Waragai M, et al. Quantification of plasma phosphorylated tau to use as a biomarker for brain Alzheimer pathology: pilot case-control studies including patients with Alzheimer's disease and down syndrome. Mol Neurodegener 2017;12:63.

[43] Mielke MM, Hagen CE, Xu J, Chai X, Vemuri P, Lowe VJ, et al. Plasma phospho-tau181 increases with Alzheimer's disease clinical severity and is associated with tau- and amyloid-positron emission tomography. Alzheimers Dement 2018;14:989−97.

[44] Thijssen EH, La Joie R, Wolf A, Strom A, Wang P, Iaccarino L, et al. Diagnostic value of plasma phosphorylated tau181 in Alzheimer's disease and frontotemporal lobar degeneration. Nat Med 2020;26:387−97.

[45] Janelidze S, Mattsson N, Palmqvist S, Smith R, Beach TG, Serrano GE, et al. Plasma P-tau181 in Alzheimer's disease: relationship to other biomarkers, differential diagnosis, neuropathology and longitudinal progression to Alzheimer's dementia. Nat Med 2020;26:379−86.

[46] Jack CR, Wiste HJ, Algeciras-Schimnich A, Figdore DJ, Schwarz CG, Lowe VJ, Ramanan VK, Vemuri P, Mielke MM, Knopman DS, Graff-Radford J, Boeve BF, Kantarci K, Cogswell PM, Senjem ML, Gunter JL, Therneau TM, Petersen RC. Predicting amyloid PET and tau PET stages with plasma biomarkers. Brain February 15, 2023:awad042. https://doi.org/10.1093/brain/awad042.

[47] Palmqvist S, Janelidze S, Quiroz YT, Zetterberg H, Lopera F, Stomrud E, et al. Discriminative accuracy of plasma phospho-tau217 for Alzheimer disease vs other neurodegenerative disorders. JAMA 2020;324(8):772−81.

[48] Barthelemy NR, Horie K, Sato C, Bateman RJ. Blood plasma phosphorylated-tau isoforms track CNS change in Alzheimer's disease. J Exp Med 2020;217.

[49] Lantero Rodriguez J, Karikari TK, Suarez-Calvet M, Troakes C, King A, Emersic A, et al. Plasma p-tau181 accurately predicts Alzheimer's disease pathology at least 8 years prior to post-mortem and improves the clinical characterisation of cognitive decline. Acta Neuropathol 2020;140:267−78.

[50] O'Connor A, Karikari TK, Poole T, Ashton NJ, Lantero Rodriguez J, Khatun A, et al. Plasma phospho-tau181 in presymptomatic and symptomatic familial Alzheimer's disease: a longitudinal cohort study. Mol Psychiatr 2021;26(10):5967−76.

[51] Khalil M, Teunissen CE, Otto M, Piehl F, Sormani MP, Gattringer T, et al. Neurofilaments as biomarkers in neurological disorders. Nat Rev Neurol 2018;14:577−89.

[52] Gaetani L, Blennow K, Calabresi P, Di Filippo M, Parnetti L, Zetterberg H. Neurofilament light chain as a biomarker in neurological disorders. J Neurol Neurosurg Psychiatry 2019;90:870−81.

[53] Bridel C, van Wieringen WN, Zetterberg H, Tijms BM, Teunissen CE, et al., the NFLG. Diagnostic value of cerebrospinal fluid neurofilament light protein in neurology: a systematic review and meta-analysis. JAMA Neurol 2019;76(9):1035−48.

[54] Weston PSJ, Poole T, O'Connor A, Heslegrave A, Ryan NS, Liang Y, et al. Longitudinal measurement of serum neurofilament light in presymptomatic familial Alzheimer's disease. Alzheimer's Res Ther 2019;11:19.

[55] Preische O, Schultz SA, Apel A, Kuhle J, Kaeser SA, Barro C, et al. Serum neurofilament dynamics predicts neurodegeneration and clinical progression in presymptomatic Alzheimer's disease. Nat Med 2019;25:277−83.

[56] Mattsson N, Cullen NC, Andreasson U, Zetterberg H, Blennow K. Association between longitudinal plasma neurofilament light and neurodegeneration in patients with Alzheimer disease. JAMA Neurol 2019;76:791−9.

[57] Simrén J, Elmgren A, Blennow K, Zetterberg H. Fluid biomarkers in Alzheimer's disease. Adv Clin Chem 2023;112:249−81. https://doi.org/10.1016/bs.acc.2022.09.006. Epub 2022 Nov 4. PMID: 36642485.

[58] Shaw LM, Arias J, Blennow K, Galasko D, Molinuevo JL, Salloway S, et al. Appropriate use criteria for lumbar puncture and cerebrospinal fluid testing in the diagnosis of Alzheimer's disease. Alzheimers Dement 2018;14:1505−21.

[59] Spitzer P, Klafki HW, Blennow K, Buee L, Esselmann H, Herruka SK, et al. cNEUPRO: novel biomarkers for neurodegenerative diseases. Int J Alzheimer's Dis 2010;2010.

[60] Hansson O, Edelmayer RM, Boxer AL, Carrillo MC, Mielke MM, Rabinovici GD, Salloway S, Sperling R, Zetterberg H, Teunissen CE. The Alzheimer's Association appropriate use recommendations for blood biomarkers in Alzheimer's disease. Alzheimers Dement 2022;18(12):2669−86.

Chapter 6d

Advances in cognitive testing

David Berron[a,b]
[a]Clinical Cognitive Neuroscience Group, German Center for Neurodegenerative Diseases (DZNE), Magdeburg, Germany; [b]Clinical Memory Research Unit, Department of Clinical Sciences Malmö, Lund University, Lund, Sweden

Novel cognitive tests sensitive to early Alzheimer's disease

Recent developments in Alzheimer's disease biomarkers have significantly improved the early detection of Alzheimer's disease pathology [1–3]. However, while biomarkers for Alzheimer's disease can indicate the underlying brain pathology and have some predictive value for future cognitive decline, they do not provide information about cognitive Alzheimer's disease phenotypes, particularly in early disease stages [4], nor do they represent optimal measures of disease progression. Thus, it is of critical importance to identify, characterize, and monitor cognitive impairment in Alzheimer's disease using cognitive assessments.

Many established neuropsychological assessments rely on recall of simple word lists or figures and were not originally developed for the detection of cognitive impairment in Alzheimer's disease. Thus, these assessments do not necessarily fully leverage our current understanding of the biological substrates of neurocognitive disorders associated with Alzheimer's disease. This is in stark contrast with our increased knowledge of the spatial distribution and temporal dynamics of Alzheimer's disease pathology, as well as our deeper understanding regarding the functional architecture of cognitive functions. The aim of this first section is to give an overview of novel test paradigms focusing on functions associated with specific brain networks that are affected early on in Alzheimer's disease and might thus show enhanced sensitivity and specificity in detecting cognitive impairment.

Progression of AD pathology and the functional architecture of episodic memory and spatial navigation

In early stages of Alzheimer's disease, accumulation of beta-amyloid (Aβ) pathology can predominantly be detected in orbitofrontal and inferior temporal regions, the posterior cingulate cortex, and the precuneus; in later stages these Aβ deposits can be found in wide parts of the temporal, frontal, and parietal association cortices, and finally in primary sensory-motor cortices [8–10]. In contrast, cortical tau pathology can be detected the earliest in the transentorhinal region, the entorhinal cortex, and the hippocampus (Braak I/II), followed by more widespread accumulation throughout the posterior and lateral temporal lobe (Braak III/IV), and finally parietal, frontal and occipital brain regions (Braak V/VI) [11–13]. However, notably the accumulation of Aβ and, in particular, tau pathology can diverge from these progression patterns, often resulting in atypical clinical presentations of Alzheimer's disease [14,15].

The hippocampus, situated in the medial temporal lobe, is densely connected with neocortical memory systems via different pathways. The perirhinal cortex connects the hippocampus with an anterior-temporal memory system including the lateral orbitofrontal cortex, the amygdala, and the ventral temporopolar cortex [16]. Another pathway is the parahippocampal cortex, which represents the connection between the hippocampus and a posterior-medial memory system that includes the retrosplenial cortex, anterior thalamus, mammillary bodies, posterior cingulate, precuneus, angular gyrus, and ventromedial prefrontal cortex [16,17]. Beside memory function, brain regions in the posterior-medial system are involved in the formation of cognitive maps, which are important for spatial navigation [18,19].

The spatio-temporal progression pattern of Alzheimer's disease pathology has important implications for optimal markers used to detect and monitor cognitive impairment in the respective disease stages [20]. For example, tasks that rely on the discrimination of items with perceptual or conceptual feature overlap as well as context-free memory tasks that are associated with the perirhinal, transentorhinal, and entorhinal cortex may be more relevant measures for preclinical Alzheimer's disease [21,22]. On the other hand, associative and context-rich memory tasks that are associated with the hippocampus are likely more sensitive in later disease stages of mild cognitive impairment.

Object memory

The perirhinal cortex as well as the anterior-lateral portion of the entorhinal cortex are critically involved in memory for objects [23–26], particularly when there is high feature overlap between the memorized objects [27–30]. As described above, tau pathology can be detected relatively early in the human medial temporal lobe and is associated with thinning and atrophy of specific medial temporal subregions—in particular the neighboring transentorhinal and entorhinal regions [11,31–34]. Object mnemonic discrimination tasks that demand detailed encoding of object features with the aim of distinguishing them from similar objects presented in later test items have been studied in early Alzheimer's disease populations, as have their association with biomarkers of Alzheimer's disease pathology. Earlier studies have consistently found impairment in object mnemonic discrimination in patients with mild cognitive impairment in comparison with cognitively unimpaired older adults [35–37]. While studies in older adults have pointed toward aberrant activity in the perirhinal and anterior-lateral entorhinal cortex being associated with reduced object mnemonic discrimination performance [27,28], studies including Alzheimer's disease biomarkers have linked this deficit to the presence of tau pathology [38–40]. In these studies, higher levels of biomarkers reflecting tau pathology, measured either in levels of phosphorylated tau in cerebrospinal fluid or tau-PET signal binding in regions in the anterior temporal lobe, were associated with lower object mnemonic discrimination performance. Thus, object memory tasks demanding detailed encoding and retrieval of object features seem to be particularly suited to detect tau-related cognitive impairment in early Alzheimer's disease [21]. Future studies are still needed to determine diagnostic sensitivity and specificity for concrete test paradigms and to investigate whether these tasks can also be used to detect subtle cognitive decline over repeated assessments in early stages of Alzheimer's disease.

Feature binding and associative memory

Models of episodic memory suggest that the hippocampus is critically involved in relational memory binding [41,42], and it has been hypothesized that the hippocampus is specifically involved in high-resolution binding of features both in perception and also working and long-term memory [43]. Interestingly, recent studies suggest that conjunctive binding processes may be particularly related to the anterior-lateral entorhinal cortex [44–46].

Several approaches exist to probe impairment in *memory binding* in Alzheimer's disease. Using visual short-term memory binding tasks in a change detection task format, it has been demonstrated that patients with dementia of the Alzheimer's type are particularly impaired when maintaining bound stimulus features [47–50]. Although change detection tasks require a binary decision of whether there was a change in the array of encoded items or not, delayed reproduction tasks allow to obtain information about memory precision in a continuous manner and inform about the source of the error [51–54]. Individuals are asked to keep in mind the identity as well as the location of colored shapes on a display. After a short delay, individuals are presented with a shape that has been encoded as well as a novel foil and need to first identify the correct shape and afterward drag it to the correct location. Patients with familial as well as sporadic Alzheimer's disease are characterized by binding errors where they identify the correct object but often associate it with the wrong location [52,55]. One cause behind these binding errors is the inability to resolve interference from competing stimuli, a process in which cortical regions in the medial temporal lobe are critical to resolving this interference because of their role in creating high-fidelity stimulus displays. Future studies are needed to determine the diagnostic accuracy of such tasks in early Alzheimer's disease, as well as the relationship with markers of AD pathology.

Face-name association memory has the advantage of drawing on ecologically relevant cognitive performance. It has been used in functional MRI studies which have demonstrated that patients with mild cognitive impairment show reduced hippocampal activity during the task [56]. Furthermore, performance in the Face-Name Associative Memory Exam (FNAME) has been validated regarding its retest reliability and validity in older adults [57,58], and has been found to be related to Aβ pathology. In a similar task, the Face-Name Associative Recognition Task (FNART), task performance separated between healthy controls and individuals with subjective cognitive decline (d = 0.58) and patients with mild cognitive impairment (d = 1.8) and was associated with subjective concerns in SCD [59].

Spatial memory and navigation

While some regions that are important for spatial navigation seem to be affected in later disease stages (e.g., the thalamus or the parahippocampal cortex), the entorhinal cortex and the hippocampus are affected by early tau pathology and gray matter atrophy, in addition to being crucial for spatial navigation [11]. Thus, measures of spatial cognition are interesting candidates to assess early cognitive impairment in Alzheimer's disease [18]. Several approaches have been investigated in clinical populations. For example, the *Four Mountains task* examines spatial memory and has been shown to tax regions in the medial temporal lobe, including the hippocampus [60]. In this task, participants view a sample image and are then required to choose a target image from four alternatives, which show the same place from a different viewpoint [61]. This task has been shown to be sensitive to detect biomarker-positive MCI and mild AD dementia populations where it showed between 50% and 80% specificity to detect early AD (MCI Aβ-positive and mild AD dementia vs. healthy controls) at 100% sensitivity [62] and has also outperformed traditional neuropsychological assessments regarding its prediction of conversion toward dementia [63].

Assessments of spatial abilities (such as navigation) using virtual reality have also been assessed in early AD populations and have been shown to differentiate patients with mild cognitive impairment at low and high risk of developing dementia. Such tasks might also be well suited to detect the earliest cognitive deficits in preclinical Alzheimer's disease [64,65]. *SeaHeroQuest* is a gamified navigation task implemented in a mobile app which allows to assess spatial abilities in the general population at scale [66,67] and has been shown to be able to differentiate individuals at-high risk for preclinical AD (i.e., ApoE-ε4 carriers) [68].

Comparison to current clinical standards

Taken together, cognitive testing is moving away from rather simple memory tests of word or figure recall and recognition toward novel assessments driven by recent findings on the spatio-temporal progression pattern of AD pathology with the aim of probing memory processes that are specifically impaired in AD. However, such novel tests should only replace the currently established tests following accurate validation and comparison with current clinical standards [69]. While the studies outlined above provide evidence for increased sensitivity by targeting anatomical networks vulnerable to early AD pathology, only few of them have been directly compared with and have shown to outperform established assessment batteries. To that end, more studies providing head-to-head comparisons in terms of sensitivity and specificity as well as relationships with Alzheimer's disease biomarkers are needed. Furthermore, it is important to consider how these novel test paradigms compare to established paper-pencil assessments with respect to operational issues such as cost, speed, and ease of use. There is enormous potential for the tests described above given that they all come in a computerized format and could thus be performed easily and scored automatically. Most of them, however, are only available in different experimental presentation software formats that have been used by the specific research groups at the moment and would thus be difficult to implement in clinical practice right away. Future efforts should thus focus on a common platform and framework to ease clinical administration and evaluation of these novel measures.

Digital cognitive assessments in Alzheimer's disease

In-clinic supervised digital cognitive assessments using tablets and personal computers

In the current clinical diagnostic process for individuals with cognitive complaints who are referred to memory clinics, the aim is to determine the presence of objective cognitive impairment. To that end, performance in a set of neuropsychological tests is assessed in a face-to-face setting using mostly paper-pencil tests. These paper-pencil tests need to be performed and scored manually, which can result in increased interrater variability and even errors in scoring. Supervised computerized web- and tablet-based implementations of traditional neuropsychological assessments have thus significant advantages.

Computerized assessments follow automated administration and scoring of test results thereby reducing potential errors and variability between raters. The data are stored on the respective platform and undergo usually automatic and standardized quality assessment. In addition, computerized testing enables the use of many different stimulus domains (e.g., pictures, short videos, game environments, standardized sound samples, etc.), as well as a variety of test and response formats (e.g., button presses, drag-and-drop, touch patterns, digitized pens) that also allow for rich, continuous outcomes. Importantly, computerized approaches also enable the collection of metrics that cannot be recorded in paper-pencil assessments, such as trial-specific response times, length and number of breaks, and inputs from different mobile device sensors or the duration of touching the screen, to name a few [70–72].

There are currently several providers of supervised cognitive testing platforms. Most of them provide supervised computerized web- and tablet-based implementations of traditional neuropsychological assessments (see Refs. [73,74]). Cogstate Ltd. provides the *Cogstate Brief Battery (CBB)* consisting of four cognitive tests that use playing cards as stimulus material: a simple reaction time task to measure psychomotor function (Detection task), a choice reaction time task to measure visual attention (Identification task), a continuous recognition learning task to assess visual learning (One Card Learning task), and a one-back task to measure working memory and attention (One-Back task; [75,76]). As an exploratory outcome for the A4 trial (ClinicalTrials.gov identifier: NCT02008357), the CBB has been extended with two cognitive test paradigms: the Behavioral Pattern Separation Object task (BPS-O) and the Face-Name Associative Memory task (FNAME; [77–80]). Cambridge Cognition provides the *Cambridge Neuropsychological Test Automated Battery (CANTAB)*, which is available on personal computers and provides measures of working memory, learning and executive functioning, attention as well as visual, verbal and episodic memory [81]. The Paired Associates Learning (PAL) task has been shown to be sensitive for mild cognitive impairment and mild probable AD dementia [82,83]. *The National Institute of Health Toolbox—Cognition Battery (NIH-TB)* is available via a web-browser as well as an iPad app and offers various tests for supervised use covering executive function, attention, episodic memory, language, processing speed, and working memory [84–87]. It was developed with the aim of providing researchers a harmonized set of cognitive tasks to compare findings across a wide range of studies and populations, and it has also been used in Alzheimer's disease research [88–90]. The *Brain Health Assessment (BHA)* is a supervised battery including measures of memory, executive function and processing speed, visuospatial skills, and language [91]. The included tests have been shown to be associated with Alzheimer's disease pathology in MCI and dementia patients [92] and a composite score created from three subtests, the BHA-Cognitive Score, has recently been found to be sensitive to longitudinal decline [93].

Remote unsupervised assessments using mobile devices in the home environment

Regarding the state of neuropsychological assessment within the current clinical diagnostic process, cognitive testing is almost exclusively performed face-to-face in a hospital setting. While this has significant advantages, most importantly by providing a standardized and private testing environment, it also comes with several limitations. First of all, individuals are required to travel to a memory clinic, which is a deviation from their daily routine and additionally constitutes an unfamiliar environment. This can cause nervousness and tension before and during cognitive assessment and might even affect cognitive performance [94,95]. Furthermore, daily cognitive functioning is influenced by many factors in everyday life such as stress [96,97] and sleep quality [98]. This results in potentially high day-to-day variability, which can obscure individual test results. Thus, face-to-face cognitive assessments represent one snapshot in time which may or may not be representative of the patient's neurocognitive status. Another critical limitation in a clinical setting is time; usually there is only a certain time period reserved for cognitive assessment during a clinic visit, which means that individuals perform a number of cognitive assessments in a row while potentially getting increasingly fatigued and thus performing more poorly than they might otherwise. Furthermore, this setting does not allow for a number of test formats that show promise in the assessment of Alzheimer's disease, such as delayed recall after longer time delays (e.g., several hours or days), or repeated learning over several days or weeks. In the following, I will outline the potential of digital cognitive assessments regarding these limitations (also see Refs. [73,152] and [151] for recent in-depth reviews on remote assessments in preclinical disease stages).

High-frequency assessments to increase the diagnostic signal

In general, unsupervised and remote digital cognitive assessments enable individuals to complete cognitive tests from their home environment using mobile devices, doing so at their own pace thereby preventing effects of fatigue and increasing ecological validity. Furthermore, digital cognitive assessments allow more frequent tests, which can result in higher diagnostic signal due to improved reliability of cognitive measures through averaging across several short assessments [99,100]. At the same time, increased frequency of testing enables the collection of measures of variability between sessions that are not possible to collect in paper-pencil tests (i.e., potential fluctuations). Fluctuations in cognitive performance itself might carry diagnostic information such as for example in Lewy body dementia [99,101,102].

Repeated assessments and novel test regimes

The measurement of cognitive functions such as learning through the repeated study of stimuli as well as long-term memory and forgetting relies on repeated assessments. While such assessments have been difficult in clinical settings, digital technologies now allow for easier repeated assessments and thus enable more complex test approaches. Several recent studies highlighted the potential for such cognitive assessments in early Alzheimer's disease.

Practice effects

While practice effects in longitudinal cognitive assessments have been considered a disadvantage, with efforts to prevent them by employing parallel task versions with differing test materials, recent studies have suggested that diminished practice effects in older individuals could be an indicator of preclinical Alzheimer's disease [103]. An attenuation of practice effects has also been associated with amyloid positivity [104–107]. Using web-based assessments of learning Chinese characters over 6 days, a recent study showed that learning deficits across such a short time frame allowed for differentiation between amyloid-positive and -negative participants with higher effect sizes than decline in episodic memory over the last 6 years [108]. Similarly, another recent study suggests that these findings can be extended to measures of associative memory (i.e., Face-Name association), and that they are valid when assessed using an iPad in the participants' home environment [109].

Long-term memory and forgetting

In order to assess long-term memory function and forgetting, an encoding session and one or several retrieval sessions are necessary. Although this is difficult to achieve in the clinical context, studies suggest that accelerated long-term forgetting can predict cognitive decline in older adults better than traditional delayed recall tests [110], and that presymptomatic autosomal dominant Alzheimer's disease is characterized by lower 7-day but not 30-min recall performance [111]. Digital cognitive assessments using smartphones allow administration of several remote encoding and retrieval sessions, and a recent study has shown that it is feasible to record forgetting curves for up to 13 days in a sample of middle-aged adults [112,113].

Longitudinal trajectories and decline: The NIA-AA research framework for Alzheimer's disease suggests that individuals in preclinical disease stages can be identified by signs of transitional cognitive decline, in terms of subjective reports of decline, or subtle worsening of cognition on longitudinal objective testing [114]. While yearly cognitive assessments are limited in their ability to detect such decline for the aforementioned reasons, frequent remote and unsupervised digital cognitive assessments in the participant's home environment might have higher sensitivity to detect long-term change over several months or years [99].

Several platforms for remote and unsupervised cognitive testing already exist. For example, the *Mezurio* smartphone app contains a collection of novel cognitive tasks measuring episodic memory, language, and executive function targeted at the detection of early Alzheimer's disease [112,113]. The *neotiv* digital platform [39,115–117] includes a mobile app that contains novel cognitive test paradigms optimized for episodic memory networks including mnemonic discrimination for objects and scenes [27,38,40], Object-in-Room-Recall [116] and Complex Scene Recognition [118,119]. The *Boston Remote Assessment for Neurocognitive Health (BRANCH)* was designed to assess learning rates using smartphone-based measures of associative memory, pattern separation, and semantically facilitated learning and recall [120–122]. The *Ambulatory Research in Cognition (ARC)* smartphone app is available for mobile devices and includes tests on working memory (Grids Test), processing speed (Symbols Test), and associative memory (Prices Test) [123].

Need for validation of novel digital assessments

While digital cognitive assessments have high potential to extend the possibilities for the assessment of cognitive impairment and decline in Alzheimer's disease, these novel solutions need to be rigorously validated [69,124,125]. As mentioned above, the biggest difference between digital and supervised in-clinic assessments is the lack of control over the standardized test environment. For example, individuals might be disturbed by a noisy environment, a distraction (such as a ringing doorbell or an incoming call), or may not even be the person who should complete the assessment in the first place. There might also be problems during cognitive assessment such as misunderstanding of the instructions or minor sensory and motor symptoms that participants may not report or even be aware of. While such issues would have been noticed by an experienced clinician, they could go undetected in a remote and unsupervised setting, thus interfering with the cognitive test and appearing like cognitive impairment. Furthermore, different mobile devices might differ in critical features such as screen size, accuracy and precision of storing response time, and similar metrics which can influence cognitive outcomes [126–128]. Thus, it is important to compare performance outcomes derived from supervised in-clinic and unsupervised at-home assessments. Several earlier studies reported evidence that it is feasible to record valid cognitive data in an unsupervised setting [129,130], however, there are also reports showing meaningful differences between supervised in-clinic and unsupervised assessments in the home environment [131]. In addition, it is important to ensure that instructions are easily understood, to record as many of the potential confounds as possible, and to make sure that there are control measures in place that can detect other problems that could result in low test performance.

Furthermore, digital unsupervised assessments need to be validated with respect to psychometric criteria. In order to use digital cognitive assessments to monitor longitudinal trajectories, they need to be validated regarding their retest reliability and potential practice effects. Their construct validity needs to be determined by comparing test scores with traditional neuropsychological assessment batteries. Recent studies using unsupervised and remote digital assessments in populations of older adults and patients with Alzheimer's disease have indeed demonstrated satisfactory construct validity and retest reliability of several digital cognitive testing approaches [116,117,120,123,132]. However, given that one of the advantages of repeated digital cognitive assessments is their increased reliability, one should be careful when selecting the appropriate validation measures and should not overestimate the relationship with paper-and-pencil assessments that are potentially less reliable.

Additionally, while digital cognitive assessments have been shown to overcome several limitations of paper-pencil assessments, only few have been validated in biomarker-characterized samples. Thus, more future studies are needed to demonstrate that digital assessments are related to Alzheimer's disease pathology and to determine whether they can identify subtle cognitive impairment cross-sectionally and longitudinally in preclinical Alzheimer's disease.

Another limitation of digital cognitive assessments is that, while they can enable frequent remote assessments, several studies have reported severe difficulties in engaging participants over longer time periods [133]. Future efforts are needed to identify critical factors associated with long-term attrition and to implement efficient counter measures including engaging and motivating test design.

Finally, it should be noted that remote digital approaches carry challenges with respect to data protection and security given that sensitive personal information needs to be regularly transferred and electronically stored.

Other approaches to infer cognitive status

Beside active digital cognitive assessments where participants are prompted to complete cognitive tests, mobile and wearable digital consumer technology can provide passively collected data—even without participants' attention [7,134]. One of the big potentials of passive data collection thus is that it allows data collection at a very high frequency or even continuous recordings that come with very low patient burden. This might result in higher adherence compared to active data collection and thus richer and more detailed datasets. In addition, such measures have high ecological validity due to their recording in everyday settings.

Speech-based markers for example can be assessed both actively (e.g., reading scripted passages, or picture description) and passively (e.g., recorded conversations or interviews), and advances in natural language-processing and machine-learning techniques allow for the extraction meaningful features

[135]. Studies have shown speech-based abnormalities in neurodegenerative diseases including Alzheimer's disease [136,137], making speech-based markers interesting candidates to potentially support dementia diagnosis, measuring disease severity, or monitoring the disease progression over time in the future. While there is some work presenting evidence for clinical validity in Alzheimer's disease by differentiating patients with mild cognitive impairment and Alzheimer's dementia patients from healthy controls [138−142], further validation is needed, especially by comparing these novel measures with current clinical standards [69,125] and assessing their potential for longitudinal monitoring.

A variety of other potential digital biomarkers is available for the detection of changes in Alzheimer's disease that can be accessed via *passive sensors in mobile devices* such as geopositioning, fine motor control assessed via touch screens, eye movements assessed via built-in cameras, keyboard interaction, or sleep patterns [7]. Similarly, sensor technology can be embedded in the home environment or in cars [143]. Using such data collection approaches, it might be possible to infer information about the cognitive performance of an individual through digital phenotyping, meaning the identification of a set of passively recorded measures while interacting with a mobile device that are able to represent neuropsychological domains [144]. Recent approaches have shown first evidence that continuous data collection from 1 week of smartphone interactions could predict performance in established neuropsychological batteries of memory, executive function, and verbal fluency [145]. Other examples similarly suggest that it might be possible to infer domains such as alertness from app usage data [146] or task-switching abilities from time costs in frequent tasks due to naturally appearing interruptions (e.g., an incoming phone call; [147]). However, while such approaches have the potential to provide important measures with increased ecological validity and less vulnerability to attrition, there are important limitations. First of all, there is still limited evidence that passive measures can provide better outcomes in the clinical setting and study samples [69]. Rigorous studies, including large and longitudinal studies, are needed to investigate how sensitive digital phenotypes are to changes over time and how they are related to cognitive decline. First efforts along those lines are underway [148]. Importantly, one major unknown is whether individuals will trust such approaches and whether they will accept continuous collection of sensitive and private data from their personal devices, cars, or home environments.

In conclusion, there are several exciting roads ahead, spanning novel anatomically- and theory-guided cognitive test paradigms, advanced digital testing solutions and approaches extending to the patient's home environment and everyday life. While all these approaches have the potential to significantly improve or extend the detection and characterization of cognitive impairment and decline as well as their clinical meaningfulness, they should not replace other currently established measures before rigorous validation and comparison with the current standard. Frameworks to validate novel markers are already available, including more specific ones regarding the validation of digital markers [69,124,125]. It will be critical to evaluate these novel approaches in the primary care and specialist setting [149,150].

References

[1] Hansson O, Blennow K, Zetterberg H, Dage J. Blood biomarkers for Alzheimer's disease in clinical practice and trials. Nat Aging 2023;3:506−19. https://doi.org/10.1038/s43587-023-00403-3.

[2] Ossenkoppele R, Hansson O. Towards clinical application of tau PET tracers for diagnosing dementia due to Alzheimer's disease. Alzheimer's Dement 2021;17(12):1998−2008. https://doi.org/10.1002/alz.12356. PMID: 33984177.

[3] Zetterberg H, Blennow K. Moving fluid biomarkers for Alzheimer's disease from research tools to routine clinical diagnostics. Mol Neurodegener 2021;16:10. https://doi.org/10.1186/s13024-021-00430-x.

[4] Dubois B, Villain N, Frisoni GB, Rabinovici GD, Sabbagh M, Cappa S, Bejanin A, Bombois S, Epelbaum S, Teichmann M, Habert M-O, Nordberg A, Blennow K, Galasko D, Stern Y, Rowe CC, Salloway S, Schneider LS, Cummings JL, Feldman HH. Clinical diagnosis of Alzheimer's disease: recommendations of the international working group. Lancet Neurol 2021. https://doi.org/10.1016/s1474-4422(21)00066-1.

[5] Weintraub S, Carrillo MC, Farias S, Goldberg TE, Hendrix JA, Jaeger J, Knopman D, Langbaum JB, Park DC, Ropacki MT, Sikkes S, Welsh-Bohmer KA, Bain LJ, Brashear R, Budur K, Graf A, Martenyi F, Storck M, Randolph C. Measuring cognition and function in the preclinical stage of Alzheimer's disease. Alzheimer's Dement 2018. https://doi.org/10.1016/j.trci.2018.01.003.

[6] Koo BM, Vizer LM. Mobile technology for cognitive assessment of older adults: a scoping review. Innov Aging 2019;3. https://doi.org/10.1093/geroni/igy038.

[7] Kourtis LC, Regele OB, Wright JM, Jones GB. Digital biomarkers for Alzheimer's disease: the mobile/wearable devices opportunity. Npj Digit Med 2019;2:9. https://doi.org/10.1038/s41746-019-0084-2.

[8] Grothe MJ, Barthel H, Sepulcre J, Dyrba M, Sabri O, Teipel SJ, Initiative A. In vivo staging of regional amyloid deposition. Neurology 2017;89:2031−8. https://doi.org/10.1212/wnl.0000000000004643.

[9] Mattsson N, Palmqvist S, Stomrud E, Vogel J, Hansson O. Staging β-amyloid pathology with amyloid positron emission tomography. JAMA Neurol 2019;76:1319. https://doi.org/10.1001/jamaneurol.2019.2214.

[10] Palmqvist S, Schöll M, Strandberg O, Mattsson N, Stomrud E, Zetterberg H, Blennow K, Landau S, Jagust W, Hansson O. Earliest accumulation of β-amyloid occurs within the default-mode network and concurrently affects brain connectivity. Nat Commun 2017;8:1214. https://doi.org/10.1038/s41467-017-01150-x.

[11] Berron D, Vogel JW, Insel PS, Pereira JB, Xie L, Wisse LEM, et al. Early stages of tau pathology and its associations with functional connectivity, atrophy and memory. Brain 2021b;144(9):2771−83. https://doi.org/10.1093/brain/awab114. PMID: 33725124; PMCID: PMC8557349.

[12] Braak H, Braak E. Staging of alzheimer's disease-related neurofibrillary changes. Neurobiol Aging 1995;16:271−8. https://doi.org/10.1016/0197-4580(95)00021-6.

[13] Pascoal TA, Therriault J, Benedet AL, Savard M, Lussier FZ, Chamoun M, Tissot C, Qureshi MNI, Kang MS, Mathotaarachchi S, Stevenson J, Hopewell R, Massarweh G, Soucy J-P, Gauthier S, Rosa-Neto P. 18F-MK-6240 PET for early and late detection of neurofibrillary tangles. Brain 2020;143:awaa180. https://doi.org/10.1093/brain/awaa180.

[14] Ossenkoppele R, Schonhaut DR, Schöll M, Lockhart SN, Ayakta N, Baker SL, et al. Tau PET patterns mirror clinical and neuroanatomical variability in Alzheimer's disease. Brain 2016;139(Pt 5):1551−67. https://doi.org/10.1093/brain/aww027. PMID: 26962052; PMCID: PMC5006248.

[15] Vogel JW, Young AL, Oxtoby NP, Smith R, Ossenkoppele R, Strandberg OT, Joie RL, Aksman LM, Grothe MJ, Iturria-Medina Y, Weiner M, Aisen P, Petersen R, Jack CR, Jagust W, Trojanowki JQ, Toga AW, Beckett L, Green RC, Saykin AJ, Morris J, Shaw LM, Liu E, Montine T, Thomas RG, Donohue M, Walter S, Gessert D, Sather T, Jiminez G, Harvey D, Bernstein M, Fox N, Thompson P, Schuff N, DeCArli C, Borowski B, Gunter J, Senjem M, Vemuri P, Jones D, Kantarci K, Ward C, Koeppe RA, Foster N, Reiman EM, Chen K, Mathis C, Landau S, Cairns NJ, Householder E, Reinwald LT, Lee V, Korecka M, Figurski M, Crawford K, Neu S, Foroud TM, Potkin S, Shen L, Kelley F, Kim S, Nho K, Kachaturian Z, Frank R, Snyder PJ, Molchan S, Kaye J, Quinn J, Lind B, Carter R, Dolen S, Schneider LS, Pawluczyk S, Beccera M, Teodoro L, Spann BM, Brewer J, Vanderswag H, Fleisher A, Heidebrink JL, Lord JL, Mason SS, Albers CS, Knopman D, Johnson K, Doody RS, Meyer JV, Chowdhury M, Rountree S, Dang M, Stern Y, Honig LS, Bell KL, Ances B, Morris JC, Carroll M, Leon S, Mintun MA, Schneider S, Oliver A, Griffith R, Clark D, Geldmacher D, Brockington J, Roberson E, Grossman H, Mitsis E, deToledo-Morrell L, Shah RC, Duara R, Varon D, Greig MT, Roberts P, Albert M, Onyike C, D'Agostino D, Kielb S, Galvin JE, Pogorelec DM, Cerbone B, Michel CA, Rusinek H, Leon MJ de, Glodzik L, Santi SD, Doraiswamy PM, Petrella JR, Wong TZ, Arnold SE, Karlawish JH, Wolk D, Smith CD, Jicha G, Hardy P, Sinha P, Oates E, Conrad G, Lopez OL, Oakley M, Simpson DM, Porsteinsson AP, Goldstein BS, Martin K, Makino KM, Ismail MS, Brand C, Mulnard RA, Thai G, Ortiz CM, Womack K, Mathews D, Quiceno M, Arrastia RD, King R, Weiner M, Cook KM, DeVous M, Levey AI, Lah JJ, Cellar JS, Burns JM, Anderson HS, Swerdlow RH, Apostolova L, Tingus K, Woo E, Silverman DHS, Lu PH, Bartzokis G, Radford NRG, Parfitt F, Kendall T, Johnson H, Farlow MR, Hake AM, Matthews BR, Herring S, Hunt C, Dyck CH van, Carson RE, MacAvoy MG, Chertkow H, Bergman H, Hosein C, Black S, Stefanovic B, Caldwell C, Hsiung GYR, Feldman H, Mudge B, Past MA, Kertesz A, Rogers J, Trost D, Bernick C, Munic D, Kerwin D, Mesulam MM, Lipowski K, Wu CK, Johnson N, Sadowsky C, Martinez W, Villena T, Turner RS, Johnson K, Reynolds B, Sperling RA, Johnson KA, Marshall G, Frey M, Yesavage J, Taylor JL, Lane B, Rosen A, Tinklenberg J, Sabbagh MN, Belden CM, Jacobson SA, Sirrel SA, Kowall N, Killiany R, Budson AE, Norbash A, Johnson PL, Obisesan TO, Wolday S, Allard J, Lerner A, Ogrocki P, Hudson L, Fletcher E, Carmichael O, Olichney J, DeCarli C, Kittur S, Borrie M, Lee TY, Bartha R, Johnson S, Asthana S, Carlsson CM, Potkin SG, Preda A, Nguyen D, Tariot P, Reeder S, Bates V, Capote H, Rainka M, Scharre DW, Kataki M, Adeli A, Zimmerman EA, Celmins D, Brown AD, Pearlson GD, Blank K, Anderson K, Santulli RB, Kitzmiller TJ, Schwartz ES, Sink KM, Williamson JD, Garg P, Watkins F, Ott BR, Querfurth H, Tremont G, Salloway S, Malloy P, Correia S, Rosen HJ, Miller BL, Mintzer J, Spicer K, Bachman D, Finger E, Pasternak S, Rachinsky I, Drost D, Pomara N, Hernando R, Sarrael A, Schultz SK, Ponto LLB, Shim H, Smith KE, Relkin N, Chaing G, Raudin L, Smith A, Fargher K, Raj BA, Pontecorvo MJ, Devous MD, Rabinovici GD, Alexander DC,

Lyoo CH, Evans AC, Hansson O. Four distinct trajectories of tau deposition identified in Alzheimer's disease. Nat Med 2021:1−11. https://doi.org/10.1038/s41591-021-01309-6.
[16] Ranganath C, Ritchey M. Two cortical systems for memory-guided behaviour. Nat Rev Neurosci 2012;13:713−26. https://doi.org/10.1038/nrn3338.
[17] Ritchey M, Cooper RA. Deconstructing the posterior medial episodic network. Trends Cogn Sci 2020;24(6):451−65. https://doi.org/10.1016/j.tics.2020.03.006. PMID: 32340798.
[18] Coughlan G, Laczó J, Hort J, Minihane AM, Hornberger M. Spatial navigation deficits—overlooked cognitive marker for preclinical Alzheimer disease? Nat Rev Neurol 2018;14(8):496−506. https://doi.org/10.1038/s41582-018-0031-x. PMID: 29980763.
[19] Epstein RA, Patai E, Julian JB, Spiers HJ. The cognitive map in humans: spatial navigation and beyond. Nat Neurosci 2017;20:1504−13. https://doi.org/10.1038/nn.4656.
[20] Jutten RJ, Sikkes SAM, Amariglio RE, Buckley RF, Properzi MJ, Marshall GA, Rentz DM, Johnson KA, Teunissen CE, Berckel BNMV, Flier WMV der, Scheltens P, Sperling RA, Papp KV. Identifying sensitive measures of cognitive decline at different clinical stages of Alzheimer's disease. J Int Neuropsychol Soc 2020:1−13. https://doi.org/10.1017/s1355617720000934.
[21] Bastin C, Delhaye E. Targeting the function of the transentorhinal cortex to identify early cognitive markers of Alzheimer's disease. Cognit Affect Behav Neurosci 2023:1−11. https://doi.org/10.3758/s13415-023-01093-5.
[22] Parra MA, Calia C, Pattan V, Sala SD. Memory markers in the continuum of the Alzheimer's clinical syndrome. Alzheimer's Res Ther 2022;14:142. https://doi.org/10.1186/s13195-022-01082-9.
[23] Libby LA, Hannula DE, Ranganath C. Medial temporal lobe coding of item and spatial information during relational binding in working memory. J Neurosci 2014;34:14233−42. https://doi.org/10.1523/jneurosci.0655-14.2014.
[24] Sheldon S, Levine B. The medial temporal lobes distinguish between within-item and item-context relations during autobiographical memory retrieval. Hippocampus 2015;25(12):1577−90. https://doi.org/10.1002/hipo.22477. PMID: 26032447.
[25] Staresina BP, Cooper E, Henson RN. Reversible information flow across the medial temporal lobe: the Hippocampus links cortical modules during memory retrieval. J Neurosci 2013;33:14184−92. https://doi.org/10.1523/jneurosci.1987-13.2013.
[26] Staresina BP, Duncan KD, Davachi L. Perirhinal and parahippocampal cortices differentially contribute to later recollection of object- and scene-related event details. J Neurosci 2011;31:8739−47. https://doi.org/10.1523/jneurosci.4978-10.2011.
[27] Berron D, Neumann K, Maass A, Schütze H, Fliessbach K, Kiven V, Jessen F, Sauvage M, Kumaran D, Düzel E. Age-related functional changes in domain-specific medial temporal lobe pathways. Neurobiol Aging 2018;65:86−97. https://doi.org/10.1016/j.neurobiolaging.2017.12.030.
[28] Reagh ZM, Noche JA, Tustison NJ, Delisle D, Murray EA, Yassa MA. Functional imbalance of anterolateral entorhinal cortex and hippocampal dentate/CA3 underlies age-related object pattern separation deficits. Neuron 2018;97:1187−1198.e4. https://doi.org/10.1016/j.neuron.2018.01.039.
[29] Reagh ZM, Yassa MA. Object and spatial mnemonic interference differentially engage lateral and medial entorhinal cortex in humans. Proc Natl Acad Sci 2014;111. https://doi.org/10.1073/pnas.1411250111.

[30] Ryan L, Cardoza JA, Barense, Kawa KH, Wallentin-Flores J, Arnold WT, Alexander GE. Age-related impairment in a complex object discrimination task that engages perirhinal cortex. Hippocampus 2012;22:1978−89. https://doi.org/10.1002/hipo.22069.

[31] Das SR, Xie L, Wisse L, Vergnet N, Ittyerah R, Cui S, et al. In vivo measures of tau burden are associated with atrophy in early Braak stage medial temporal lobe regions in amyloid-negative individuals. Alzheimers Dement 2019;15(10):1286−95. https://doi.org/10.1016/j.jalz.2019.05.009. PMID: 31495603; PMCID: PMC6941656.

[32] Wolk DA, Das SR, Mueller SG, Weiner MW, Yushkevich PA. Medial temporal lobe subregional morphometry using high resolution MRI in Alzheimer's disease. Neurobiol Aging 2017;49:204−13. https://doi.org/10.1016/j.neurobiolaging.2016.09.011.

[33] Xie L, Das SR, Wisse L, Ittyerah R, Yushkevich PA, Wolk DA. Early tau burden correlates with higher rate of atrophy in transentorhinal cortex. J Alzheim Dis 2018;62:85−92. https://doi.org/10.3233/jad-170945.

[34] Xie L, Wisse LEM, Das SR, Vergnet N, Dong M, Ittyerah R, et al. Longitudinal atrophy in early Braak regions in preclinical Alzheimer's disease. Hum Brain Mapp 2020;41(16):4704−17. https://doi.org/10.1002/hbm.25151. PMID: 32845545; PMCID: PMC7555086.

[35] Bakker A, Albert MS, Krauss G, Speck CL, Gallagher M. Response of the medial temporal lobe network in amnestic mild cognitive impairment to therapeutic intervention assessed by fMRI and memory task performance. Neuroimage 2015;7:688−98. https://doi.org/10.1016/j.nicl.2015.02.009.

[36] Bakker A, Krauss GL, Albert MS, Speck CL, Jones LR, Stark CE, Yassa MA, Bassett SS, Shelton AL, Gallagher M. Reduction of hippocampal hyperactivity improves cognition in amnestic mild cognitive impairment. Neuron 2012;74:467−74. https://doi.org/10.1016/j.neuron.2012.03.023.

[37] Yassa MA, Stark SM, Bakker A, Albert MS, Gallagher M, Stark C. High-resolution structural and functional MRI of hippocampal CA3 and dentate gyrus in patients with amnestic Mild Cognitive Impairment. Neuroimage 2010;51:1242−52. https://doi.org/10.1016/j.neuroimage.2010.03.040.

[38] Berron D, Cardenas-Blanco A, Bittner D, Metzger CD, Spottke A, Heneka MT, et al. Higher CSF tau levels are related to hippocampal hyperactivity and object mnemonic discrimination in older adults. J Neurosci 2019;39(44):8788−97. https://doi.org/10.1523/jneurosci.1279-19.2019. PMID: 31541019; PMCID: PMC6820211.

[39] Berron D, Olsson E, Andersson F, Janelidze S, Tideman P, Düzel E, et al. Remote and unsupervised digital memory assessments can reliably detect cognitive impairment in Alzheimer's disease. Alzheimers Dement 2024b;20(7):4775−91. https://doi.org/10.1002/alz.13919. PMID: 38867417; PMCID: PMC11247711.

[40] Maass A, Berron D, Harrison TM, Adams JN, Joie RL, Baker S, Mellinger T, Bell RK, Swinnerton K, Inglis B, Rabinovici GD, Düzel E, Jagust WJ. Alzheimer's pathology targets distinct memory networks in the ageing brain. Brain 2019;142:2492−509. https://doi.org/10.1093/brain/awz154.

[41] Davachi L. Item, context and relational episodic encoding in humans. Curr Opin Neurobiol 2006;16:693700. https://doi.org/10.1016/j.conb.2006.10.012.

[42] Ranganath C. Binding items and contexts the cognitive neuroscience of episodic memory. Curr Dir Psychol Sci 2010;19:131−7. https://doi.org/10.1177/0963721410368805.

[43] Yonelinas AP. The hippocampus supports high-resolution binding in the service of perception, working memory and long-term memory. Behav Brain Res 2013;254:34−44. https://doi.org/10.1016/j.bbr.2013.05.030.

[44] Besson G, Simon J, Salmon E, Bastin C. Familiarity for entities as a sensitive marker of antero-lateral entorhinal atrophy in amnestic mild cognitive impairment. Cortex 2020;128:61−72. https://doi.org/10.1016/j.cortex.2020.02.022.

[45] Delhaye E, Bahri MA, Salmon E, Bastin C. Impaired perceptual integration and memory for unitized representations are associated with perirhinal cortex atrophy in Alzheimer's disease. Neurobiol Aging 2018;73:135−44. https://doi.org/10.1016/j.neurobiolaging.2018.09.021.

[46] Yeung L-K, Olsen RK, Bild-Enkin HE, D'Angelo MC, Kacollja A, McQuiggan DA, Keshabyan A, Ryan JD, Barense MD. Anterolateral entorhinal cortex volume predicted by altered intra-item configural processing. J Neurosci 2017;37:5527−38. https://doi.org/10.1523/jneurosci.3664-16.2017.

[47] Parra MA, Sala SD, Abrahams S, Logie RH, Méndez LG, Lopera F. Specific deficit of colour−colour short-term memory binding in sporadic and familial Alzheimer's disease. Neuropsychologia 2011;49:1943−52. https://doi.org/10.1016/j.neuropsychologia.2011.03.022.

[48] Parra MA, Abrahams S, Logie RH, Méndez LG, Lopera F, Sala SD. Visual short-term memory binding deficits in familial Alzheimer's disease. Brain 2010;133:2702−13. https://doi.org/10.1093/brain/awq148.

[49] Parra MA, Abrahams S, Fabi K, Logie R, Luzzi S, Sala SD. Short-term memory binding deficits in Alzheimer's disease. Brain 2009;132:1057−66. https://doi.org/10.1093/brain/awp036.

[50] Sala SD, Parra MA, Fabi K, Luzzi S, Abrahams S. Short-term memory binding is impaired in AD but not in non-AD dementias. Neuropsychologia 2012;50:833−40. https://doi.org/10.1016/j.neuropsychologia.2012.01.018.

[51] Bays PM, Catalao RFG, Husain M. The precision of visual working memory is set by allocation of a shared resource. J Vis 2009;9(7):1−11. https://doi.org/10.1167/9.10.7.

[52] Liang Y, Pertzov Y, Nicholas JM, Henley S, Crutch S, Woodward F, et al. Visual short-term memory binding deficit in familial Alzheimer's disease. Cortex 2016;78:150−64. https://doi.org/10.1016/j.cortex.2016.01.015. PMID: 27085491; PMCID: PMC4865502.

[53] Richter FR, Cooper RA, Bays PM, Simons JS. Distinct neural mechanisms underlie the success, precision, and vividness of episodic memory. Elife 2016;5. https://doi.org/10.7554/elife.18260.

[54] Zhang W, Luck SJ. Discrete fixed-resolution representations in visual working memory. Nature 2008;453:233−5. https://doi.org/10.1038/nature06860.

[55] Zokaei N, Sillence A, Kienast A, Drew D, Plant O, Slavkova E, Manohar SG, Husain M. Different patterns of short-term memory deficit in Alzheimer's disease, Parkinson's disease and subjective cognitive impairment. Cortex 2020;132:41−50. https://doi.org/10.1016/j.cortex.2020.06.016.

[56] Sperling R, Bates J. fMRI studies of associative encoding in young and elderly controls and mild Alzheimer's disease. 2003.

[57] Amariglio RE, Frishe K, Olson LE, Wadsworth LP, Lorius N, Sperling RA, Rentz DM. Validation of the face name associative memory Exam in cognitively normal older individuals. J Clin Exp Neuropsychol 2012;34:580−7. https://doi.org/10.1080/13803395.2012.666230.

[58] Rentz DM, Amariglio RE, Becker JA, Frey M, Olson LE, Frishe K, Carmasin J, Maye JE, Johnson KA, Sperling RA. Face-name associative memory performance is related to amyloid burden in normal elderly. Neuropsychologia 2011;49:2776−83. https://doi.org/10.1016/j.neuropsychologia.2011.06.006.

[59] Polcher A, Frommann I, Koppara A, Wolfsgruber S, Jessen F, Wagner M. Face-name associative recognition deficits in subjective cognitive decline and mild cognitive impairment. J Alzheimers Dis Park 2017:1—13. https://doi.org/10.3233/jad-160637.

[60] Hartley T, Bird CM, Chan D, Cipolotti L, Husain M, Vargha-Khadem F, Burgess N. The hippocampus is required for short-term topographical memory in humans. Hippocampus 2007;17:34—48. https://doi.org/10.1002/hipo.20240.

[61] Chan D, Gallaher L, Moodley K, Minati L, Burgess N, Hartley T. The 4 Mountains test: a short test of spatial memory with high sensitivity for the diagnosis of pre-dementia Alzheimer's disease. JoVE 2016;(116):54454. https://doi.org/10.3791/54454. PMID: 27768046; PMCID: PMC5092189.

[62] Moodley K, Minati L, Contarino V, Prioni S, Wood R, Cooper R, D'Incerti L, Tagliavini F, Chan D. Diagnostic differentiation of mild cognitive impairment due to Alzheimer's disease using a hippocampus-dependent test of spatial memory. Hippocampus 2015;25:939—51. https://doi.org/10.1002/hipo.22417.

[63] Wood RA, Moodley KK, Lever C, Minati L, Chan D. Allocentric spatial memory testing predicts conversion from mild cognitive impairment to dementia: an initial proof-of-concept study. Front Neurol 2016;7:215. https://doi.org/10.3389/fneur.2016.00215.

[64] Allison SL, Fagan AM, Morris JC, Head D. Spatial navigation in preclinical alzheimer's disease. J Alzheim Dis 2016;52:77—90. https://doi.org/10.3233/jad-150855.

[65] Howett D, Castegnaro A, Krzywicka K, Hagman J, Marchment D, Henson R, et al. Differentiation of mild cognitive impairment using an entorhinal cortex-based test of virtual reality navigation. Brain 2019;142(6):1751—66. https://doi.org/10.1093/brain/awz116. PMID: 31121601; PMCID: PMC6536917.

[66] Coughlan G, Puthusseryppady V, Lowry E, Gillings R, Spiers H, Minihane A-M, Hornberger M. Test-retest reliability of spatial navigation in adults at-risk of Alzheimer's disease. PLoS One 2020;15:e0239077. https://doi.org/10.1371/journal.pone.0239077.

[67] Coutrot A, Schmidt S, Coutrot L, Pittman J, Hong L, Wiener JM, Hölscher C, Dalton RC, Hornberger M, Spiers HJ. Virtual navigation tested on a mobile app is predictive of real-world wayfinding navigation performance. PLoS One 2019;14:e0213272. https://doi.org/10.1371/journal.pone.0213272.

[68] Coughlan G, Coutrot A, Khondoker M, Minihane A-M, Spiers H, Hornberger M. Toward personalized cognitive diagnostics of at-genetic-risk Alzheimer's disease. Proc National Acad Sci 2019;116:201901600. https://doi.org/10.1073/pnas.1901600116.

[69] Frisoni GB, Boccardi M, Barkhof F, Blennow K, Cappa S, Chiotis K, Démonet J-F, Garibotto V, Giannakopoulos P, Gietl A, Hansson O, Herholz K, Jack CR, Nobili F, Nordberg A, Snyder HM, Kate M, Varrone A, Albanese E, Becker S, Bossuyt P, Carrillo MC, Cerami C, Dubois B, Gallo V, Giacobini E, Gold G, Hurst S, Lönneborg A, Lovblad K-O, Mattsson N, Molinuevo J-L, Monsch AU, Mosimann U, Padovani A, Picco A, Porteri C, Ratib O, Saint-Aubert L, Scerri C, Scheltens P, Schott JM, Sonni I, Teipel S, Vineis P, Visser P, Yasui Y, Winblad B. Strategic roadmap for an early diagnosis of Alzheimer's disease based on biomarkers. Lancet Neurol 2017;16:661—76. https://doi.org/10.1016/s1474-4422(17)30159-x.

[70] Dahmen J, Cook D, Fellows R, Schmitter-Edgecombe M. An analysis of a digital variant of the Trail Making Test using machine learning techniques. Technol Health Care Preprint 2016:1—14. https://doi.org/10.3233/thc-161274.

[71] Fellows RP, Dahmen J, Cook D, Schmitter-Edgecombe M. Multicomponent analysis of a digital trail making test. Clin Neuropsychol 2016;31:1—14. https://doi.org/10.1080/13854046.2016.1238510.

[72] Müller S, Preische O, Heymann P, Elbing U, Laske C. Diagnostic value of a tablet-based drawing task for discrimination of patients in the early course of Alzheimer's disease from healthy individuals. J Alzheim Dis 2017;55:1463−9. https://doi.org/10.3233/jad-160921.

[73] Öhman F, Hassenstab J, Berron D, Schöll M, Papp KV. Current advances in digital cognitive assessment for preclinical Alzheimer's disease. Alzheimers Dement Diagnosis Assess Dis Monit 2021;13:e12217. https://doi.org/10.1002/dad2.12217.

[74] Staffaroni AM, Tsoy E, Taylor J, Boxer AL, Possin KL. Digital Cognitive Assessments for Dementia: digital assessments may enhance the efficiency of evaluations in neurology and other clinics. Pr. Neurol. 2020;2020:24−45.

[75] Maruff P, Lim YY, Darby D, Ellis KA, Pietrzak RH, Snyder PJ, Bush AI, Szoeke C, Schembri A, Ames D, Masters CL, Group AR. Clinical utility of the cogstate brief battery in identifying cognitive impairment in mild cognitive impairment and Alzheimer's disease. BMC Psychol 2013;1:30. https://doi.org/10.1186/2050-7283-1-30.

[76] Maruff P, Thomas E, Cysique L, Brew B, Collie A, Snyder P, Pietrzak RH. Validity of the CogState brief battery: relationship to standardized tests and sensitivity to cognitive impairment in mild traumatic brain injury, schizophrenia, and AIDS dementia complex. Arch Clin Neuropsychol 2009;24:165−78. https://doi.org/10.1093/arclin/acp010.

[77] Edgar CJ, Siemers E, Maruff P, Petersen RC, Aisen PS, Weiner MW, Albala B. Pilot evaluation of the unsupervised, at-home cogstate brief battery in ADNI-2. J Alzheim Dis 2021;83:915−25. https://doi.org/10.3233/jad-210201.

[78] Papp KV, Rentz DM, Maruff P, Sun C-K, Raman R, Donohue MC, Schembri A, Stark C, Yassa MA, Wessels AM, Yaari R, Holdridge KC, Aisen PS, Sperling RA. The computerized cognitive composite (C3) in A4, an alzheimer's disease secondary prevention trial. J Prev Alzheimers Dis 2021;8:59−67. https://doi.org/10.14283/jpad.2020.38.

[79] Rentz DM, Dekhtyar M, Sherman J, Burnham S, Blacker D, Aghjayan SL, Papp KV, Amariglio RE, Schembri A, Chenhall T, Maruff P, Aisen P, Hyman BT, Sperling RA. The feasibility of at-home iPad cognitive testing for use in clinical trials. J Prev Alzheimers Dis 2016;3:8−12. https://doi.org/10.14283/jpad.2015.78.

[80] Stark SM, Yassa MA, Lacy JW, Stark C. A task to assess behavioral pattern separation (BPS) in humans: data from healthy aging and mild cognitive impairment. Neuropsychologia 2013;51:2442−9. https://doi.org/10.1016/j.neuropsychologia.2012.12.014.

[81] Fray PJ, Robbins TW, Sahakian BJ. Neuorpsychiatyric applications of CANTAB. Int J Geriatr Psychiatr 1996;11:329−36. https://doi.org/10.1002/(sici)1099-1166(199604) 11:4<329::aid-gps453>3.0.co;2-6.

[82] Barnett JH, Blackwell AD, Sahakian BJ, Robbins TW. Translational neuropsychopharmacology. Curr Top Behav Neurosci 2015;28:449−74. https://doi.org/10.1007/7854_2015_5001.

[83] Junkkila J, Oja S, Laine M, Karrasch M. Applicability of the CANTAB-PAL computerized memory test in identifying amnestic mild cognitive impairment and alzheimer's disease. Dement Geriatr Cogn 2012;34:83−9. https://doi.org/10.1159/000342116.

[84] Gershon RC, Sliwinski MJ, Mangravite L, King JW, Kaat AJ, Weiner MW, Rentz DM. The Mobile Toolbox for monitoring cognitive function. Lancet Neurol 2022;21:589−90. https://doi.org/10.1016/s1474-4422(22)00225-3.

[85] Gershon RC, Wagster MV, Hendrie HC, Fox NA, Cook KF, Nowinski CJ. NIH Toolbox for assessment of neurological and behavioral function. Neurology 2013;80:S2−6. https://doi.org/10.1212/wnl.0b013e3182872e5f.

[86] Weintraub S, Dikmen SS, Heaton RK, Tulsky DS, Zelazo PD, Slotkin J, Carlozzi NE, Bauer PJ, Wallner-Allen K, Fox N, Havlik R, Beaumont JL, Mungas D, Manly JJ, Moy C,

Conway K, Edwards E, Nowinski CJ, Gershon R. The cognition battery of the NIH Toolbox for assessment of neurological and behavioral function: validation in an adult sample. J Int Neuropsychol Soc 2014;20:567−78. https://doi.org/10.1017/s1355617714000320.

[87] Weintraub S, Dikmen SS, Heaton RK, Tulsky DS, Zelazo PD, Bauer PJ, Carlozzi NE, Slotkin J, Blitz D, Wallner-Allen K, Fox NA, Beaumont JL, Mungas D, Nowinski CJ, Richler J, Deocampo JA, Anderson JE, Manly JJ, Borosh B, Havlik R, Conway K, Edwards E, Freund L, King JW, Moy C, Witt E, Gershon RC. Cognition assessment using the NIH Toolbox. Neurology 2013;80:S54−64. https://doi.org/10.1212/wnl.0b013e3182872ded.

[88] Buckley RF, Sparks KP, Papp KV, Dekhtyar M, Martin C, Burnham S, Sperling RA, Rentz DM. Computerized cognitive testing for use in clinical trials: a comparison of the NIH Toolbox and cogstate C3 batteries. J Prev Alzheimers Dis 2017;4:3−11. https://doi.org/10.14283/jpad.2017.1.

[89] Hackett K, Krikorian R, Giovannetti T, Melendez-Cabrero J, Rahman A, Caesar EE, Chen JL, Hristov H, Seifan A, Mosconi L, Isaacson RS. Utility of the NIH Toolbox for assessment of prodromal Alzheimer's disease and dementia. Alzheimers Dement Diagnosis Assess Dis Monit 2018;10:764−72. https://doi.org/10.1016/j.dadm.2018.10.002.

[90] Snitz BE, Tudorascu DL, Yu Z, Campbell E, Lopresti BJ, Laymon CM, Minhas DS, Nadkarni NK, Aizenstein HJ, Klunk WE, Weintraub S, Gershon RC, Cohen AD. Associations between NIH Toolbox Cognition Battery and in vivo brain amyloid and tau pathology in non-demented older adults. Alzheimers Dement Diagnosis Assess Dis Monit 2020;12:e12018. https://doi.org/10.1002/dad2.12018.

[91] Possin KL, Moskowitz T, Erlhoff SJ, Rogers KM, Johnson ET, Steele NZ, Higgins JJ, Stiver J, Alioto AG, Farias ST, Miller BL, Rankin KP. The brain health assessment for detecting and diagnosing neurocognitive disorders. J Am Geriatr Soc 2018;66:150−6. https://doi.org/10.1111/jgs.15208.

[92] Tsoy E, Strom A, Iaccarino L, Erlhoff SJ, Goode CA, Rodriguez A-M, Rabinovici GD, Miller BL, Kramer JH, Rankin KP, Joie RL, Possin KL. Detecting Alzheimer's disease biomarkers with a brief tablet-based cognitive battery: sensitivity to Aβ and tau PET. Alzheimer's Res Ther 2021;13:36. https://doi.org/10.1186/s13195-021-00776-w.

[93] Tsoy E, Erlhoff SJ, Goode CA, Dorsman KA, Kanjanapong S, Lindbergh CA, Joie RL, Strom A, Rabinovici GD, Lanata SC, Miller BL, Farias SET, Kramer JH, Rankin KP, Possin KL. BHA-CS: a novel cognitive composite for Alzheimer's disease and related disorders. Alzheimers Dement Diagnosis Assess Dis Monit 2020;12:e12042. https://doi.org/10.1002/dad2.12042.

[94] Bo M, Mario B, Massaia M, Massimiliano M, Merlo C, Chiara M, Sona A, Alessandro S, Canadè A, Antonella C, Fonte G, Gianfranco F. White-coat effect among older patients with suspected cognitive impairment: prevalence and clinical implications. Int J Geriatr Psychiatr 2009;24:509−17. https://doi.org/10.1002/gps.2145.

[95] Schlemmer M, Desrichard O. Is medical environment detrimental to memory? A test of A white coat effect on older people's memory performance. Clin Gerontol 2017;41:77−81. https://doi.org/10.1080/07317115.2017.1307891.

[96] Hyun J, Sliwinski MJ, Smyth JM. Waking up on the wrong side of the bed: the effects of stress anticipation on working memory in daily life. J Gerontol B Psychol Sci Soc Sci 2018;74:38−46. https://doi.org/10.1093/geronb/gby042.

[97] Sliwinski M, Smyth J, Hofer S, Stawski R. Intraindividual coupling of daily stress and cognition. 2006.

[98] Wild CJ, Nichols ES, Battista ME, Stojanoski B, Owen AM. Dissociable effects of self-reported daily sleep duration on high-level cognitive abilities. Sleep 2018;41(12):zsy182. https://doi.org/10.1093/sleep/zsy182. PMID: 30212878; PMCID: PMC6289236.

[99] Sliwinski MJ. Measurement-burst designs for social health research: longitudinal measurement-burst design. Soc Personality Psychol Compass 2008;2:245−61. https://doi.org/10.1111/j.1751-9004.2007.00043.x.

[100] Sliwinski MJ, Mogle JA, Hyun J, Munoz E, Smyth JM, Lipton RB. Reliability and validity of ambulatory cognitive assessments. Assessment 2018;25:14−30.

[101] Giil LM, Aarsland D. Greater variability in cognitive decline in Lewy body dementia compared to alzheimer's disease. J Alzheim Dis 2020;73:1321−30. https://doi.org/10.3233/jad-190731.

[102] Matar E, Shine JM, Halliday GM, Lewis SJG. Cognitive fluctuations in Lewy body dementia: towards a pathophysiological framework. Brain 2019;143:31−46. https://doi.org/10.1093/brain/awz311.

[103] Hassenstab J, Ruvolo D, Jasielec M, Xiong C, Grant E, Morris JC. Absence of practice effects in preclinical Alzheimer's disease. Neuropsychology 2015;29:940−8. https://doi.org/10.1037/neu0000208.

[104] Duff K, Hammers D, et al. Short-term practice effects and amyloid deposition: providing information above and beyond baseline cognition. J Prev Alzheimers Dis 2017;4.

[105] Duff K, Foster N, et al. Practice effects and amyloid deposition: preliminary data on a method for enriching samples in clinical trials. Alzheimer Dis Assoc Disord 2014;28.

[106] Jutten RJ, Rentz DM, Fu JF, Maybrium DV, Amariglio RE, Buckley RF, Properzi MJ, Maruff P, Stark CE, Yassa MA, Johnson KA, Sperling RA, Papp KV. Monthly at-home computerized cognitive testing to detect diminished practice effects in preclinical alzheimer's disease. Front Aging Neurosci 2022;13:800126. https://doi.org/10.3389/fnagi.2021.800126.

[107] Jutten RJ, Grandoit E, Foldi NS, Sikkes SAM, Jones RN, Choi S-E, Lamar ML, Louden DKN, Rich J, Tommet D, Crane PK, Rabin LA. Lower practice effects as a marker of cognitive performance and dementia risk: a literature review. Alzheimers Dement 2020;12:e12055. https://doi.org/10.1002/dad2.12055.

[108] Lim YY, Baker JE, Bruns L, Mills A, Fowler C, Fripp J, Rainey-Smith SR, Ames D, Masters CL, Maruff P. Association of deficits in short-term learning and Aβ and hippoampal volume in cognitively normal adults. Neurology 2020. https://doi.org/10.1212/WNL.0000000000010728.

[109] Samaroo A, Amariglio RE, Burnham S, Sparks P, Properzi M, Schultz AP, Buckley R, Johnson KA, Sperling RA, Rentz DM, Papp KV. Diminished learning over repeated Exposures (LORE) in preclinical Alzheimer's disease. Alzheimers Dement Diagnosis Assess Dis Monit 2020;12:e12132. https://doi.org/10.1002/dad2.12132.

[110] Wearn AR, Saunders-Jennings E, Nurdal V, Hadley E, Knight MJ, Newson M, Kauppinen RA, Coulthard EJ. Accelerated long-term forgetting in healthy older adults predicts cognitive decline over 1 year. Alzheimer's Res Ther 2020;12:119. https://doi.org/10.1186/s13195-020-00693-4.

[111] Weston PSJ, Nicholas JM, Henley SMD, Liang Y, Macpherson K, Donnachie E, Schott JM, Rossor MN, Crutch SJ, Butler CR, Zeman AZ, Fox NC. Accelerated long-term forgetting in presymptomatic autosomal dominant Alzheimer's disease: a cross-sectional study. Lancet Neurol 2018;17:123−32. https://doi.org/10.1016/s1474-4422(17)30434-9.

[112] Lancaster C, Koychev I, Blane J, Chinner A, Wolters L, Hinds C. Evaluating the feasibility of frequent cognitive assessment using the Mezurio smartphone app: observational and

interview study in adults with elevated dementia risk. Jmir Mhealth Uhealth 2020b;8:e16142. https://doi.org/10.2196/16142.

[113] Lancaster C, Koychev I, Blane J, Chinner A, Chatham C, Taylor K, Hinds C. Gallery Game: smartphone-based assessment of long-term memory in adults at risk of Alzheimer's disease. J Clin Exp Neuropsychol 2020a;42:329−43. https://doi.org/10.1080/13803395.2020.1714551.

[114] Jack C, Bennett D, et al. NIA-AA Research Framework: toward a biological definition of Alzheimer's disease. Alzheimers Dement 2018.

[115] Berron D, Glanz W, Clark L, Basche K, Grande X, Güsten J, Billette OV, Hempen I, Naveed MH, Diersch N, Butryn M, Spottke A, Buerger K, Perneczky R, Schneider A, Teipel S, Wiltfang J, Johnson S, Wagner M, Jessen F, Düzel E. A remote digital memory composite to detect cognitive impairment in memory clinic samples in unsupervised settings using mobile devices. Npj Digit Med 2024a;7:79. https://doi.org/10.1038/s41746-024-00999-9.

[116] Berron D, Ziegler G, Vieweg P, Billette O, Güsten J, Grande X, Heneka MT, Schneider A, Teipel S, Jessen F, Wagner M, Düzel E. Feasibility of digital memory assessments in an unsupervised and remote study setting. Front Digit Heal 2022;4:892997. https://doi.org/10.3389/fdgth.2022.892997.

[117] Öhman F, Berron D, Papp KV, Kern S, Skoog J, Bodin TH, Zettergren A, Skoog I, Schöll M. Unsupervised mobile app-based cognitive testing in a population-based study of older adults born 1944. Front Digital Heal 2022;4:933265. https://doi.org/10.3389/fdgth.2022.933265.

[118] Düzel E, Berron D, Schütze H, Cardenas-Blanco A, Metzger C, Betts M, Ziegler G, Chen Y, Dobisch L, Bittner D, Glanz W, Reuter M, Spottke A, Rudolph J, Brosseron F, Buerger K, Janowitz D, Fliessbach K, Heneka M, Laske C, Buchmann M, Nestor P, Peters O, Diesing D, Li S, Priller J, Spruth EJ, Altenstein S, Ramirez A, Schneider A, Kofler B, Speck O, Teipel S, Kilimann I, Dyrba M, Wiltfang J, Bartels C, Wolfsgruber S, Wagner M, Jessen F. CSF total tau levels are associated with hippocampal novelty irrespective of hippocampal volume. Alzheimers Dement 2018;10:782−90. https://doi.org/10.1016/j.dadm.2018.10.003.

[119] Düzel E, Schütze H, Yonelinas AP, Heinze H. Functional phenotyping of successful aging in long-term memory: preserved performance in the absence of neural compensation. Hippocampus 2011;21:803−14. https://doi.org/10.1002/hipo.20834.

[120] Papp KV, Samaroo A, Chou H, Buckley R, Schneider OR, Hsieh S, Soberanes D, Quiroz Y, Properzi M, Schultz A, García-Magariño I, Marshall GA, Burke JG, Kumar R, Snyder N, Johnson K, Rentz DM, Sperling RA, Amariglio RE. Unsupervised mobile cognitive testing for use in preclinical Alzheimer's disease. Alzheimers Dement Diagnosis Assess Dis Monit 2021;13. https://doi.org/10.1002/dad2.12243.

[121] Papp KV, Jutten RJ, Soberanes D, Weizenbaum E, Hsieh S, Molinare C, Buckley R, Betensky RA, Marshall GA, Johnson KA, Rentz DM, Sperling R, Amariglio RE. Early detection of amyloid-related changes in memory among cognitively unimpaired older adults with daily digital testing. Ann Neurol 2023. https://doi.org/10.1002/ana.26833.

[122] Weizenbaum EL, Soberanes D, Hsieh S, Molinare CP, Buckley RF, Betensky RA, et al. Capturing learning curves with the multiday Boston remote assessment of neurocognitive health (BRANCH): feasibility, reliability, and validity. Neuropsychology 2023;38(2):198−210. https://doi.org/10.1037/neu0000933. PMID: 37971862; PMCID: PMC10841660.

[123] Nicosia J, Aschenbrenner AJ, Balota DA, Sliwinski MJ, Tahan M, Adams S, Stout SS, Wilks H, Gordon BA, Benzinger TLS, Fagan AM, Xiong C, Bateman RJ, Morris JC, Hassenstab J. Unsupervised high-frequency smartphone-based cognitive assessments are reliable, valid, and feasible in older adults at risk for Alzheimer's disease. J Int Neuropsychol Soc 2022a;29:459−71. https://doi.org/10.1017/s135561772200042x.

[124] Goldsack JC, Coravos A, Bakker JP, Bent B, Dowling AV, Fitzer-Attas C, Godfrey A, Godino JG, Gujar N, Izmailova E, Manta C, Peterson B, Vandendriessche B, Wood WA, Wang KW, Dunn J. Verification, analytical validation, and clinical validation (V3): the foundation of determining fit-for-purpose for Biometric Monitoring Technologies (BioMeTs). Npj Digit Med 2020;3:55. https://doi.org/10.1038/s41746-020-0260-4.

[125] Robin J, Harrison JE, Kaufman LD, Rudzicz F, Simpson W, Yancheva M. Evaluation of speech-based digital biomarkers: review and recommendations. Digit Biomark 2020;4:99−108. https://doi.org/10.1159/000510820.

[126] Germine L, Reinecke K, Chaytor NS. Digital neuropsychology: challenges and opportunities at the intersection of science and software. Clin Neuropsychol 2019:1−16. https://doi.org/10.1080/13854046.2018.1535662.

[127] Nicosia J, Wang B, Aschenbrenner AJ, Sliwinski MJ, Yabiku ST, Roque NA, Germine LT, Bateman RJ, Morris JC, Hassenstab J. To BYOD or not: are device latencies important for bring-your-own-device (BYOD) smartphone cognitive testing? Behav Res Methods 2022:1−13. https://doi.org/10.3758/s13428-022-01925-1.

[128] Passell E, Strong RW, Rutter LA, Kim H, Scheuer L, Martini P, Grinspoon L, Germine L. Cognitive test scores vary with choice of personal digital device. Behav Res Methods 2021:1−14. https://doi.org/10.3758/s13428-021-01597-3.

[129] Mackin RS, Insel PS, Truran D, Finley S, Flenniken D, Nosheny R, Ulbright A, Comacho M, Bickford D, Harel B, Maruff P, Weiner MW. Unsupervised online neuropsychological test performance for individuals with mild cognitive impairment and dementia: results from the Brain Health Registry. Alzheimers Dement Diagnosis Assess Dis Monit 2018;10:573−82. https://doi.org/10.1016/j.dadm.2018.05.005.

[130] Mielke MM, Machulda MM, Hagen CE, Edwards KK, Roberts RO, Pankratz VS, Knopman DS, Jack CR, Petersen RC. Performance of the CogState computerized battery in the Mayo clinic study on aging. Alzheimer's Dementia 2015;11:1367−76. https://doi.org/10.1016/j.jalz.2015.01.008.

[131] Stricker NH, Lundt ES, Alden EC, Albertson SM, Machulda MM, Kremers WK, Knopman DS, Petersen RC, Mielke MM. Longitudinal comparison of in clinic and at home administration of the cogstate brief battery and demonstrated practice effects in the Mayo clinic study of aging. J Prev Alzheimers Dis 2020;7:21−8. https://doi.org/10.14283/jpad.2019.35.

[132] Berron D, Glanz W, Billette O, Grande X, Guesten J, Hempen I, Naveed MH, Butryn M, Spottke A, Buerger K, Perneczky R, Schneider A, Teipel S, Wiltfang J, Wagner M, Jessen F, Duezel E. A remote digital memory composite to detect cognitive impairment in memory clinic samples in unsupervised settings using mobile devices. NPJ Digit Med 2021. https://doi.org/10.1101/2021.11.12.21266226.

[133] Pratap A, Neto EC, Snyder P, Stepnowsky C, Elhadad N, Grant D, Mohebbi MH, Mooney S, Suver C, Wilbanks J, Mangravite L, Heagerty PJ, Areán P, Omberg L. Indicators of retention in remote digital health studies: a cross-study evaluation of 100,000 participants. NPJ Digit Med 2020;3:21. https://doi.org/10.1038/s41746-020-0224-8.

[134] Gold M, Amatniek J, Carrillo MC, Cedarbaum JM, Hendrix JA, Miller BB, Robillard JM, Rice JJ, Soares H, Tome MB, Tarnanas I, Vargas G, Bain LJ, Czaja SJ. Digital technologies

as biomarkers, clinical outcomes assessment, and recruitment tools in Alzheimer's disease clinical trials. Alzheimer's Dement 2018;4:234−42. https://doi.org/10.1016/j.trci.2018.04.003.
[135] König A, Satt A, Sorin A, et al. Automatic speech analysis for the assessment of patients with predementia and Alzheimer's disease. Alzheimer's Dement 2015.
[136] Boschi V, Catricalà E, Consonni M, Chesi C, Moro A, Cappa SF. Connected speech in neurodegenerative language disorders: a review. Front Psychol 2017;8:269. https://doi.org/10.3389/fpsyg.2017.00269.
[137] Szatloczki G, Hoffmann I, Vincze V, Kalman J, Pakaski M. Speaking in alzheimer's disease, is that an early sign? Importance of changes in language abilities in alzheimer's disease. Front Aging Neurosci 2015;7:195. https://doi.org/10.3389/fnagi.2015.00195.
[138] Asgari M, Kaye J, Dodge H. Predicting mild cognitive impairment from spontaneous spoken utterances. Alzheimers Dement Transl Res Clin Inter 2017;3:219−28. https://doi.org/10.1016/j.trci.2017.01.006.
[139] Fraser KC, Meltzer JA, Rudzicz F. Linguistic features identify alzheimer's disease in narrative speech. J Alzheim Dis 2016;49:407−22. https://doi.org/10.3233/jad-150520.
[140] Schäfer S, Mallick E, Schwed L, König A, Zhao J, Linz N, Bodin TH, Skoog J, Possemis N, Huurne D ter, Zettergren A, Kern S, Sacuiu S, Ramakers I, Skoog I, Tröger J. Screening for mild cognitive impairment using a machine learning classifier and the remote speech biomarker for cognition: evidence from two clinically relevant cohorts. J Alzheim Dis 2022;91:1165−71. https://doi.org/10.3233/jad-220762.
[141] Themistocleous C, Eckerström M, Kokkinakis D. Voice quality and speech fluency distinguish individuals with mild cognitive impairment from healthy controls. PLoS One 2020;15:e0236009. https://doi.org/10.1371/journal.pone.0236009.
[142] Toth L, Hoffmann I, Gosztolya G, Vincze V, Szatloczki G, Banreti Z, Pakaski M, Kalman J. A speech recognition-based solution for the automatic detection of mild cognitive impairment from spontaneous speech. Curr Alzheimer Res 2018;15:130−8. https://doi.org/10.2174/1567205014666171121114930.
[143] Piau A, Wild K, Mattek N, Kaye J. Current state of digital biomarker technologies for real-life, home-based monitoring of cognitive function for mild cognitive impairment to mild alzheimer disease and implications for clinical care: systematic review. J Med Internet Res 2019;21:e12785. https://doi.org/10.2196/12785.
[144] Insel TR. Digital phenotyping: technology for a new science of behavior. JAMA 2017;318:1215−6. https://doi.org/10.1001/jama.2017.11295.
[145] Dagum P. Digital biomarkers of cognitive function. NPJ Digit Med 2018;1:10. https://doi.org/10.1038/s41746-018-0018-4.
[146] Murnane EL, Väänänen K, Church K, Kay M, Krüger A, Choudhury T, Gay G, Cosley D. Mobile manifestations of alertness: connecting biological rhythms with patterns of smartphone app use. In: Mobile HCI '16: Proceedings of the 18th International Conference on human-Computer interaction with mobile devices and Services; 2016. p. 465−77. https://doi.org/10.1145/2935334.2935383.
[147] Leiva L, Böhmer M, Gehring S, Krüger A. Back to the app: the costs of mobile application interruptions. In: Proc 14th Int Conf human-computer Interact Mob devices Serv - Mobilehci, vol 12; 2012. p. 291−4. https://doi.org/10.1145/2371574.2371617.
[148] Butler P, Au R, Becker A, Bianchi M, Brown R, Cosne G, Demircioglu G, Erkkinen M, Gabelle A, Hobbs M, Hughes R, Juraver A, Langbaum J, Lenyoun H, Lingler J, Penalver-Andres J, Pham H, Porsteinsson A, Price P, Quiroz Y, Roggen D, Saha-Chaudhuri P, Scotland A, Sha S, Sliwinski M, Song H, Yang J, Belachew S. Intuition Brain Health

Study: a smartphone- and smartwatch-based virtual, observational study using multimodal mobile sensing to classify and detect mild cognitive impairment. 2024. https://doi.org/10.21203/rs.3.rs-4173311/v1.

[149] Duzel E, Schöttler M, Sommer H, Griebe M. Protocol: prospective evaluation of feasibility, added value and satisfaction of remote digital self-assessment for mild cognitive impairment in routine care with the neotivCare app. BMJ Open 2024;14:e081159. https://doi.org/10.1136/bmjopen-2023-081159.

[150] Malzbender K, Barbarino P, Ferrell PB, Bradshaw A, Brookes AJ, Díaz C, Flier WM van der, Georges J, Hansson O, Hartmanis M, Jönsson L, Krishnan R, MacLeod T, Mangialasche F, Mecocci P, Minguillon C, Middleton L, Pla S, Sardi SP, Schöll M, Suárez-Calvet M, Weidner W, Visser PJ, Zetterberg H, Bose N, Solomon A, Kivipelto M. Validation, deployment, and real-world implementation of a modular Toolbox for alzheimer's disease detection and dementia risk reduction: the AD-RIDDLE project. J Prev Alzheimers Dis 2024:1–10. https://doi.org/10.14283/jpad.2024.32.

[151] Sabbagh MN. Early detection of Mild Cognitive Impairment (MCI) in an at-home setting. J Prev Alzheimers Dis 2020;7:171–8.

[152] Polk SE, Öhman F, Hassenstab J, König A, Papp KV, Schöll M, et al. A systematic review of remote and unsupervised digital cognitive assessments in preclinical Alzheimer's disease. medRxiv 2024.

Chapter 7

Challenges for the new approach

Chapter 7a

Ethical challenges associated with the early detection of Alzheimer's disease

Richard Milne[a,b], Alessia Costa[a,b] and Carol Brayne[c]
[a]Engagement and Society, Wellcome Connecting Science, Hinxton, United Kingdom; [b]Kavli Centre for Ethics, Science, and the Public, University of Cambridge, Cambridge, United Kingdom; [c]Cambridge Public Health, University of Cambridge, Cambridge, United Kingdom

Introduction

The reduction of risk of developing dementia for the greatest number of people is a pressing and important concern around the world. Despite the demonstration of declining incidence of dementia in many high income countries, the number of people with the condition is still expected to increase dramatically as the global population ages [1,2]. As the other chapters in this book demonstrate, one strategy that has been embraced enthusiastically by policymakers looking to reduce the impact of dementia has been the early detection of pathologies associated with dementia, most notably those that characterize Alzheimer's disease (AD).

In this chapter, we introduce and review the ethical challenges associated with the move to early detection and dementia risk prediction. These discussions reflect the engagement of clinicians, scientists, bioethicists and the wider public with the question of how to deal with the predictive information about dementia provided by the identification of dominantly inherited genetic forms of Alzheimer's disease [3], ApoE genotype [4], amyloid biomarkers [5] and more recently digital and blood-based markers of disease status [6].

For much of the last 4 decades, this discussion has been dominated by questions of utility and actionability—what is the value of predicting or detecting early in the absence of effective interventions? Over recent years, the discourse has begun to shift - regardless of their efficacy, the approval of the first drugs that may modify the disease course refocuses, but does not resolve, ethical debates around early detection [7].

Our aim is to set out the major ethical and societal challenges that the implementation of early detection continues to face. Some of these, particularly those concentrated around the impact of predictive information on individual psychology and wellbeing, have received systematic research attention, and we draw attention to this literature. Other challenges, particularly those encountered outside the contexts of clinical trials and specialist clinical practice, are under-studied. We consider whether and how these questions have been reshaped by technological developments—not least the regulatory approval (and rejection) of new drugs, and the novel and scalable forms of testing enabled by blood-based biomarkers and digital tools.

Our discussion is structured in three stages, starting from the individual's encounter with testing in the context of the clinic or clinical trial. Here the emphasis of bioethical discussion has been on questions of autonomy along with the potential benefits and harms associated with a prediction of future dementia. It then considers the wider societal context of early detection, through the question of stigma and discrimination associated with risk states and a discussion of social justice and equity. Finally, we address the ethics of future care, considering the emerging questions associated with recently approved drugs and the expanded availability of technologies for early detection.

Early what?

At the outset, it is crucial to delineate the areas of overlap and distinction between early detection and early diagnosis. While these terms are frequently conflated in discourse among policymakers, funders, and even within patient organizations and research communities, they have nuanced differences that bear significant implications for patient care and healthcare systems. Both early detection and early diagnosis focus on the future, on providing a prognosis and a prediction of the likelihood of future cognitive decline and dementia [8]. They also relate to practices of classification [9]. Classification is a fundamental role of diagnosis, linked to categories set out in guidelines including the International Classification of Diseases (ICD) or the Diagnostic and Statistical Manual of Mental Disorders (DSM) [9,10].

Diagnosis, as a practice of clinical medicine, encompasses both the assignment of a disease label at a specific juncture and the temporal process leading to this designation [11]. In the context of a biomarker-driven field where endophenotypic signs often predominate over symptoms, there exists a propensity to conceptualize diagnosis as a linear trajectory from detection to post-diagnostic support [12]. However, empirical evidence suggests that the diagnosis of dementia is, in practice, a complex, temporal, and non-linear process [11,13,14].

As work in the sociology of medicine has demonstrated, the origin of diagnostic meaning is "the social activity of conversation and not the interpretive application of knowledge" [15]. Diagnosis is both interactive and, to an extent, negotiated, a "site of contest and compromise"9(p279), in which the outcome is "collective, cumulative, [and] contingent" [16]. This understanding is critical to efforts to nuance discussions of early diagnosis by emphasizing *timely* diagnosis. Here a diagnosis is not necessarily made at the "the earliest stage possible" [17], but timed to when individuals and families notice changes in cognitive function, recognize that there may be developing disease and can use diagnostic information to make sense of what is happening [18–20].

While diagnosis is a clinical practice, detection is a technical one that involves the "discovery of what is unknown or hidden" [21]. While it constitutes an element of diagnosis, it is not historically synonymous with it, given the complex meanings of diagnosis elucidated above. Furthermore, as will be explored in subsequent sections, the technical act of detection is not confined to diagnostic settings but can be integrated into diverse medical practices, notably screening. Consequently, in these contexts, the technical and ethical considerations pertaining to detection may diverge from those relevant to clinical diagnostic settings. As such, in this chapter, we focus on early *detection* and risk prediction, and their implications for clinical and population health. We do not touch upon the wider, and important discussion of timely diagnosis. These discussions are covered comprehensively elsewhere [16,19,20,22].

The right to know and the consequences of early detection for the individual

For much of the last 3 decades, from contemporary discussions about the use of biomarkers to longer-standing debates about the use and wider societal value of testing for the ApoE e4 susceptibility allele [23], the starting point for discussion of the ethics of early detection has been arguments around the use of this information in the clinic and by and with patients [24,25]. These discussions start from discussion of the autonomy of the individual being tested, their right to know (or not) information about their current and future health. These conversations are informed by studies of public attitudes that suggest interest in, and support for, the communication of information at early stages of disease, particularly among those with a family history or a higher perceived personal susceptibility [26–30].

The value of knowing (and knowing what)

Information about early disease states can only be produced through the deliberate application of testing. Before one arrives at the "right to know", there is a requirement for research that identifies measures and establishes their validity and predictive value including uncertainty for individuals, and the value, benefit and harms for them [31]. For those who have early symptoms, detection may enable timely diagnosis and access to services, or provide patients and their doctors with greater certainty [32]. The benefits of communicating information about early detection for people without symptoms have, in general, been described in terms of the personal utility of this information [24]—even where its clinical value and prognostic ability has been contested [33]. To date, however, no studies have prospectively tested personal utility in practice. Retrospective and hypothetical studies suggest that it may include an increased ability to plan for the future, or to take action to reduce risk [24,32]. Studies with members of the public and patients describe these potential benefits and perceived personal utility as important, including the ability to be "monitored to diagnosis" [26,34].

However, it is still not clear whether these presumptive benefits are attained as a result of early detection—and thus, whether personal utility exists, and what the potential consequences are of medicalization for both individuals and the wider health service [33]. Whereas a predictive test associated with heart disease, such as elevated blood pressure or cholesterol levels, may inform treatment and ultimately improve patient outcomes in a clear manner as demonstrated convincingly by many trials, it is not currently clear that the early detection of changes associated with dementia does result in future benefit [35]. In terms of behavior change, the interview study by Largent et al. found that many participants did report undertaking health behavior changes and making revised future plans—but that participants also denied that these changes would be initiated by testing, but were rather initiated due to increasing age [36]. In another interview study with cognitively unimpaired adults who had received amyloid biomarker results, those with elevated biomarker results were significantly more likely than those with non-elevated results to want to change their lifestyle [37].

While people often express an intention to change behavior as a result of information about their disease risk, reviews of work in other clinical areas has been that the individualized provision of information on its own does not have has strong or consistent effects on health-related behaviors [26,38,39]. For instance, genetic risk information alone has also been shown to have no or very limited effects on health behaviors, although ApoE disclosure has been associated with some behavior change, particularly increased vitamin intake [40,41]. Where lifestyle interventions have had some impact on risk status, this has been associated with more intensive intervention, and active support and motivation for healthy lifestyle changes [42].

The potential harms of predictive information

In addition to uncertainty about benefits, attention has been drawn to the potential harms associated with the early detection of a late-onset disorder. A central focus has been the psychological impact of learning this information in terms of depression and anxiety, including the possibility of an increase in suicidal behavior and requests for euthanasia associated with early detection [43–45].

Concerns about the use of predictive information about dementia were first raised in discussions of genetic testing, and particularly following the discovery of the ApoE e4 allele in the early 1990s, which confers an increased risk of Alzheimer's disease. In response to these concerns a series of randomized clinical trials of ApoE disclosure were conducted under the auspices of the REVEAL studies [46]. These concluded that for first generation relatives of people with dementia in a controlled research environment, learning one's Alzheimer's disease risk does not result in significant short-term psychological harm for most [46,47]. However, the generalizability of these findings is unclear given the bias in the study population, and it is important to note that some people did experience significant psychological impact [48]. Further studies have suggested the possibility that genetic information may have a negative impact on both subjective and objective memory performance among older adults [49]. Longer-term qualitative follow-up suggests that many people come to terms with learning ApoE e4 homozygote status, but that for a few, the information remains life-changing [50].

A growing number of studies has extended work on the impact of risk information to studies of biomarker disclosure to both asymptomatic individuals and those with mild cognitive impairment [32]. Here also, the currently available evidence suggests that the communication of risk information to individuals who do not currently have symptoms of cognitive impairment does not lead to short-term psychological effects for the majority [29,51]. A 2023 meta-analysis of seven studies that present quantitative data about the impact of communicating biomarker information found that it did not result in short-term psychological harm [29].

Across qualitative studies, those who learned their amyloid levels were elevated felt sadness, worry, or despair, although often reported that they appreciated knowing the cause of the cognitive complaints, the greater certainty about prognosis, and the potential for more active monitoring of their symptoms [29]. The SOKRATES study, which followed-up participants in the A4 clinical trial of solanezumab in asymptomatic individuals, found that participants who received an elevated amyloid PET scan result did view the result as more serious and sensitive than other medical test results, primarily because of its unique implications for identity, self-determination, and stigma, which we consider in the following section [36].

It seems, based on the available data, that psychological harm does not result for *most* people who have taken part in such studies as a result of learning they are at increased risk of developing dementia. However, one key characteristic and important limitation of this research to date is that it involves disclosure to volunteer research participants using standardized disclosure protocols and educational sessions [36,52,53]. These protocols, developed with reference to those for the disclosure of genetic information [54], involve detailed and in-depth discussions with participants about what biomarkers can and cannot tell them about their future health. As methods and tools for early detection proliferate, and as they are incorporated into clinical practice, case finding or screening programmes, it will be necessary to better understand whether and how these pre-disclosure preparatory sessions can be incorporated. This includes understanding what cost and opportunity cost will be incurred by systems provide them when set against the benefit/harm for those receiving the information.

Discussions of both benefits and harms presuppose the existence of meaningful information. This takes us beyond "would you want to know" and why, to questions of *what* you would come to know. Such questions relate to the validity of risk models [25], which in turn affects the potential benefits. In the case of blood-based biomarkers, for example, the population accuracy of tools developed in specialist secondary care settings needs to be evaluated before they are implemented in population settings, where the average age of dementia onset is later and levels of blood biomarkers are affected by the presence of chronic conditions [55]. Overall, the continued development of models and tools for early detection developed with the populations intended to receive them is needed to ensure the accuracy and consistency of testing, and to minimize the potential for over-diagnosis and overtreatment among individuals who may be erroneously identified as at-risk of future dementia [56]. Transparency about the accuracy of prediction, and the uncertainties associated with early detection technologies are also essential to ensure reasonable expectations from patients and clinicians about their value [27,34].

The social consequences of early detection

Ethical considerations associated with early detection extend beyond individual autonomy and the psychological impact of risk information to incorporate the place of early detection in society. Specifically, the ethical introduction of early detection tools requires effort to address questions of stigma and discrimination, equity and social justice.

Stigma and discrimination

An individual who receives information about a biomarker associated with risk of dementia (and accompanying uncertainties) is confronted with the question of whether, how and when to share it with their family and beyond. Such sharing of information may be essential to realize what are perceived to be potential benefits of early detection, but will be influenced by people's expectations of how others will treat them as a result of this information, and how such information may be used.

Dementia has long been associated with stigma, particularly in the context of our contemporary "hypercognitive society" [43]. This stigma is a recognized barrier to timely diagnosis, both on the part of patients, who may fear being labeled with a condition associated with lack of autonomy and independence, and on behalf of clinicians, who may assume a patient does not want a diagnosis [18]. Such stigma may also extend to early disease states, with studies in Europe and the USA suggesting that people who receive biomarker results are concerned about how they would be treated by others, or would be concerned about how family members and friends would see them [36,57,58]. This question has been explored in detail in the SOKRATES I and II studies with participants in the A4 Study clinical trial and the API Generation Program [59]. While all participants shared information about their amyloid or APOE results with at least one person - most often their spouse, sibling, adult child, or friend, they described the surprising burden of the decision-making about whether to share with others. Among the reasons for not sharing were both the emotional impact on recipients and fear about being treated differently and stigmatized. In survey research with members of the US public about hypothetical biomarker results, positive biomarker results were associated with stigma, even when not accompanied by symptoms of dementa—and this was not affected by the availability of a disease-modifying therapy [60]. However, it may be that expectations of stigma are greater than that experienced [61].

The disclosure of information regarding current and future brain health carries significant implications for insurance and employment. Studies on ApoE e4 genotype disclosure in the United States have revealed that one of the most notable consequences of learning about increased risk was an uptick in long-term care insurance acquisition - ApoE e4 carriers were more than twice as likely to change their long-term insurance coverage than non-carriers [62]. Furthermore, a U.S. study on perceptions of early Alzheimer's disease diagnosis found that nearly half of the respondents anticipated health insurance limitations for affected individuals, while over half expected employment discrimination [58].

The potential for early detection to result in employment discrimination is potentially considerable. It is therefore critical that early detection is not conflated with incapacity, that it is recognized that people with "early" stages of disease may never have symptoms, are able to continue to work, and that provisions are made for appropriate accommodations if cognitive impairment progresses. This requires particular emphasis if digital tools such as those discussed elsewhere in this book become widely used—as discussed below. It is essential in such contexts to protect employee's ability to opt-out of such programmes without penalty, both to protect them from potential discrimination and to preserve their right not to know.

Justice and equity considerations

The social context of early detection of dementia extends beyond questions of stigma and discrimination to include ensuring that early detection and associated clinical and "precision" public health practices are socially just and equitable [63]. Addressing the equity challenge for early detection is critically important given social, economic and ethnic disparities in the distribution of the risk of developing dementia and access to healthcare services, and the uneven impacts on population health of existing early detection programmes [64–66].

"Early detection" represents an approach focused on targeting "high risk" individuals for prevention strategies. However, at a population level, such strategies may have less impact that those aimed at achieving relatively smaller reductions in dementia risk among a greater number of people [67]. They focus resources on individuals most likely to seek clinical assessment at stages of disease, and may thereby reinforce an "inverse care law", whereby those with most access to healthcare resources benefit disproportionately [63,68]. The overall benefit to societies of such individualized approaches has yet to be demonstrated. The single trial of a screening program for dementia in practice failed to show any benefits [69]. A retrospective analysis of a large UK cohort suggests that individuals identified as "high risk" on the basis of validated dementia risk prediction models accounted for only three in 10 cases of dementia, while eight in every 10 people at high risk did not develop dementia [70].

Further, as an approach to risk reduction, early detection places an emphasis on risk factors that are amenable to individual behavior change and lifestyle approaches. However, again, interventions that require an individual to draw on their own resources—whether social, economic or psychological—tend to disproportionately benefit those with more of these resources [65]. Moreover, many factors which are often considered as "individual" and modifiable by individuals are not equally distributed across society, and are closely associated with poverty, deprivation and social and economic conditions that run through the lifecourse [66,71]. In contrast, population level interventions come with the advantage of equity [67]. Such interventions may focus on fiscal, marketing/advertising, availability, and legislative approaches that address modifiable risk factors such as levels of education, smoking or air pollution [72].

Questions of accuracy and bias also remain critical for both tools for early detection and drugs associated with them. As the former are developed and used to determine access or eligibility for a treatment, it is essential that researchers give attention to potential sources of inaccuracy, including the potential for ethnic and gender bias in clinical and research datasets, and the over-representation of European and US populations in dementia risk model development [25,73]. For example, cognitive assessments developed on the basis of European or North American data are often not validated in, and may have limited applicability to, low and middle income settings, raising the possibility of education, literacy, and cultural biases in their applicability [74]. In the case of genomic or imaging data, predictive models are potentially affected by reliance on a small number non-diverse data sources, such as ADNI [75]. Similar attention needs to be paid to ensure as far as possible that the drugs that target early disease stages are informed by clinical trials with populations that reasonably represent the population in which such drugs would be applied [76]. To date, however, clinical trials of the recently approved disease modifying drugs have been non-diverse, presenting concerns about the wider validity of the results, particularly among older, socially marginalized and underrepresented populations [77–79]. This reflects multiple barriers to participation, including attitudes and beliefs about dementia, recruitment methods and access to research sites, as well as concerns about invasive procedures and mistrust of investigators [80,81].

While growing attention and effort has been directed toward tackling the stigma associated with dementia, further efforts are needed to engage with questions of social justice. One potential advantage of digital tools in relation to dementia, is their potential to expand access to diagnoses, monitoring and support with lifestyle changes—within high income countries and in global health, not least in the wake of the COVID-19 pandemic [82]. Experience to date suggests that uptake of such approaches is highly biased to the advantaged [83]. A clear and concerted effort will be required to realize genuine equity of access and care [63].

The future of early detection

In this final section, we consider future directions for early detection and the emerging ethical considerations associated with it. We focus on three specific developments: the implications of recent approvals of monoclonal antibody therapies for discussions of risk communication; of these approvals and the growing availability of less invasive blood-based and digital biomarkers for discussions of screening; and potential for scientific, clinical and above all, public and patient hopes around early detection to become the ground for speculation and exploitation.

Early detection and genetic risk

After 2 decades of unsuccessful clinical trials, there is hope that a therapeutic breakthrough, for Alzheimer's disease at least, has been achieved with the approval by the US FDA of aducanumab/Aduhelm, donanemab/Kisunia, and lecanemab/Leqembi. A disease modifying drug has often been seen as presenting a definitive rupture between the present and future of clinical practice, for example in proposals such as the Edinburgh Consensus and recent work on

Brain Health Clinics, which orient themselves toward a future of clinical practice that looks quite different to the present [84,85].

In particular, the existence of these approved drugs, it would seem, provides actionability, the lack of which has long been one of the central ethical arguments against the early detection of changes associated with later dementia [24,25]. It is possible that the ability to intervene to affect levels of biomarkers associated with dementia will reframe questions about the right to know and the psychological impact of risk information—albeit not, in the short term, those around stigma. However, other considerations may become more salient.

A treatment aimed toward individuals whom diagnostic technologies suggest are in early stages of disease, would still present significant challenges [25]. The first is again highlighted by the case of ApoE testing, particularly its history as a potential pharmacogenetic marker [86]. In the late 1990s, the treatment outcome of therapy with tacrine—now banned but the first centrally acting cholinesterase inhibitor to be approved for treating Alzheimer's disease—was found to depend on the apolipoprotein genotype and gender of patients [87]. However, clinical hesitation in adopting ApoE testing persisted because of concerns about using genetic testing to prescribe tacrine to those with milder cognitive impairment, given the wider potential for harm associated with the psychological impact of testing and implications for insurance and employment [88]. Moreover, given the absence of any other treatments, clinicians often felt that a drug with even a minimal chance of working would have more value than no drug [86].

In the case of the new monoclonal antibodies, efficacy but also side effects, are again associated with ApoE e4 carrier status [89]. In the case of lecanemab, the rate of ARIA in APOE e4 noncarriers in the pivotal CLARITY AD trial was 5.4% (1.4% symptomatic), while in APOE e4 heterozygotes it was 10.9% (1.7%), and in APOE e4 homozygotes 32.6% (9.2%). This presents a double-bind for patients and healthcare services, and both revives and layers complexity onto pharmacogenetic testing discussions—in that those who are at greatest risk and may have most to gain are also those who are most likely to suffer serious consequences. As a result, ApoE testing may become a more routine part of diagnostic workups, and are recommended in appropriate use criteria [90]. This has potential consequences for both patients themselves, but also for family members, who may as a result learn unsolicited information about their own genetic status, and in turn for genetic counseling services and the viability of long-term care insurance [91].

In practice, it is likely that significant clinician and patient demand will exist for potentially disease-modifying drugs regardless of testing for ApoE status, and that the risk of being excluded from treatment, as in the case of the US Veteran's Affairs Pharmacy criteria for lecanemab use, will have a chilling effect on genetic testing. As these criteria recognize, the decision *not* to have genetic testing in such circumstances comes with potential risks for both individuals and health systems. However, they highlight there is a need to revisit discussions and guidance about the pharmacogenetic role of ApoE testing.

Expanding early detection and screening

Combined clinical, pharmacological and commercial drives toward early detection are facilitated by the increasing availability of tools for early detection that are less burdensome and invasive than imaging and CSF analysis. This includes, for example, blood-based biomarkers, the use of routine health data for detection, and digital markers that draw on either active assessments or the analysis of passively collected data [6,55].

The development of these new tools and their implementation, renews existing questions. For example, the development of early detection and prediction models based on clinical data will necessarily reflect inequalities in access to diagnosis and services associated with age, gender, education, socioeconomic status, and ethnicity including income [92–95]. The most clinically vulnerable people and those on the lowest household incomes are among those most likely to be digitally excluded and report that they do not have access to either broadband or smartphones, presenting challenges for equitable digital health programmes.

Digital biomarkers also bring the early detection of dementia into dialog with wider questions associated with digital health—including those raised around commercialization above, and the privacy of health information. The latter is particularly acute in the case of applications that draw on the "data exhaust" of everyday activities to make "passive" assessments of cognitive state—as, for example, in Google's patent for the Google Home device's capacity to "infer a higher probability that the household occupant has [Alzheimer's] disease" based on monitoring users' movement patterns [96]. Such applications present challenges in terms of ensuring awareness of, and continuing consent for, cognitive and other relevant assessments.

However, increasing technical ease and the changing environment for detection also lead to discussions of the role of early detection in population health. While much of the discussion introduced above has taken place in the context of memory clinic care or clinical trials, there is a growing push for large-scale population testing or screening programmes [97]. Proponents of such calls argue that the availability of new drugs, and the increased possibility of testing at scale change the evidence base for screening that has underpinned the negative evaluations of the UK National Screening Committee (NSC) and equivalent bodies in the USA and Canada and reiterated by the 2024 Lancet Commission on Dementia Prevention, Intervention, and Care [98–100].

Whereas the introduction of screening programmes often requires the involvement of bodies such as the NSC to justify the investment in equipment, personnel and technology required, the potential—and perils—of digital screening lies in the ability to reduce this investment. Digital tools have the potential to be applied to large numbers of people to sort those who are well from those who may have early or incipient disease, but are, to the best of their knowledge, asymptomatic. They also have the potential to be implemented as widely and for as long as the ubiquitous digital technologies on which they rely. The scale up of testing carries the potential for a sufficiently widely adopted technology to become screening by stealth. What little evidence is available suggests limited effectiveness of screening programmes [69], while it remains unclear that tools for early detection, including digital biomarkers, meet standard criteria for the implementation of screening programmes and thus cannot be ethically nor practically justified [6]. The danger the remains that we risk "sleepwalking into the roll out of 'exciting' new technological screening or detection opportunities without proper planning and resourcing" [6].

The risk of premature adoption is compounded by the social and cultural context of Alzheimer's disease. Dementia is among the most feared conditions among older adults [101], creating a large potential market for products and services that claim to assess and reduce risk. The desperation of those living with, at risk of, or affected by dementia and their willingness "to accept the trade-off of some uncertainty about clinical benefit in exchange for earlier access to a potentially effective drug" [102] played an influential role in the FDA's contested decision to grant accelerated approval to aducanumab [103].

Similarly, despite the limitations of detection tools and current preventative and therapeutic interventions, there are a burgeoning number of early detection platforms that cater to the concerned consumer. The potential for consumer exploitation as a result of these developments has already spurred regulatory attention in the case of digital health. The "brain training" company Lumos Labs, for example, was fined $2 million by the US Federal Trade Commission for misleading statements about their ability to delay cognitive decline. While cost and closer regulation of testing equipment limits the marketing of CSF or imaging biomarkers direct to consumers, blood-based biomarkers are also likely to be available outside clinical settings, reflecting the longstanding commercial appeal of predictive testing for dementia—from the marketing of ApoE tests to clinicians within months of the identification of the ApoE susceptibility allele in the early 1990s through to the offerings of consumer genetics companies [43,104].

Conclusion

The move to early detection of Alzheimer's disease presents a range of ethical challenges, across multiple scales. The challenges span from deeply personal considerations of individual autonomy and psychological well-being to broader societal concerns of equity, stigma, and responsible innovation.

The advent of new diagnostic technologies and potentially disease-modifying therapies has not resolved these ethical dilemmas but, rather, reshaped and in some cases intensified them. While early detection offers the tantalizing prospect of timely intervention and improved outcomes, it also risks exacerbating health inequalities, generating undue anxiety, and blurring the lines between health, risk, and disease.

Moving forward, it is imperative that the development and implementation of early detection strategies be guided by a holistic ethical framework that balances individual rights with societal benefits, addresses issues of access and equity, and remains flexible enough to accommodate rapidly evolving scientific understanding and technological capabilities.

The scale and scope of these considerations emphasize the now urgent need for a wide range of expertize and experience to continue to be brought to discussions of early detection. Answering them requires addressing the question of "under which conditions [could] emerging diagnostic technologies for AD be considered a *responsible innovation*" [105], (p2) and what other possible directions for the future of dementia care are de-emphasized in the pursuit of early detection. but also to question the framing of future clinical practice that drives innovation, and whose priorities and concerns this reflects.

References

[1] Wolters FJ, Chibnik LB, Waziry R, et al. 27-year time trends in dementia incidence in Europe and the US: the Alzheimer Cohorts Consortium. Neurology 2020. https://doi.org/10.1212/WNL.0000000000010022.

[2] Mukadam N, Wolters FJ, Walsh S, et al. Changes in prevalence and incidence of dementia and risk factors for dementia: an analysis from cohort studies. Lancet Public Health 2024;9(7):e443−60. https://doi.org/10.1016/S2468-2667(24)00120-8.

[3] Pollen DA. Hannah's Heirs: the quest for the genetic origins of Alzheimer's disease. USA: Oxford University Press; 1996.

[4] Post SG. Future scenarios for the prevention and delay of Alzheimer disease onset in high-risk groups: an ethical perspective. Am J Prev Med 1999;16(2):105−10. https://doi.org/10.1016/S0749-3797(98)00139-1.

[5] Karlawish J. Addressing the ethical, policy, and social challenges of preclinical Alzheimer disease. Neurology 2011;77(15):1487−93. https://doi.org/10.1212/WNL.0b013e318232ac1a.

[6] Ford E, Milne R, Curlewis K. Ethical issues when using digital biomarkers and artificial intelligence for the early detection of dementia. WIREs Data Min Knowl Discov 2023:e1492. https://doi.org/10.1002/widm.1492.

[7] Walsh S, Merrick R, Milne R, Nurock S, Richard E, Brayne C. Considering challenges for the new Alzheimer's drugs: clinical, population, and health system perspectives. Alzheimer's Dement 2024. https://doi.org/10.1002/alz.14108.

[8] Boenink M, van der Molen L. The biomarkerization of Alzheimer's disease: from (early) diagnosis to anticipation?. In: A pragmatic approach to conceptualization of health; 2024. p. 141.

[9] Jutel A. Sociology of diagnosis: a preliminary review. Sociol Health Illness 2009;31(2):278−99. https://doi.org/10.1111/j.1467-9566.2008.01152.x.
[10] Geoffrey CB, Star SL. Sorting things out: classification and its consequences. MIT Press; 1999.
[11] Hofmann B. Temporal uncertainty in disease diagnosis. Med Health Care Philos 2023;26(3):401−11. https://doi.org/10.1007/s11019-023-10154-y.
[12] Krolak-Salmon P, Maillet A, Vanacore N, et al. Toward a sequential strategy for diagnosing neurocognitive disorders: a consensus from the "act on dementia" European joint action. J Alzheim Dis 2019;72(2):363−72. https://doi.org/10.3233/JAD-190461.
[13] Bailey C, Dooley J, McCabe R. 'How do they want to know?' Doctors' perspectives on making and communicating a diagnosis of dementia. Dementia 2019;18(7−8):3004−22. https://doi.org/10.1177/1471301218763904.
[14] Dooley J, Bass N, McCabe R. How do doctors deliver a diagnosis of dementia in memory clinics? Br J Psychiatr 2018;212(04):239−45. https://doi.org/10.1192/bjp.2017.64.
[15] Lynch M. "Turning up signs" in neurobehavioral diagnosis. Symbolic Interact 1984;7(1):67−86. https://doi.org/10.1525/si.1984.7.1.67.
[16] Dhedhi SA, Swinglehurst D, Russell J. 'Timely' diagnosis of dementia: what does it mean? A narrative analysis of GPs' accounts. BMJ Open 2014;4(3). https://doi.org/10.1136/bmjopen-2013-004439.
[17] Prince MJ. World Alzheimer Report 2011: the benefits of early diagnosis and intervention. September 13, 2011. https://www.alz.co.uk/research/world-report-2011.
[18] Vernooij-Dassen MJFJ, Moniz-Cook ED, Woods RT, et al. Factors affecting timely recognition and diagnosis of dementia across Europe: from awareness to stigma. Int J Geriatr Psychiatr 2005;20(4):377−86. https://doi.org/10.1002/gps.1302.
[19] Woods B, Arosio F, Diaz A, et al. Timely diagnosis of dementia? Family carers' experiences in 5 European countries. Int J Geriatr Psychiatr 2019;34(1):114−21. https://doi.org/10.1002/gps.4997.
[20] Brooker D, La Fontaine J, Evans S, Bray J, Saad K. Public health guidance to facilitate timely diagnosis of dementia: ALzheimer's COoperative Valuation in Europe recommendations. Int J Geriatr Psychiatr 2014;29(7):682−93. https://doi.org/10.1002/gps.4066.
[21] Oxford English Dictionary. Detection. Oxford University Press; 2023. https://doi.org/10.1093/OED/8603378359.
[22] Nuffield Council on Bioethics. Dementia: ethical issues. Nuffield Council on Bioethics; 2009. https://www.nuffieldbioethics.org/publications/dementia.
[23] Roses AD. Apolipoprotein E genotyping in the differential diagnosis, not prediction, of Alzheimer's disease. Ann Neurol 1995;38(1):6−14. https://doi.org/10.1002/ana.410380105.
[24] Smedinga M, Tromp K, Schermer MH, Richard E. Ethical arguments concerning the use of Alzheimer's disease biomarkers in individuals with No or mild cognitive impairment: a systematic review and framework for discussion. J Alzheim Dis 2018;66(4):1309−22.
[25] Angehrn Z, Sostar J, Nordon C, et al. Ethical and social implications of using predictive modeling for Alzheimer's disease prevention: a systematic literature review. J Alzheim Dis 2020;76(3):923−40. https://doi.org/10.3233/JAD-191159.
[26] Caselli RJ, Langbaum J, Marchant GE, et al. Public perceptions of presymptomatic testing for Alzheimer's disease. Mayo Clin Proc 2014;89(10):1389−96. https://doi.org/10.1016/j.mayocp.2014.05.016.
[27] Gooblar J, Roe CM, Selsor NJ, Gabel MJ, Morris JC. Attitudes of research participants and the general public regarding disclosure of Alzheimer disease research results. JAMA Neurol 2015;72(12):1484−90. https://doi.org/10.1001/jamaneurol.2015.2875.

[28] Alzheimers Research UK. Detecting and diagnosing Alzheimer's disease. 2019.
[29] van der Schaar J, Visser LNC, Ket JCF, et al. Impact of sharing Alzheimer's disease biomarkers with individuals without dementia: a systematic review and meta-analysis of empirical data. Alzheimer's Dementia 2023;19(12):5773−94. https://doi.org/10.1002/alz.13410.
[30] Lingler JH, Roberts JS, Kim H, et al. Amyloid positron emission tomography candidates may focus more on benefits than risks of results disclosure. Alzheimer's Dementia 2018. https://doi.org/10.1016/j.dadm.2018.05.003.
[31] Chadwick R, Levitt M, Shickle D. The right to know and the right not to know: genetic privacy and responsibility. Cambridge University Press; 2014.
[32] Lingler JH, Sereika SM, Butters MA, et al. A randomized controlled trial of amyloid positron emission tomography results disclosure in mild cognitive impairment. Alzheimer's Dementia 2020. https://doi.org/10.1002/alz.12129.
[33] Bunnik EM, Richard E, Milne R, Schermer MHN. On the personal utility of Alzheimer's disease-related biomarker testing in the research context. J Med Ethics 2018. https://doi.org/10.1136/medethics-2018-104772.
[34] Milne R, Bunnik E, Diaz A, et al. Perspectives on communicating biomarker-based assessments of Alzheimer's disease to cognitively healthy individuals. J Alzheim Dis 2018;62(2):487−98. https://doi.org/10.3233/JAD-170813.
[35] Brayne C. Evidence still lacking for recommendation of screening for cognitive impairment in older adults. JAMA Intern Med 2020. https://doi.org/10.1001/jamainternmed.2019.7522.
[36] Largent EA, Harkins K, Dyck CH van, Hachey S, Sankar P, Karlawish J. Cognitively unimpaired adults' reactions to disclosure of amyloid PET scan results. PLoS One 2020;15(2):e0229137. https://doi.org/10.1371/journal.pone.0229137.
[37] Ketchum FB, Erickson CM, Basche KE, et al. Informing Alzheimer's biomarker communication: concerns and understanding of cognitively unimpaired adults during amyloid results disclosure. J Prev Alzheimers Dis 2024. https://doi.org/10.14283/jpad.2024.151.
[38] Usher-Smith JA, Silarova B, Schuit E, Moons KG, Griffin SJ. Impact of provision of cardiovascular disease risk estimates to healthcare professionals and patients: a systematic review. BMJ Open 2015;5(10):e008717. https://doi.org/10.1136/bmjopen-2015-008717.
[39] French DP, Cameron E, Benton JS, Deaton C, Harvie M. Can communicating personalised disease risk promote healthy behaviour change? A systematic review of systematic reviews. Ann Behav Med 2017;51(5):718−29. https://doi.org/10.1007/s12160-017-9895-z.
[40] Hollands GJ, French DP, Griffin SJ, et al. The impact of communicating genetic risks of disease on risk-reducing health behaviour: systematic review with meta-analysis. BMJ 2016;352:i1102. https://doi.org/10.1136/bmj.i1102.
[41] Chao S, Roberts JS, Marteau TM, Silliman R, Cupples LA, Green RC. Health behavior changes after genetic risk assessment for alzheimer disease: the REVEAL study. Alzheimer Dis Assoc Disord 2008;22(1):94−7. https://doi.org/10.1097/WAD.0b013e31815a9dcc.
[42] Solomon A, Stephen R, Altomare D, et al. Multidomain interventions: state-of-the-art and future directions for protocols to implement precision dementia risk reduction. A user manual for Brain Health Services—part 4 of 6. Alz Res Therapy 2021;13(1):171. https://doi.org/10.1186/s13195-021-00875-8.
[43] Post DSG. The moral challenge of Alzheimer disease: ethical issues from diagnosis to dying. second edition. The Johns Hopkins University Press; 2000.
[44] Draper B, Peisah C, Snowdon J, Brodaty H. Early dementia diagnosis and the risk of suicide and euthanasia. Alzheimer's Dementia 2010;6(1):75−82. https://doi.org/10.1016/j.jalz.2009.04.1229.

[45] Largent EA, Terrasse M, Harkins K, Sisti DA, Sankar P, Karlawish J. Attitudes toward physician-assisted death from individuals who learn they have an Alzheimer disease biomarker. JAMA Neurol 2019;76(7):864−6.
[46] Green RC, Roberts JS, Cupples LA, et al. Disclosure of APOE genotype for risk of Alzheimer's disease. N Engl J Med 2009;361(3):245−54.
[47] Green RC, Christensen KD, Cupples LA, et al. A randomized noninferiority trial of condensed protocols for genetic risk disclosure of Alzheimer's disease. Alzheimer's Dement 2015;11(10):1222−30. https://doi.org/10.1016/j.jalz.2014.10.014.
[48] Bemelmans S, Tromp K, Bunnik EM, et al. Psychological, behavioral and social effects of disclosing Alzheimer's Disease biomarkers to research participants - a systematic review. Alzheimer's Res Ther 2016;8(46):46. https://doi.org/10.1186/s13195-016-0212-z.
[49] Lineweaver TT, Bondi MW, Galasko D, Salmon DP. Effect of knowledge of APOE genotype on subjective and objective memory performance in healthy older adults. Am J Psychiatr 2014;171(2):201−8. https://doi.org/10.1176/appi.ajp.2013.12121590.
[50] Zallen DT. "Well, good luck with that": reactions to learning of increased genetic risk for Alzheimer disease. Genet Med 2018;20(11):1462−7. https://doi.org/10.1038/gim.2018.13.
[51] Grill JD, Raman R, Ernstrom K, et al. Short-term psychological outcomes of disclosing amyloid imaging results to research participants who do not have cognitive impairment. JAMA Neurol 2020. https://doi.org/10.1001/jamaneurol.2020.2734.
[52] Wake T, Tabuchi H, Funaki K, et al. Disclosure of amyloid status for risk of alzheimer disease to cognitively normal research participants with subjective cognitive decline: a longitudinal study. Am J Alzheimers Dis Other Demen 2020;35. https://doi.org/10.1177/1533317520904551.
[53] Hartz SM, Goswami S, Oliver A, et al. Returning research results that indicate risk of alzheimer disease dementia to healthy participants in longitudinal studies (WeSHARE). 2024. https://doi.org/10.1101/2024.07.01.24309801.
[54] Harkins K, Sankar P, Sperling R, et al. Development of a process to disclose amyloid imaging results to cognitively normal older adult research participants. Alzheimer's Res Ther 2015;7(1). https://doi.org/10.1186/s13195-015-0112-7.
[55] Mielke MM, Fowler NR. Alzheimer disease blood biomarkers: considerations for population-level use. Nat Rev Neurol 2024;20(8):495−504. https://doi.org/10.1038/s41582-024-00989-1.
[56] Langa KM, Burke JF. Preclinical alzheimer disease—early diagnosis or overdiagnosis? JAMA Intern Med 2019;179(9):1161−2. https://doi.org/10.1001/jamainternmed.2019.2629.
[57] Milne R, Diaz A, Badger S, Bunnik E, Fauria K, Wells K. At, with and beyond risk: expectations of living with the possibility of future dementia. Sociol Health Illness 2018;40(6):969−87. https://doi.org/10.1111/1467-9566.12731.
[58] Stites SD, Rubright JD, Karlawish J. What features of stigma do the public most commonly attribute to Alzheimer's disease dementia? Results of a survey of the U.S. general public. Alzheimer's Dement 2018;14(7):925−32. https://doi.org/10.1016/j.jalz.2018.01.006.
[59] Largent EA, Stites SD, Harkins K, Karlawish J. "That would be dreadful": the ethical, legal, and social challenges of sharing your Alzheimer's disease biomarker and genetic testing results with others. J Law Biosci 2021;8(1):lsab004. https://doi.org/10.1093/jlb/lsab004.
[60] Stites SD, Gill J, Largent EA, et al. The relative contributions of biomarkers, disease modifying treatment, and dementia severity to Alzheimer's stigma: a vignette-based experiment. Soc Sci Med 2022;292:114620. https://doi.org/10.1016/j.socscimed.2021.114620.

[61] Stites S, Dedhia M, Harkins K, et al. Learning about a heightened genetic risk for dementia: anticipated stigma is greater than experienced stigma. 2024. https://doi.org/10.2139/ssrn.4764784.
[62] Taylor DH, Cook-Deegan RM, Hiraki S, Roberts JS, Blazer DG, Green RC. Genetic testing for Alzheimer's and long-term care insurance. Health Aff 2010;29(1):102–8. https://doi.org/10.1377/hlthaff.2009.0525.
[63] Milne R, Altomare D, Ribaldi F, et al. Societal and equity challenges for brain health services. A user manual for brain health services—part 6 of 6. Alzheimer's Res Ther 2021;13(1):173. https://doi.org/10.1186/s13195-021-00885-6.
[64] Krogsbøll LT, Jørgensen KJ, Gøtzsche PC. General health checks in adults for reducing morbidity and mortality from disease. Cochrane Database Syst Rev 2019;(1). https://doi.org/10.1002/14651858.CD009009.pub3.
[65] Capewell S, Graham H. Will cardiovascular disease prevention widen health inequalities? PLoS Med 2010;7(8):e1000320. https://doi.org/10.1371/journal.pmed.1000320.
[66] Livingston G, Huntley J, Sommerlad A, et al. Dementia prevention, intervention, and care: 2020 report of the Lancet Commission. Lancet 2020;0(0). https://doi.org/10.1016/S0140-6736(20)30367-6.
[67] Rose G. Sick individuals and sick populations. Int J Epidemiol 1985;14(1):32–8. https://doi.org/10.1093/ije/14.1.32.
[68] Hart JT. The inverse care law. Lancet 1971;297(7696):405–12. https://doi.org/10.1016/S0140-6736(71)92410-X.
[69] Fowler NR, Perkins AJ, Gao S, Sachs GA, Boustani MA. Risks and benefits of screening for dementia in primary care: the Indiana University cognitive health outcomes investigation of the comparative effectiveness of dementia screening (IU CHOICE)trial. J Am Geriatr Soc 2020;68(3):535–43. https://doi.org/10.1111/jgs.16247.
[70] Walsh S, Wallace L, Merrick R, et al. How many future dementia cases would be missed by a high-risk screening program? A retrospective cohort study in a population-based cohort. Alzheimer's Dement 2024. https://doi.org/10.1002/alz.14113.
[71] Russ TC, Stamatakis E, Hamer M, Starr JM, Kivimäki M, Batty GD. Socioeconomic status as a risk factor for dementia death: individual participant meta-analysis of 86 508 men and women from the UK. Br J Psychiatr 2013;203(1):10–7. https://doi.org/10.1192/bjp.bp.112.119479.
[72] Walsh S, Wallace L, Kuhn I, et al. Population-level interventions for the primary prevention of dementia: a complex evidence review. eClinicalMedicine 2024;70. https://doi.org/10.1016/j.eclinm.2024.102538.
[73] Magklara E, Stephan BCM, Robinson L. Current approaches to dementia screening and case finding in low- and middle-income countries: research update and recommendations. Int J Geriatr Psychiatr 2019;34(1):3–7. https://doi.org/10.1002/gps.4969.
[74] Stephan BCM, Pakpahan E, Siervo M, et al. Prediction of dementia risk in low-income and middle-income countries (the 10/66 Study): an independent external validation of existing models. Lancet Global Health 2020;8(4):e524–35. https://doi.org/10.1016/S2214-109X(20)30062-0.
[75] Goerdten J, Čukić I, Danso SO, Carrière I, Muniz-Terrera G. Statistical methods for dementia risk prediction and recommendations for future work: a systematic review. Alzheimer's Dement 2019;5:563–9. https://doi.org/10.1016/j.trci.2019.08.001.
[76] Reed C, Belger M, Dell'Agnello G, et al. Representativeness of European clinical trial populations in mild Alzheimer's disease dementia: a comparison of 18-month outcomes

with real-world data from the GERAS observational study. Alz Res Therapy 2018;10(1):36. https://doi.org/10.1186/s13195-018-0360-4.

[77] Manly JJ, Glymour MM. What the aducanumab approval reveals about alzheimer disease research. JAMA Neurol 2021;78(11):1305–6. https://doi.org/10.1001/jamaneurol.2021.3404.

[78] Babulal GM, Quiroz YT, Albensi BC, et al. Perspectives on ethnic and racial disparities in Alzheimer's disease and related dementias: update and areas of immediate need. Alzheimer's Dement 2019;15(2):292–312. https://doi.org/10.1016/j.jalz.2018.09.009.

[79] Raman R, Quiroz YT, Langford O, et al. Disparities by race and ethnicity among adults recruited for a preclinical alzheimer disease trial. JAMA Netw Open 2021;4(7):e2114364. https://doi.org/10.1001/jamanetworkopen.2021.14364.

[80] Waheed W, Mirza N, Waheed MW, et al. Recruitment and methodological issues in conducting dementia research in British ethnic minorities: a qualitative systematic review. Int J Methods Psychiatr Res 2020;29(1):e1806. https://doi.org/10.1002/mpr.1806.

[81] Salazar CR, Hoang D, Gillen DL, Grill JD. Racial and ethnic differences in older adults' willingness to be contacted about Alzheimer's disease research participation. Alzheimer's Dement 2020;6(1). https://doi.org/10.1002/trc2.12023.

[82] Knapp M, Shehaj X, Wong G. Digital interventions for people with dementia and carers: effective, cost-effective and equitable? Neurodegener Dis Manag 2022;12(5):215–9. https://doi.org/10.2217/nmt-2022-0025.

[83] Coley N, Andre L, Hoevenaar-Blom MP, et al. Factors predicting engagement of older adults with a coach-supported eHealth intervention promoting lifestyle change and associations between engagement and changes in cardiovascular and dementia risk: secondary analysis of an 18-month multinational randomized controlled trial. J Med Internet Res 2022;24(5):e32006. https://doi.org/10.2196/32006.

[84] Ritchie CW, Russ TC, Banerjee S, et al. The Edinburgh Consensus: preparing for the advent of disease-modifying therapies for Alzheimer's disease. Alzheimer's Res Ther 2017;9(1):85. https://doi.org/10.1186/s13195-017-0312-4.

[85] Frisoni GB, Molinuevo JL, Altomare D, et al. Precision prevention of Alzheimer's and other dementias: anticipating future needs in the control of risk factors and implementation of disease-modifying therapies. Alzheimer's Dement 2020. https://doi.org/10.1002/alz.12132.

[86] Hedgecoe A. The politics of personalised medicine: pharmacogenetics in the clinic. Cambridge University Press; 2004. https://doi.org/10.1017/CBO9780511489136.

[87] Farlow MR, Lahiri DK, Poirier J, Davignon J, Schneider L, Hui SL. Treatment outcome of tacrine therapy depends on apolipoprotein genotype and gender of the subjects with Alzheimer's disease. Neurology 1998;50(3):669–77. https://doi.org/10.1212/WNL.50.3.669.

[88] Issa AM, Keyserlingk EW. Apolipoprotein E genotyping for pharmacogenetic purposes in Alzheimer's disease: emerging ethical issues. Can J Psychiatry 2000;45(10):917–22. https://doi.org/10.1177/070674370004501007.

[89] Evans CD, Sparks J, Andersen SW, et al. APOE ε4's impact on response to amyloid therapies in early symptomatic Alzheimer's disease: analyses from multiple clinical trials. Alzheimer's Dement 2023;19(12):5407–17. https://doi.org/10.1002/alz.13128.

[90] Cummings J, Apostolova L, Rabinovici GD, et al. Lecanemab: appropriate use recommendations. J Prev Alzheimers Dis 2023;10(3):362–77. https://doi.org/10.14283/jpad.2023.30.

[91] Thambisetty M, Howard R. Lecanemab and APOE genotyping in clinical practice—navigating uncharted Terrain. JAMA Neurol 2023;80(5):431–2. https://doi.org/10.1001/jamaneurol.2023.0207.

[92] Savva GM, Arthur A. Who has undiagnosed dementia? A cross-sectional analysis of participants of the Aging, Demographics and Memory Study. Age Ageing 2015;44(4):642−7. https://doi.org/10.1093/ageing/afv020.

[93] Lang L, Clifford A, Wei L, et al. Prevalence and determinants of undetected dementia in the community: a systematic literature review and a meta-analysis. BMJ Open 2017;7(2):e011146. https://doi.org/10.1136/bmjopen-2016-011146.

[94] Amjad H, Roth DL, Sheehan OC, Lyketsos CG, Wolff JL, Samus QM. Underdiagnosis of dementia: an observational study of patterns in diagnosis and awareness in US older adults. J Gen Intern Med 2018;33(7):1131−8. https://doi.org/10.1007/s11606-018-4377-y.

[95] Qian W, Schweizer TA, Fischer CE. Impact of socioeconomic status on initial clinical presentation to a memory disorders clinic. Int Psychogeriatr 2014;26(4):597−603. https://doi.org/10.1017/S1041610213002299.

[96] Fadell AM, Matsuoka Y, Sloo D, Veron M. Monitoring and reporting household activities in the smart home according to a household policy. 2018. https://patents.google.com/patent/US9872088B2/en?q=monitoring+and+reporting&inventor=fadell&oq=fadell+monitoring+and+reporting.

[97] Reiman EM, Cummings JL, Langbaum JB, Mattke S, Alexander RC. A chance to prevent Alzheimer's disease sooner than you think. Lancet Neurol 2024;23(2):144−5. https://doi.org/10.1016/S1474-4422(23)00508-2.

[98] Livingston G, Huntley J, Liu KY, et al. Dementia prevention, intervention, and care: 2024 report of the Lancet standing Commission. Lancet 2024;404(10452):572−628. https://doi.org/10.1016/S0140-6736(24)01296-0.

[99] Ismail Z, Black SE, Camicioli R, et al. Recommendations of the 5th Canadian Consensus Conference on the diagnosis and treatment of dementia. Alzheimers Dement 2020;16(8):1182−95. https://doi.org/10.1002/alz.12105.

[100] Owens DK, Davidson KW, Krist AH, et al. Screening for cognitive impairment in older adults: US preventive services task force recommendation statement. JAMA 2020;323(8):757−63. https://doi.org/10.1001/jama.2020.0435.

[101] Alzheimers Research UK. Dementia attitudes monitor: Wave 3. Alzheimer's Research UK; 2023. https://www.dementiastatistics.org/attitudes/.

[102] Dunn B, Stein P, Cavazzoni P. Approval of aducanumab for alzheimer disease—the FDA's perspective. JAMA Intern Med 2021;181(10):1276−8. https://doi.org/10.1001/jamainternmed.2021.4607.

[103] Largent EA, Peterson A, Lynch HF. FDA drug approval and the ethics of desperation. JAMA Intern Med 2021. https://doi.org/10.1001/jamainternmed.2021.6045.

[104] Annas GJ, Elias S. 23 and Me and the FDA. N Engl J Med 2014;370(11):985−8. https://doi.org/10.1056/NEJMp1316367.

[105] Boenink M, Van Lente H, Moors E. Emerging technologies for diagnosing Alzheimer's disease: innovating with care. In: Emerging technologies for diagnosing Alzheimer's disease. Springer; 2016. p. 1−17.

Chapter 7b

Privacy

Abhirup Ghosh[a], Jagmohan Chauhan[b] and Cecilia Mascolo[c]

[a]*School of Computer Science, University of Birmingham, Birmingham, United Kingdom;*
[b]*Electronics and Computer Science, University of Southampton, Southampton, United Kingdom;*
[c]*Department of Computer Science and Technology, University of Cambridge, Cambridge, United Kingdom*

Introduction

Widely available devices, such as smartphones and smart home appliances, have the ability to capture a wide range of signals spanning physiological (e.g., heart rate), motor (e.g., acceleration), environmental (e.g., air temperature, humidity), contextual (e.g., location, device interactions), and social (e.g., interpersonal voice or electronic communication) measures. This ability to collect multiple sets of data representing direct and indirect measures of physical and mental health has massive implications for the usage of such devices for medical practice, both in terms of diagnosis and management, including evaluation of treatment on digital outcome measures that are reflective of real life activity.

However, the acquisition of personal data in this way, particularly when encompassing a range of behaviors and contexts, necessarily raises the risk of privacy loss. This can be illustrated by the example of an individual with early stage Alzheimer's disease whose spatial navigational skills are affected by disease-related disruption to the medial temporal and medial parietal lobe function. The real life consequences of this impaired navigational ability will be a reduction of time spent away from home, fewer places visited and changes in driving behavior with an increased tendency to keep to familiar locations. These changes can be detected by tracking that individual's everyday movements using a location tracking app in their smartphone, which periodically records the individual's location. Without appropriate modifications, such location data could expose potentially sensitive information such as the identities of locations visited which may have negligible clinical utility but be of high personal value. This will be used as the running example to explain various topics as and where relevant to the discussion.

The risk of leaking private information goes beyond the issues tackled by secure systems, which guarantees that only authorized personnel have access to the data. However, data might have multiple dimensions and monitored individuals may not like to expose all of them to the authorized entities. This issue becomes more complex with inferences powered by pattern analysis in literature. For example, finger tapping speed and accuracy have been studied as potential markers of disease severity in dementia. As such, one could track touch events during everyday mobile phone usage in order to extract digital equivalents of these measures. However, the key press patterns (e.g., time in between key presses) can reveal what keys were pressed and synchronized events from multiple people could reveal their communication via a messaging application. Moreover, even secure communication can reveal sensitive information. For example, an application that sends notification to a server upon detecting a health crisis. A third party individual or organization acting independently to the monitored person without their consent and potentially with separate objectives not necessarily aligned to that person's best interests (known henceforth as an adversary) only knowing the timing of the messages would know the sensitive health condition. Therefore, while designing a mobile health system, the issue of privacy needs special consideration.

The types of privacy risks vary with applications and adversaries. Suppose the above location tracking study is conducted on asymptomatic people at risk of developing dementia. Thus, the fact that a person participated in the study is sensitive. In another application where the participation is voluntary, location information remains sensitive. For example, an unscrupulous employer, knowing that an employee regularly visited a hospital, could use such knowledge to make a decision potentially deleterious to the employee. Several types of such privacy issues are discussed along with different types of adversaries in section Privacy issues.

Privacy preserving data analysis and private data publication constitute a large body of research in computer science. The major challenge in developing privacy protection methods lies in achieving a balance between the utility of the sanitized data and the level of privacy protection. For instance, using the prior example of the location monitoring application, we can reduce the spatial resolution of the collected data (e.g., to the level of cities visited) so that it would be impossible to derive personal privacy sensitive inferences, but such coarse-grained data may fail to capture information of potential value, such as the frequency of going out, thus rendering the data useless for clinical purposes. Choice of resolution is therefore crucial to the balance between privacy and utility.

Privacy issues

In a typical mobile health setting, several devices (*data contributors*) sense personal and environmental factors and send the collected data to a *server* over the internet for processing (Fig. 1). The server processes the aggregate data

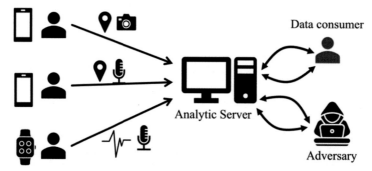

FIG. 1 A typical system design in a mobile health setting with data contributors, analytic server and consumers. Multiple devices at the data contributors collect and send sensing data to the server such as location, audio, temperature, heart rate, photos, etc. The consumers interact with the server either through queries or receiving published data.

through pattern analysis, e.g., fitting analytical functions (called models). The *data consumers* either pose queries to the server or receive the aggregate dataset. Data contributed by individual devices often have multiple dimensions or *attributes*, for example, name, age, gender, address, location or photoplethysmography (PPG) signals. For simplicity, here we concentrate on protecting the demographic attributes at first and consider that a subset of such attributes is sensitive. Indeed, the sensor signals, such as the location tracked in the application discussed earlier need protection against privacy leakage, they are discussed later in sections Privacy leaks in apps and Other privacy preserving system

Depending on the application, several of the system components, such as the server or communication medium, can be susceptible to privacy leaks. A *data curator* (server) applies privacy protection methods to remove personal traits from the data prior to sending the data out to an untrusted medium.

In the following, we describe different types of adversaries followed by the types of privacy vulnerabilities.

Types of adversaries

An adversary aims to know sensitive attributes of certain data contributors (*the target*). An important question toward protecting the sensitive information of the target is to characterize the adversary in terms of background knowledge, capability, and objective.

Adversaries with access to additional data sources can significantly reduce privacy. For example, suppose we decide to publish anonymous outdoor mobility traces (i.e., traces are not marked with identifiers like names, emails, etc.) collected by the above location tracking application. Even then an adversary can re-identify a participant's trace knowing home location; simply by finding where people have spent their nights. Thus, it is crucial to estimate an adversary's background knowledge prior to designing a privacy preserving mechanism. This often is difficult to estimate exactly in practice and different privacy preserving methods have different assumptions.

Adversaries can have varied levels of access to the data through a query, published dataset or side channel information. In a query-based system, the server exposes a set of data operations to the consumers. The above location tracking system can expose an interface to support queries such as whether, on the basis of mobility traces, an individual is manifesting early features of dementia. Publishing datasets differ from query-based systems as data consumers can use them for any query. While publication increases utility of the data it also enhances privacy risks. The privacy vulnerability through side channel information is more subtle, for example, an adversary could infer sensitive texts like passwords typed in a mobile phone using the actigraphy data from a wrist worn device.

In turn, approaches to preserve privacy differ depending on which system component is vulnerable. In the location tracking application, if either the central data processing server or the communication channel is untrusted, then the data contributor devices need to sanitize the traces before sending. If both the communication channel and the central server are trusted then the aggregated data can be sanitized at the server.

Privacy attacks and mitigation techniques

All privacy leaks are targeted toward inferring sensitive attributes of the data contributors. Depending on the capability and objective of the adversary, there exist multiple types of privacy attacks. The rest of this section discusses several types of attacks and sanitization methods to protect against such attacks.

Re-identification attack—anonymization frameworks

The aim of the adversary here is to re-identify a contributor from a supposedly anonymous dataset. For simplicity, let us consider tabular data, e.g., consisting of Electronic Health Record data. We assume each patient contributes a row containing multiple attributes divided into three categories: (i) attributes identifying the individual exactly (e.g., email, phone number, driving license number), (ii) quasi-identifiers (QID) (provides a subtle way of identification), and (iii) sensitive attributes. Removing exact identifiers is trivial but do not eradicate the privacy risks because of the QIDs as we discuss in detail below.

Let us consider that the adversary has access to the data as in Fig. 2A and is interested in knowing the diagnosis of a target person. The exact identifiers are removed in this table to make the data anonymous. However, if the adversary knew the target's age (a QID in this case) then can narrow down a few possible

Age	Job	Diagnosis
45	Engineer	Mild
47	Lawyer	Severe
53	Dancer	Mild
62	Writer	Severe
60	Engineer	Severe

(A)

Age	Job	Diagnosis
(36-45)	Engineer	Mild
(46-55)	Lawyer	Severe
(46-55)	Dancer	Mild
(56-65)	Writer	Severe
(56-65)	Engineer	Severe

(B)

FIG. 2 (A) is an anonymized dataset with each row describing an individual. Age and Job constitute the QID. The sensitive attribute, Diagnosis describes the severity of a disease in that individual. In the sanitized dataset (B) the age is generalized using intervals. However, all records in the age range (56–65) have severe disease and thus the anonymized data reveals the diagnosis if the target is one of the them (and even the adversary doesn't know target's job). This shows a major drawback in the k-anonymity framework.

diagnosis results (exact diagnosis in this example). Naturally, the more information an adversary has regarding the target (e.g., profession along with age), the more precise sensitive information can be extracted.

A privacy protection framework called, k-anonymity framework prevents such privacy leak by modifying the QID values of the individuals to ensure that each group of records with the same QID is of size at least k. Therefore, an adversary knowing the QID value of the target cannot localize the target's record beyond a set of size k. However, this method has several limitations, for example as shown in Fig. 2.

Statistical inference attack

In many applications, it is sufficient to publish statistics on the data aggregated from all the contributors. For example, to find a correlation between smoking and cancer, a survey asks the participants whether they smoke and if they are diagnosed with cancer. Only publishing aggregate statistics such as the fraction of smokers who have cancer suffices to show correlation, however let's consider the follwoing to see that such statistics can still reveal if a participant has cancer, thus uncovering personally sensitive information. If an adversary knows both the survey answers from all contributors except only the target individual and knows that the target is a smoker then publishing the fraction of smokers who have cancer reveals whether the target has cancer.

The most popular framework to prevent this attack is known as differential privacy, in which a random number is added to the results prior to releasing it to ensure that the same aggregate conclusions can be reached irrespective of whether an individual participated in the study. This foils efforts to infer a target individual's data. The amount of randomization determines the privacy and in-turn utility of the data, with greater noise ensuring greater privacy but poorer data quality. An important property of this framework is that it is impossible to reverse the effect of sanitization after the data is sanitized. This property makes the framework useful for designing more complex analysis pipelines. An example of real-world application of this approach is Apple's use of local differential privacy while collecting user data such as what users type on their keyboards and websites they visit.

Privacy leaks in apps

Monitoring apps running on contributors' devices raise privacy issues through malicious exploitation of permissions or side channels. In most phones user needs to grant permissions to apps to access certain sensors such as the camera. However, many apps request extra permissions which can lead to collecting sensitive information or installation of malware on the user device. Studies on analyzing communication patterns and code of apps have also shown that many apps send private information such as location, phone numbers, and audio to third parties without consent. Separate to this, side channels refer to attacks aimed at exposing private information through interfaces that are not designed for that purpose. For example, an app monitoring motion sensors for tracking activities can infer keystrokes on a smartphone's keypad, and thus can potentially guess typed passwords.

Analyzing permissions and behaviors of apps through software tools can prevent privacy leaks. In depth analysis of an app's runtime behavior and its code can reveal how a permission is used and if sensitive information is being sent to third-party vendors. This information can be used by marketplaces such as Google Play to inform their app publishing, thereby preventing users from installing malicious apps. Marketplaces can also enforce compliance to certain privacy standards for the app developers. Side channel attacks can, to an extent, be prevented by appropriate app permissions and using other privacy frameworks in conjunction. For example, an audio recording app can protect privacy by sending derived features instead of raw audio to the server.

Other privacy preserving systems

Wearable and mobile devices generate a significant amount of health data which are typically aggregated and stored at a server, to be made available for access by multiple authorized parties. However, in this sort of arrangement the server represent a single point of potential failure—if the server fails then the data become inaccessible—as well as a point of vulnerability where sensitive health data may be compromised.

These risks may be mitigated by recent advances in hardware and computation technologies in smartphones and wearable devices which permit partial or full processing of recorded data using the devices' own CPUs. For example, in the location tracking application discussed previously, given a diagnostic classifier—built previously from prior datasets acquired from multiple individuals with the condition of interest—an individual's own phone can run the inference on its own sensitive location data without the need for sharing with a third party.

Other privacy preserving options include encryption-based privacy protection mechanisms such as homomorphic encryption (HE), representing an algebraic system that allows functions to be performed on the encrypted data without accessing the raw data.

Conclusion and future directions

Privacy of data contributors is an issue of paramount importance for mobile health applications and the failure to preserve privacy would fatally compromise future efforts to use such applications for clinical purposes. Given its importance, this chapter has outlined for the non-expert reader the various challenges to privacy and the potential solutions to the problems posed. However, beyond computer science solutions, any future use of mobile sensing for AD diagnostic purposes will need to be couched within a robust regulatory environment, ranging from adherence to privacy preserving principles by device manufacturers, app developers and data scientists through to regulations in international law such as GDPR (General Data Protection Regulation). Ongoing clinician input and feedback will be required to ensure the correct balance between the clinical utility of captured data and privacy concerns. Finally, and arguably most important of all, the clinical rationale for the acquisition of such data, the potential risks to privacy and the risk mitigating solutions all need to be communicated to patients and the general public, to ensure that any consent given to proceed with these diagnostic approaches is made with the appropriate prior information.

Chapter 8

An eye to the future

8a. Beyond diagnosis: how to use such approaches for intervention

Chapter 8a(i)

Clinical trials

Sarah Gregory[a,b], Stina Saunders[a,c] and Craig W. Ritchie[a,b,d]
[a]Centre for Clinical Brain Sciences, University of Edinburgh, Edinburgh, United Kingdom; [b]Scottish Brain Sciences, Edinburgh, United Kingdom; [c]Linus Health Europe, Edinburgh, United Kingdom; [d]Mackenzie Institute, University of St Andrews, St Andrews, United Kingdom

Introduction

In the years that followed Alois Alzheimer's first description of the (to become) eponymous condition of Alzheimer's disease in 1906, clinicians have had to rely on a syndromal approach to the diagnosis. This reliance was inevitable and acceptable until the late 20th century as, until then, there was no clinically appropriate means to identify the core pathological lesions of misfolded and then aggregated amyloid beta and tau or the cortical impact of this on neurodegeneration. Accordingly, clinical symptoms like memory loss formed a prominent and then core part of the diagnosis of Alzheimer's disease which could only be made with any degree of certainty in older people when function was also affected. This became Alzheimer's dementia. In essence, Alzheimer's disease transitioned from being a brain disease in 1906 to a *cognitive disorder* by the end of the 20th Century.

This legacy, that is proving very resilient, hampers both clinical practice and clinical trials as services focus until recently on older people with overt and intractable symptoms at the very late or dementia phase of the disease. Clinical trials require a cognitive outcome to demonstrate efficacy which is impossible in early phases of disease when cognitive symptoms are absent unless trials last for several years if not decades. Moreover, conducting clinical trials needs access to a pool of patients for approach and recruitment and if the source of patients is predominantly memory clinics diagnosing people with dementia, then trials to target disease modifying therapies become exceptionally difficult to undertake. Consequently, clinical trials of disease modifying therapies are undertaken in very late-stage disease because they can be done—not because they should be done—in that population.

The development of increasingly reliable, valid and standardized metabolic and structural imaging, protein biomarkers in spinal fluid and plasma, risk prediction algorithms under development and Brain Health Services for early disease detection and risk profiling will ensure that as we go deeper into the 21st century we can reconsider Alzheimer's disease and other neurodegenerative diseases as brain diseases that can be detected and managed decades before the irreversible *symptomatic* stages of disease where people are diagnosed today. This though is not a trivial task, there is much invested in "dementia"; a common understanding between the public and the clinician, regulatory perspectives, an overwhelming amount of research and knowledge relative to that available in preclinical and prodromal disease stages. Massive investments in many different ways, such as investment in clinics, nursing homes, development pipelines, hospital beds, medical and other clinical specialties, third sector organizations and national and international government agencies all act as impediments to change despite the strong scientific and indeed ethical impetus to intervene decades before the dementia syndrome.

Early disease identification and the predictive value of biomarkers

As clinical trials of either symptomatic or disease modifying treatments advance in targeting new mechanisms causing Alzheimer's disease, so does our understanding of the earliest pathological changes in the disease process. The underlying characteristics of Alzheimer's disease start developing years before any symptomatic manifestations of dementia occur, offering an opportunity to identify individuals who are at a high risk of future dementia and intervening in the disease process to either prevent deterioration entirely or delay it significantly. However, the correct identification of individuals who are truly at risk of future dementia (true positive cases) relies on any tests being not only sensitive in picking up the populations at risk, but also specific, meaning these tests need to have strong predictive value. In the next section of the chapter, we discuss the early identification of Alzheimer's disease, detail the latest scientific evidence around Alzheimer's disease pathology and describe biomarker tests currently available.

A shift from late-stage clinical presentation to early disease detection

Historically, Alzheimer's disease was primarily described by the late-stage clinical manifestation of dementia [1] and even then, a diagnosis would be confirmed by post-mortem brain examination. However, with advances in medicine over the last 100 years since this condition was first described, evidence now suggests Alzheimer's disease is a neurodegenerative condition which develops over a long time. In fact, the earliest pathological changes may happen as early as 20–30 years before an individual is diagnosed with dementia [2–5]. Accordingly, the Alzheimer's disease continuum *ends* with the dementia syndrome, when life expectancy is generally expected to be around 8 years if diagnosed at the age of 65 [6]. We now also differentiate between Early Onset Alzheimer's disease for those diagnosed before the age of 65 and Late Onset (also *sporadic*) Alzheimer's disease for those diagnosed after the age of 65 [7]. Early Onset AD is rare and is widely reported to occur between the ages of 30—60 or 65 years [8]. Familial or autosomal dominant AD (FAD or ADAD) is a rare condition, affecting less than 1% of the Alzheimer's disease community, with age of onset typically before 65 years. As detailed in the following section of the chapter, despite years of research, there are currently four treatments globally available for the dementia stage of the disease process and these only provide symptomatic relief for about a year, with a handful of new treatments approved for earlier stages of the disease process which offer a modest delay in impairment. It is crucial to examine what has contributed to the decades of unsuccessful drug development as not to repeat the same disappointing results over and over. With increasing evidence supporting a long *silent* period in the Alzheimer's disease pathology, some of the futility of drug development may be explained by recruiting individuals into clinical trials who are too far advanced in the disease continuum, thus attempting to intervene in the disease process too late.

The Alzheimer's disease continuum

When Alzheimer's disease starts developing, in the *preclinical* stage, there are no outwardly symptoms but underlying pathological changes are already present; the earliest symptoms may be noticed in the *prodromal* stage of the illness where there is mild impairment but activities of daily living are not affected [3,9—12]. Differentiating between an early and late stage in the Alzheimer's disease pathology is reflected in the 2011 NIA-AA guidelines [13] which for the first-time incorporate neuroimaging biomarkers and have recently been updated with recommendations from the International Working Group [13]. However, as individuals may have some biological evidence consistent with Alzheimer's disease pathology, but this may never deteriorate to dementia [14,15], it is vital that we are able to correctly identify people who are in the pre-dementia stages of the illness continuum. To this end, there are major global initiatives aimed at detecting the very earliest changes in Alzheimer's disease, such as the European cohorts like the PREVENT study in the UK and Ireland [11], ALFA study in Spain [12] and the cross Europe EPAD study [16]. The US based GAP who are facilitating trial recruitment [17].

The neuropathology of Alzheimer's disease

As other chapters have mentioned, on a macro level, the Alzheimer's disease pathology involves brain shrinkage with cortical thinning and atrophy [3,7], typically starting in the medial temporal lobe (hippocampus and entorhinal cortex) [18]. One of the earliest markers in the disease process is vascular dysregulation [19], resulting in reduced cerebral blood flow. Reduced cerebral blood flow is associated with neurodegeneration through two processes, the upregulation of the BACE1 enzyme that produces amyloid-β and the build-up of tau [20], the two hallmark proteins associated with Alzheimer's disease. Interestingly, post mortem studies have found that on a cellular level, the aggregations of tau may precede the deposition of amyloid-β by approximately a decade [21] though this is the opposite direction of what we measure in individuals at the earliest stages of Alzheimer's disease where amyloid build up is considered the first detectable sign of disease pathology. These pathological tau structures spread to medial and basal temporal lobes, then into neocortical associative regions and finally into the unimodal sensory and motor cortex [22]. Additionally, there is increasing evidence that the presence of a sustained immune response in the brain is associated with oxidative stress, neuroinflammation and glial activation [23,24] though the direction of these events remains uncertain and subject to investigations in research studies as described earlier in the chapter involving individuals in the earliest stages of the disease process. The early pathological changes associated with amyloid and tau are succeeded by synaptic loss, initially localized in the entorhinal cortex though over time, synaptic loss spreads to other regions of the brain [25]. Synaptic loss is a critical event in the disease pathology as synaptic function mediates cognitive performance [26] and as such, decades after the first pathological changes occur, this is the point in time where an individual may be exhibiting outwardly observable cognitive decline for the first time. The disease pathology continues with neuronal loss and structural brain atrophy [27,28]. Evidence also suggests early disruption of the blood-brain barrier is involved in the Alzheimer's disease process [29,30], leading to neurovascular dysfunction [31]. However, as evidenced by the remaining uncertainties, there is a lot we do not yet understand about the underlying disease mechanisms and the very earliest processes to be able to detect pathology and intervene years in advance when any intervention may have the most effect.

Methods for investigating Alzheimer's disease pathology

As the previously described neuropathological changes progress over decades and Alzheimer's disease finally manifests with outwardly clinical signs (of emerging dementia), earlier disease identification needs to detect change not expressed by any clinical symptoms. The clinical manifestations of dementia are well recognized (and described in the following section of this chapter), primarily including progressive memory loss and subsequent decline in cognitive abilities such as language, understanding and ultimately carrying out daily activities [32]. However, these symptoms are not applicable to the much earlier disease stage years before. At the earliest stage of the disease continuum, detecting pathological changes characteristic of Alzheimer's disease relies principally on biomarker assessments. The primary methods for assessing biomarkers in Alzheimer's disease involve brain imaging techniques such as MRI, CT scan, positron emission tomography (PET) or single-photon emission computerized tomography (SPECT) imaging or obtaining bodily fluids such as cerebrospinal fluid (CSF), blood, saliva or urine and digital biomarkers such as changes in speech [33].

Genetics

Genetically, the presence of the APOE-e4 allele is an important risk factor for Alzheimer's disease [21] and is associated with a lower age of disease onset [34]. A recent study comparing the likelihood of developing Alzheimer's disease with people with different APOE genotypes found that while the mean age at diagnosis in the whole study was 72.8, for individuals carrying the APOE-e4 allele, the diagnosis was between 65 and 70 [35]. The effect of the APOE-e4 allele has a stronger risk effect in men [36] and studies have shown higher risk in white populations as African Americans and Hispanics have an increased frequency of Alzheimer's disease regardless of their APOE genotype [37]. Interestingly, it has also been shown that healthy individuals who carry the APOE-e4 allele exhibit deficits in pattern separation, indicating an early marker for Alzheimer's disease related neuronal dysfunction [38]. Other than the APOE-e4 allele, no other gene locus has been found to have a consistently strong association with Late Onset Alzheimer's disease [39]. However, not everyone who carries the APOE-e4 allele develops Alzheimer's disease and the disease may occurs in people who do not have an APOE e4 gene [40].

Conclusion

There have been major shifts in our understanding of the silent phase and the pre-dementia stages of the illness. This move toward early disease identification offers hope for intervening in the disease process at a time when maximum benefit could be yielded. The last 20 years have seen an enormous expansion in research on biomarkers [41] and there are now many potential biomarkers which may deliver new insights on underlying disease mechanisms, though for now many biomarker tests are still only appropriate for research purposes rather than have diagnostic utility [42]. Any good biomarker should be stable across different populations, easy to use and standardize, cost effective and sensitive and specific [43]. With all the advances in neuroimaging techniques and biomarker analysis, it is vital to consider these criteria to develop truly useful and accessible markers of early disease which would benefit patient populations across the globe.

Current drugs development and past successes/failures

At the time of writing this chapter, four drugs have been fully licensed and are available for the treatment of Alzheimer's disease (AD) dementia. These drugs fall into two classes; acetylcholinesterase inhibitors (donepezil, galantamine, rivastigmine-note tacrine was previously licensed in the USA but has since been withdrawn due to pharmacokinetic and pharmacodynamic problems) and one N-Methyl-D-aspartate (NMDA) receptor antagonist (memantine). Some of the recently approved medications are discussed in more detail later in the chapter. For instance, Aducanumab was given accelerated approval in June 2021 by the Food and Drug Administration (FDA) with a requirement to complete a post-approval confirmatory phase four trial [44], though has since been discontinued by Biogen. Aducunamab failed to gain marketing approval from the European Medicines Agency, who cited a lack of evidence to support any clinical improvement following removal of amyloid beta in the brain, the conflicting results reported from the main studies and safety concerns regarding the compound's side effects [45]. Other similar drugs followed, e.g., Lecanemab was approved in 2023, intended for use in mild cognitive impairment due to AD to mild AD dementia with confirmed brain amyloid pathology [46] and donanemab most recently which is likely to receive approval in the near future [47]. Uptake in clinical practice has been limited so far for these recent treatments, with financial considerations for aducanumab cited by the manufacturer as a major barrier, leading to the cost being reduced from $56,000 to $28,200 per year [48]. Health systems readiness [49] is another major concern with the dementia (late care) model being widely used globally and expensive monitoring of new therapies in combination with case finding costs to identify the right people at early stages AD causing immense stress in the real world implementation of these treatments.

As our understanding of AD and AD dementia has developed, the number of potential treatable targets has increased. This has resulted in a diversity in the mechanisms of action targeted by molecules and biologics currently undergoing evaluation [50–53]. In this section we will provide an overview of previous and current drug trials alongside commentary on where the future gains are likely to be. Drug development efforts for AD and AD dementia can broadly be divided into two overarching strategies: symptomatic and disease modifying treatments. Additionally, we tend to separate the types of compounds into small molecules and biologics. Small molecules are organic compounds with low molecular weight. Most therapeutics would be considered small molecules. Small molecules can be administered orally which is often easier for the patient and available at a lower cost for the health care system. Biologics are compounds which are isolated, in full or part, from living systems such microorganisms, plants, animals or humans. Biologics may be administered by infusion, subcutaneous, intramuscular or intrathecal injection. They tend to require more specialist settings for preparing and administering the treatment and therefore can be more expensive than small molecules and more demanding for patients to engage with.

Symptomatic treatments: Past, present and future

Symptomatic treatments aim to treat and improve the symptoms of AD dementia. Modalities that have been targeted for pharmacological treatments range from cognition and everyday function to sleep and agitation. It should be noted here that the regulatory bodies still require cognition to be measured as an outcome in most trials involvement people with AD or AD dementia, which limits proper investigation of the important non-cognitive symptoms. Symptomatic treatments do not target the underlying biological causes of the disease and do not aim to have any long-term impact on the overall neurodegenerative disease process. All symptomatic treatments previously and currently in development are small molecules. We anticipate that this will remain so in the future, as it would be likely that the cost and treatment regimens of biologics are prohibitive when considering symptom management without disruption of disease pathology.

Cognition and everyday function

Cognitive impairment is the most commonly discussed symptom of AD dementia. Ability to function well in daily life is clearly an important goal for us all as we age. As such it is not surprising that much of the symptomatic drug development effort to date has focused on cognition and daily function as the primary target. To assess cognitive function, researchers use rating scales with which to monitor change from baseline performance. The most common of these cognitive outcome measures in clinical trials is the ADAS-Cog, a brief battery of tasks measuring performance in a number of cognitive domains [54]. Everyday function is typically measured using scales assessing a person's ability to complete a variety of activities of daily living (such as making a cup of tea, getting dressed, driving a car) [55,56].

Acetylcholinesterase inhibitors

Acetylcholine supports inter-neuronal communication and is broken down to an inactive form by acetylcholinesterase. Levels of acetylcholine have been established as significantly reduced in AD dementia [57,58]. Acetylcholinesterase inhibitors (AChEIs) block the action of acetylcholinesterase, increasing the concentration of acetylcholine [58]. The licensed AChEIs introduced at the start of this section were all approved based on their demonstrated effectiveness in phase three trials for improvement in ADAS-Cog scores compared to placebo as well as superiority on a measure of function [59–64]. A considerable number of studies have built on these original trials to understand more about the therapeutic potential of these molecules. Whilst original efficacy studies focused on mild-to-moderate AD dementia, the DOMINO-AD study demonstrated that continued treatment with donepezil in people with moderate-to-severe AD dementia resulted in clinically important cognitive benefits compared to withdrawal of treatment [65]. The same trial also found it to be more cost effective to continue donepezil or start memantine treatment compared to discontinuing donepezil [66]. A recent Cochrane review looked at the data from DOMINO-AD along with six other trials and found the data suggested discontinuation of the AChEIs may result in worsening of cognition, psychiatric and functional impairment for people with AD dementia, although there continued to be a lack of evidence at a high certainty level [67]. A small number of studies have trialled AChEIs in mild cognitive impairment (MCI). A Cochrane review of donepezil for MCI identified two studies, concluding there is currently no evidence to support the use of donepezil for this indication [68]. A more recent systematic review and meta-analysis identified 17 studies enrolling 2847 participants with MCI treated with donepezil and/or placebo. This study found that short term cognitive function was improved, however there was no significant delay in disease progression and adverse reactions to treatment were commonplace [69]. A review that additionally looked at rivastigmine and galantamine found no benefit of treatment on cognition, although did find those with MCI treated with galantamine (data from two studies) has a lower incidence of progression compared to placebo [70]. Overall the clinical benefits of AChEI treatment in MCI currently appear limited and the risk of experiencing side effects is moderate. Donepezil was the first of the currently used AChEIs to be licensed (as noted earlier, tacrine was licensed a year earlier than donepezil but withdrawn from the market due to side effects), and while there are conflicting results about the most efficacious of the drug class, donepezil appears to have the lowest frequency in adverse events reported in trials [71,72]. Given the paucity of development of new AChEIs in the last decade it seems unlikely novel compounds in this drug class will be in the future pipeline. Evaluation of this drug class in other disorders causing dementia, such as Lewy Body and Parkinson's disease dementia [73], as well as identifying when and how the AChEIs are most potent [74,75] will see the continuation of research in this drug class.

NMDA receptor antagonist

Memantine is the only NMDA receptor antagonist licensed for AD dementia. It is a low affinity antagonist, meaning there is only short term blockage of the receptor thereby avoiding side effects such as impacts on learning and memory that are associated with high affinity NMDA receptor antagonists [76]. Memantine was approved in 2002 as a monotherapy for moderate-to-severe AD dementia, with symptomatic effects on both cognition and function [77]. Memantine is commonly used as adjunctive therapy to an AChEI with important impact on cognition and function over that of AChEIs alone at the moderate-to-severe AD dementia stage [78]. Very few studies have investigated the effects of memantine earlier in the disease process. One small study trialled the benefits of memantine alongside metformin compared to metformin alone in a group of participants with prediabetes and MCI. In this group the combination therapy was associated with significant oxidative indices decreases compared to monotherapy [79]. No cognitive tests were included in this study protocol, however as oxidative stress is associated with diabetic cognitive dysfunction this may be an important area for future work to explore this special population. As with AChEIs it is unlikely there will be further compounds developed within this class for AD dementia but rather efforts will continue in understanding additional indications for memantine [80].

α7 nicotinic receptor agonists and modulators

Several studies have investigated the efficacy of targeting the α7 nicotinic receptor. This target was selected as α7 subtype of nicotinic acetylcholine receptors are expressed by basal forebrain cholinergic projection neurons as well as in the hippocampus [81]. A number of small molecules acting as partial or total α7 agonists or α7 allosteric modulators have been trialled in mild-to-moderate AD dementia, however most have met futility criteria or have been stopped due to safety concerns [82]. Although galantamine has been licensed as an AChEI it is also known to be an effective positive α7 nicotinic allosteric modulator [83]. This drug class continues to be developed in animal model studies [84] and if the safety profile can be improve we may see future clinical trials of these compounds in a preclinical and prodromal population [85].

MAO-B inhibitors

Monoamine oxidase B (MOA-B) levels are known to be increased in the brains of people with AD dementia, particularly in astrocytes and pyramidal neurons. MAO-B may be important in the regulation of Aβ production in neurons via γ-secretase [86]. A number of MOA-B inhibitors are licensed for the treatment of Parkinson's disease [87]. Multiple trials of MAO-B drugs have taken place however none have met primary endpoints. A Cochrane review of selegiline identified 17 trials and found no clinically meaningful benefit, recommended no further studies should continue with selegiline and AD dementia [88]. A phase II study of ladostigil in 210 participants with mild cognitive impairment (MCI) found no differences in rates of progression to dementia compared to placebo, although there was less whole brain and hippocampal atrophy in the treatment group [89]. The Mayflower Road study, sponsored by Roche, tested sembragiline compared to placebo over 52 weeks in a group of participants with moderate AD dementia. This study found no differences in cognition or function between treatment and placebo over the study duration. Interestingly there was a treatment effect seen on the BEHAVE-AD-FW scale which measures neuropsychiatric symptoms, which was stronger in those with higher baseline behavioral symptoms, however this was not planned as a primary or secondary endpoint [90]. Rasagiline has been investigated with FDG-PET and perfusion as outcomes, both of which reported significant results suggesting an area for future development [91,92]. There are currently no ongoing trials of rasigiline, ladostigil or sembragiline in MCI due to AD or AD dementia. Re-aligning the MAO-B inhibitors with the purpose of targeting neuropsychiatric symptoms may be the best approach to continue the development of these compounds. This would build on the learnings of the effect of sembragiline on BEHAVE-AD-FW scale, which particularly identified an impact on paranoia and delusions, and the effect of rasagiline on brain regions such as the interior frontal gyrus, an area associated with delusions in dementia [93,94].

11β-HSD1 inhibitors

Higher than expected levels of cortisol have been associated with MCI and AD dementia, as well as decline in cognitive function in healthy volunteers, in a number of cross-sectional and cohort studies [95–99]. 11β-HSD1, a regular of glucocorticoids, is abundantly found in the hippocampus and central nervous system [100]. The action of 11β-HSD1 amplifies cortisol through catalyzing the conversion from cortisone to cortisol [101]. 11β-HSD1 inhibitors work to partially inhibit this conversion, thereby lowering cortisol levels [102]. 11β-HSD1 inhibitors have been studied in AD dementia to understand the impact on cognitive function [103]. Two molecules have been trialled in human clinical trials to date (ABT384 and Xanamem) however neither have demonstrated efficacy on outcome measures [104]. Xanamem remains under development. 11β-HSD1 inhibitors are likely to play a role not only in the symptomatic treatment of AD dementia, but also in the modification of risk factors such as diabetes, obesity and depression [103]. As we learn more about the role of cortisol and 11β-HSD1 in cognitive decline, we anticipate a shift to a stratified medicine approach for 11β-HSD1 inhibitors, whereby those with genetic vulnerabilities [105] or abnormal cortisol levels will be specifically targeted for clinical trials of these inhibitors.

Sleep

Changes in circadian rhythm are known to occur in AD dementia, with associated problems such as sundowning, reported by patients and their family members [106,107]. Irregular Sleep-Wake Rhythm Disorder (ISWRD) is one of the common sleep disorders experienced by people living with AD dementia. ISWRD is the absence of a clear sleep wake cycle and people experience this will typically exhibit a sporadic pattern of sleep over at least three episodes in a 24 h period [108]. A phase II study reporting on the use of lemborexant found a significant reduction in restlessness and greater distinction between night and day rest-activity rhythm in treatment compared to placebo, with treatment well tolerated by subjects [109]. Studies are continuing to further understand the potential therapeutic potential of Lemborexant for people with mild to moderate AD dementia and ISWRD. Two small trials have studied melatonin for sleep disturbances in MCI due to AD and mild AD dementia, with results indicating positive benefits for sleep latency in the treated group compared to placebo [110,111]. Sleep is an area likely to garner increased attention in the coming years as we understand more about the associations between sleep and AD dementia progression [112] and the impact of disturbed sleep on caregivers [113].

Disease modifying treatments: Past, present and future

Disease modifying treatments (DMTs) aim to disrupt the disease process to stop further progression, prevent disease development and even reverse damage caused by the disease. DMTs under development in the field include both small molecules and biologics. Typically these treatments focus on biomarker outcomes (most commonly amyloid) as well as cognitive and function outcomes. Validation of specific AD biomarkers against the gold standard of neuropathology has accelerated this field of research.

Treatments targeting amyloid

Development of immunotherapies for AD and AD dementia began in earnest with the investigation of monoclonal antibodies (mAbs). Exogenous mAbs, as trialled in AD and AD dementia, work through the mechanism of passive immunization to bind and clear amyloid [114]. A number of mAbs have been trialled, typically in mild-to-moderate AD dementia or MCI due to AD with varying success. The introduction of mAb trials to the AD research field significantly changed the profile of participants entering trials and the screening procedures undertaken. Due to the target of the mAbs (our examples all focus on amyloid) participants entering trials are required to demonstrate amyloid positivity, through positron emission tomography (PET) scans or lumbar punctures. Early trials of bapineuzumab, solanezumab and gantenerumab failed to reach significance on primary outcomes in original phase III trials [115]. However many of these biologics remain under investigation, with consideration being paid to whether these might be more suitable earlier in the disease process or whether there are sub-sets of participants who would be more appropriately targeted for treatment. The A4 and Dominantly Inherited Alzheimer Network Trial (DIAN-TU) studies trialled solanezumab and gantenerumab to determine whether these compounds are beneficial in early stages of AD before symptom emergence [116,117]. Neither solanezumab or gantenerumab met the primary outcome (multivariate cognitive endpoint) in the DIAN-TU trial, however data presented at the AAT-AD/PD conference suggests gantenerumab removed amyloid-β plaques, normalized both p and t tau levels and blocked increases in NFL [118,119]. The A4 trial of solanezumab is due to present results in 2022. Additionally gantenurumab is being studied in participants with prodromal AD and results are anticipated in 2022 [120]. During the bapineuzumab trials a series of side effects of concern emerged which were termed ARIA-amyloid related imaging abnormalities, and typically are separated into two categories; microhaemorrhages (ARIA-H) and cerebral edema (ARIA-E). While these can be serious, fortunately most participants enrolled in mAb trials with these imaging findings are symptom free and the findings tend to resolve without further intervention when the mAb is ceased. Some trials have also successfully managed redosing for participants who had ARIA with no reoccurrence or worsening of findings. The ARIA side effects appear to be more common in APOE e4 carriers and as such subsequent trials have tested for this gene at baseline and allocated participants to slower treatment titration pathways which lowers risk of ARIA occurrence.

In recent years the most discussed amyloid-targeting biologic in development has been aducanumab [121] which was given accelerated approval in June 2021 by the US Food and Drug Administration (FDA) with a requirement to complete a post-approval confirmatory phase IV trial [44]. Aducanumab has been trialled in two phase 3 trials, which were stopped due to futility. However subsequent post-hoc analyses of the data led the sponsor Biogen to submit a marketing application, suggesting a significant difference between treated and placebo groups on a number of outcome measures [122]. Consensus appropriate use recommendations developed by a small expert group suggest treatment with aducanumab is restricted to the population included in the efficacy and safety studies (i.e., those with MCI due to AD and mild AD dementia), titration to 10 mg/kg occurs over a 6-month period, patients require pre-dosing and peri-titration MRIs as well as repeat scans for anyone with emergent symptoms, as well as patient-led discussions about expectations and aims of treatment [123]. There remains a level of skepticism in the field as to whether post-hoc data analyses were appropriate to support this submission [124–126]. What is likely is that the approval of the first new compound for AD, and the only disease-modifying compound, will change the AD clinical trial landscape [127]. In the wake of the accelerated approval for aducanumab, two other compounds (donanemab and lecanemab) were granted FDA breakthrough therapy status, meaning they will also be eligible to apply for accelerated approval during the licensing application process. Further development and commercialization of aducanumab was discontinued in January 2024.

A phase II trial of lecanemab, developed by Eisai Inc., did not meet its 12 month primary endpoint (cognition as measured by the Alzheimer's Disease Composite Score (ADCOMS)), however a prespecified 18 month Bayesian analysis demonstrated reduction in brain amyloid as measured by brain PET scans alongside superiority demonstrated on the ADCOMS, ADAS-Cog and Clinical Dementia Rating scale Sum of Boxes (CDR-SB) [128]. Two phase III studies of lecanemab in preclinical and early AD are underway and will be used to make a decision on submitting a licensing application [129,130]. A phase III study of lecanemab versus placebo demonstrated a significant slowing of global cognitive decline (measured by the CDR-SB) and greater reductions in brain amyloid (as measured by PET) compared to placebo over 18 months of treatment [131]. These results led to the approval of lecanemab (LEQEMBI) under the traditional approval pathway by the FDA in July 2023. Consensus-appropriate use recommendations, similar to those for aducanumab, were developed by a small expert group. These recommendations suggested that treatment with lecanemab was restricted to the population included in the efficacy and safety studies (i.e., those with MCI due to AD and mild AD dementia), no required titration period, patients requiring pre-dosing and peri-titration MRIs as well as repeat scans for anyone with emergent symptoms, as well as patient-led discussions about expectations and aims of treatment [46].

Donanemab is currently being developed by Eli Lilly. Dose escalation studies suggest donanemab is well tolerated and results in reduction of amyloid deposits in AD [132]. In a phase II study treatment with donanemab of participants with prodromal AD resulted in better cognition, as measured on a composite score, and functional activity, as measured using an activities of daily living scale, compared to those treated with placebo [133]. An 18-month, phase III study reported significantly better outcomes across 23 of the 24 pre-specific gated outcomes (including global function, cognition and amyloid burden) for participants treated with donanemab compared to placebo [47].

The accelerated approval of aducanumab and the traditional approval of lecanemab are likely to herald the arrival of a number of first-generation disease modifying mAbs for AD. As we continue to learn more about both the biology of the disease and the effects of these compounds, we anticipate second-generation mAbs delivered to high risk strata of patient populations who are most likely to respond to these treatments.

Treatments targeting tau

Tau-targeting treatments are now a significant portion of the AD research and development pipeline [52,134,135], although most remain in phase II of development and few results are reported to date. Examples of tau targeting treatments in development include LY3303560, RO7105705 and TRx0237.

LY3303560 is a humanized monoclonal antibody engineered to target aggregated tau [136], shown to be tolerated in both healthy volunteers and participants with AD in phase I trials and currently undergoing phase II investigations with participants with mild-to-moderate AD dementia [137]. Similarly RO7105705 is an IG4 antibody designed to intercept extracellular tau, which has passed safety and tolerability assessments in phase I studies and is currently under investigation in two phase II trials in mild and moderate AD [137,138]. TRx0237 is a stabilized form of the methylthioninium moiety. A phase III trial of TRx0237 investigated disease modifying effects in participants with mild-to-moderate AD dementia with no significant effects on the primary outcome measures of cognition and function [139]. An open label extension study of a second phase II/III study (LUCIDITY) and expanded access program are ongoing to further understand TRx0237.

Development of knowledge about the sequencing of tau in the AD process and expansion of access to tau testing through initiatives such as blood plasma markers [140,141] will accelerate research into tau-targeting treatments. One can forsee a future where tau and amyloid targeting treatments will be trialled in sequence to impact sequential stages of the AD process.

Treatments targeting other biomarkers

Alongside the more commonly considered targets of amyloid and tau, a raft of other treatment targets have been identified and are under investigation. Interesting avenues of investigation include synaptic plasticity (ANAVEX2-73 [142]; neflamapimod [143]), metabolism (metformin [144]), epigenetic (ORY 2001 [145]) and proteostasis (posiphen [146]; nilotinib [147]).

It is likely that as we continue to develop our knowledge of AD and AD dementia, a cocktail of treatments, likely personalized to the patient, will be the mainstay of medication treatment.

Future directions e.g., novel outcome measures

It is important to use outcome measures which capture both biological as well as clinical change in Alzheimer's disease trials. Any treatment would need to show effectiveness in halting or reversing some of the pathological change, but equally in having an effect on the clinical manifestations of disease. However, Alzheimer's disease trials commonly employ outcome measures which are validated for use in the dementia stage of the illness and may therefore have a ceiling effect (where individuals score highly as the measure captures only more advanced impairment) in the earlier outwardly healthy study population. As the scientific understanding of early development of dementia suggests intervening at a much earlier stage in the pathological process, any assessment measures should therefore be sensitive in the younger preclinical or prodromal study population. Crucially, any change must be clinically meaningful for the individual receiving treatment.

Measuring treatment success

With the move to earlier interventions in the disease process, there is a fundamental shift in what constitutes treatment success, and therefore, what outcome measures need to capture. In preclinical Alzheimer's disease, the aim is to *maintain ability* by preventing entirely or delaying significantly the development of symptoms in the first place. Accordingly, in dementia prevention trials where individuals do not have symptomatic disease, an outcome measure would need to be sensitive enough to detect change if there is a decline (in the placebo group) but also detect *stability* (in the treatment group) as the desired outcome. It is critical that participants in randomized controlled trials are phenotyped based on biomarker criteria and grouped based on the disease stage in order to use appropriate outcome measures in each stage of the disease spectrum. In addition to detecting biological changes detailed earlier in the chapter, emerging research suggests there may be areas in everyday functioning subtly impacted even years earlier than symptomatic disease manifests. Critically, these areas would need to be assessed using novel outcome measures rather than existing and widely used tests which are aimed at a later disease stage.

Clinical meaningfulness in Alzheimer's disease outcomes

Currently, Alzheimer's disease trials use a range of outcome measures, leading to difficulties with data comparison and standardization. A recent review identified 81 different outcome measures used across the different disease stages [148]. Although impairment in cognition is a fairly late-stage manifestation of Alzheimer's disease, cognition still remains a central target for measuring treatment success in clinical trials. However, there is no direct evidence how a change in score on a cognitive outcome measure translates into clinical meaningfulness for a study participant. As such, regulatory bodies which approve new medications such as the US based FDA and Europe based EMA advise that Alzheimer's disease trials include Patient Reported Outcome Measures (PROMs) which capture clinical meaningfulness for the patient i.e., any treatment benefit should be experienced by the study participant [149,150]. However, while scales measuring activities of daily living are intended to capture the disease impact on functioning, many of the commonly used measures (such as the widely used Clinical Dementia Rating Scale [151]) are not applicable before the dementia syndrome manifests. Furthermore, even in the case of symptomatic disease, a recent study demonstrated there are many domains which matter to individuals with dementia which are not captured by the CDR (In press, ref clinical ePSOM paper), such as confidence.

Electronic person specific outcome measure (ePSOM) development program

A treatment's success should therefore be evidenced by change in the disease pathology but also by how the change translates into clinical meaningfulness for the individual. The aim of the electronic Person Specific Outcome Measure (ePSOM) development program [152], is to understand what outcomes matter to the individual when developing new treatments for Alzheimer's disease. The ePSOM study uses a stepped approach to gather evidence and ultimately develop an electronic outcome measure to capture treatment/intervention outcomes which are tailored to the individual and reflect *maintenance* of ability as well as any decline. While commonly used paper-pen outcome measures may capture a limited number of domains, the ePSOM study using Natural Language Processing resulted in 184 themes of importance about brain health, demonstrating a need for a person specific approach in reflecting clinical meaningfulness [153]. Importantly, the ePSOM findings also suggest that outcomes that matter shift along the preclinical, prodromal and overt dementia continuum. This has important implications for the development of outcome measures that may be used in long term prevention studies that last several years where participants may pass through different stages of disease.

Computerized assessment methods

Although measuring change in cognition remains the primary outcome in Alzheimer's disease trials, there are promising new targets for detecting Alzheimer's disease pathology early in the disease course. Many of the novel assessment methods utilize advances in technology to assess complex behaviors or even everyday tasks, resulting in a type of *digital biomarker.*

There is emerging evidence that spatial processing may be a potential indicator and indeed manifestation of early Alzheimer's disease pathology. Findings from mid-life cohorts such as the PREVENT Dementia study suggest allocentric processing (a person's ability to understand object-to-object spatial relations, i.e., looking at images of a mountain range from different angles to identify which of the images are of the same mountain range [154]) rather than egocentric processing (a person's ability to understand self-to-object spatial relations, i.e., finding a way around a supermarket [155]) may be a potential indicator for Alzheimer's disease pathology [156]. In line with other research on the importance of spatial processing in the pre-dementia stages, a recent study concluded that driving may serve as an effective and accurate digital biomarker for identifying preclinical Alzheimer's disease among older adults [157].

Symptomatic Alzheimer's disease may also change an individual's language, especially certain elements in speech such as elaboration and attribution [158]. However, De La Fuente et al., (2020) carried out a first systematic review on the potential of using interactive AI methods analyzing speech data as Alzheimer's disease biomarker already in the pre-dementia stages of the illness pathology. The review concluded that when compared to traditional neuropsychological assessment methods, speech and language technology were found to be at least equally discriminative between different groups [159]. Indicators of disease in language may be confounded by amyloid levels as people with higher amyloid levels (both healthy and those with mild cognitive impairment), exhibit greater rate of decline in episodic memory and language [160].

Another novel outcome for assessing early Alzheimer's disease pathology is eye-tracking, with poorer performance at eye tracking potentially revealing underlying cognitive decline. An explanation may be that people with Alzheimer's disease gradually lose the efficient control of attention and develop impairments of both inhibitory control and eye movement error-correction [161] or people with Alzheimer's disease may have difficulty selecting relevant information and therefore attend more frequently to irrelevant information [162]. Eye tracking has been found to be a useful tool in differentiating between healthy stage, mild impairment and manifest dementia [163]. While research on the digital biomarkers is only beginning to emerge, it will undoubtedly play a major role in screening for and tracking preclinical Alzheimer's disease in the future [164].

Finally, digitalizing existing cognitive assessments such as the clock drawing test to analyze the process of completing the drawing-based task using artificial intelligence has proven a sensitive measure of cognition [165].

Conclusion

As we understand more about the earliest stages of disease, it is critical that clinical trials design and outcome measures accurately capture change in the earlier predementia study population. There is a fundamental shift in measuring not only decline but also relative stability or decline consistent with what is expected in age matched individuals as success of treatment. There are a range of novel outcome measures harnessing the advances in technology which, if more widely available in the future, could offer a window of biological processes through everyday activities. Outcome measures need to be sensitive, specific, but also ecologically valid and reflect clinically meaningful benefits to patients. Importantly, patients should be enrolled into clinical trials early in the disease course and alongside biological measures, robust, reliable and valid tools should be used to measure the intervention's effectiveness from the study participant's perspective (In press, ref clinical ePSOM paper).

Clinical trials designs, considerations, major initiatives

Randomized control trials

Randomized control trials

Many trials in AD still follow a traditional randomized controlled trial (RCT) approach, particularly when we look at the symptomatic trials. Inclusion criteria can often be reliant on clinical diagnoses and where evidence of previous imaging is required it is not uncommon for CT scans to be accepted.

Whilst there is a balance to maintain between keeping invasive procedures to the minimum required for study entry and managing a cost-effective research study, in some cases lack of adequate participant profiling at screening may mean the study is enrolling those without the target disease of interest. As we look at an older adult population with dementia, post-mortem data suggests it is common for multiple underlying pathologies to be at play [166,167]. The development of blood-based AD biomarkers brings new possibility to better profile participants during screening visits for research studies, to ensure only those with the disease of interest are exposed to experimental procedures whilst maximizing the chances of trials reaching their end points [41,168,169]. Additionally, digital screening tools are developed at a rapid rate to be able to perform de-centralized eligibility assessments. With the shift toward preclinical and prodromal AD studies as well as expansion to symptoms other than cognition (e.g., sleep and agitation), we have seen more of a selective approach taken in participant enrollment. As these drugs have more specific targets, only those with the biological target or symptom of interest can be included in the trials. We think it is likely that over time these will become increasingly specific as we understand more about what is still a heterogenous clinical diagnosis. A shift toward midlife interventions is also likely, given we now understand the genesis of AD during this period of life. Brain health clinics, serving younger, healthier populations, exemplified by the Scottish Brain Health Service model [170], will become integral to future clinical trials.

Database discoveries

We have also seen some interesting discoveries through harnessing the power of clinical and research databases. One such example was the discovery in a German health records database that people living with diabetes who were taking pioglitazone had significantly lower rates of dementia compared to expectations (especially interesting given diabetes is a known risk factor for dementia) [171]. Pioglitazone was trialled against placebo in the TOMMORROW study. The TOMMORROW study was an ambitious RCT which enrolled healthy volunteers and allocated them to treatment based on their genetic status-of interest was a combination of APOE e4 carriers and those with an SNP of the TOMM40 gene. Participants were followed over a number of years and primary outcome was the number of events of dementia in the treatment compared to the placebo group. Unfortunately the trial did not meet the primary endpoint however the study has laid the ground for future innovative trials with a similar design [172,173].

Platform trials

Another innovation which has begun to emerge in the field of AD research is the concept of the platform trial. With origins in cancer research, platform trials utilize Bayesian methods to fast track testing of novel compounds [174]. The need for platform trials in AD arises from the lack of translation from phase II to phase III trials. One theory behind this lack of efficacy seen in phase III AD trials, despite apparent promise in phase II trials, is that there is insufficient proof of concept of the right dose, right population and right time.

The European Prevention of Alzheimer's Dementia (EPAD) project was the first of its kind in sporadic AD to attempt to develop platform studies [16]. The objective was to develop a clinical trial platform with the capacity to test multiple interventions concurrently in multiples sites across Europe. Participants for the clinical trial platform would be highly phenotyped and deemed eligible for secondary-prevention studies. The underpinning principles of the EPAD platform trial were to offer a single operational environment, protocol and sponsor responsible for the overarching governance framework, alongside the site network and readiness cohort. One of the supporting elements was a master protocol, designed using a Bayesian adaptive approach with a shared placebo across all interventions. The operational elements of the platform trial were under a single contract research organization (CRO) who took on the management of vendors including laboratory services, imaging vendors, database systems, randomization systems, and cognitive test vendors. As mentioned the platform trial also fell under a single sponsor, who were experienced in trial sponsorship, adequately resourced, and neutral in terms of the outcomes from the trial. As part of the Bayesian adaptive design, a priori decisions on success and futility had to be articulated in the master protocol. The pre-planned analyses were planned on a limited number of variables to be carried out by the independent data monitoring committee (IDMC), with the RBANS selected as the primary outcome. The template established by EPAD provides us with an opportunity to consider alternative approaches to AD clinical trial designs.

The Dominantly Inherited Alzheimer Network Trials Unit (DIAN-TU) have also launched an adaptive platform trial [175], after evolving from a more traditional phase II/III trial of solanezumab. The next generation of DIAN-TU will test two potential disease-modifying therapies in 160 autosomal dominant AD (ADAD) mutation carriers who are either pre-symptomatic or experiencing only mild symptoms over 4 years with a cognitive composite primary endpoint. The AHEAD 3—45 platform-based study combines two phase III trials (A45 and A3 trials) of lecanemab with trials differing in dosage regime based on baseline amyloid burden. The primary outcome of these studies is amyloid measured by brain PET. The platform shares a common screening protocol, PET scanning intervals (amyloid and tau at baseline, two and 4 years) and enrollment age of 55—80 years of age [119,129]. Two tau platforms are being developed, one by Eli Lilly and one by the Alzheimer's Clinical Trial Consortium (ACTC) [119].

Having a diversity of platforms allows the development of more targeted, stratified medicine. The outcome measures and safety assessments in tau-targeting compounds may be very different to those of amyloid-targeting compounds. Similarly participants with ADAD are very different to those with sporadic AD. It is entirely appropriate therefore to develop these different research environments. However alongside this sharing of participants and data across the platforms and the potential for development of more diverse trial ready participants for non-amyloid and tau targets will also be critical to prevention and treatment trial efforts.

Stratified medicine

In the future we may see more development in the field of stratified and personalized medicine for AD dementia. Other causes of dementia are already investigating these routes, particularly the field researching frontotemporal dementia (FTD) and glutamate and γ-aminobutyric acid (GABA). The Genetic FTD Initiative (GENFI) has already facilitated a number of discoveries about the earliest changes in FTD, such as insula and temporal lobe atrophy [176], which will further supported the development of stratified medicine trials in this field. GABAergic deficits in the frontal lobes of people with FTD are well established and have been highlighted as a potential treatment target [177,178]. Given the diversity of risk factors involved in AD [9], differential trajectories of cognitive and functional decline [179] and response to current treatments [180], it is reasonable to assume that the future of AD and AD dementia treatment will involve individual prevention and treatment plans with one or more pharmacological compounds along lifestyle modifications and psychosocial interventions. As we continue to move forward it will be important to learn from approaches taken by our colleagues in other neurodegenerative diseases.

Conclusion

To paraphrase Prof Sir John Bell, University of Oxford Regius Professor of Medicine, the best time to have entered the "modern" phase of Alzheimer's disease was 30 years ago when biomarkers of this brain disease were proving of real clinical value. The second-best time is today.

Knowledge emerging from mid-life cohort studies, learnings from familial AD cohorts, much greater understanding of risk factor/mechanism interactions, public demand, the applicability of machine learning approaches to the real world "big data" coming on line for prognostication, dissatisfaction with the glacial rate of progress over the last 50 years for better outcomes following a "dementia" diagnosis and massive societal and economic cost of dementia are now being allowed to fuel the "brain health" revolution. This revolution had as its spark the irrefutable fact that Alzheimer's disease is a disease that starts in mid-life that has a very late-stage consequence as "dementia". This revolution will facilitate the symbiotic development of new clinical services with the new and right clinical trials being done in the right population with the right (combination of) drugs.

A recent publication suggested a tripling of the numbers of people with dementia by 2050. This assumes no change to the status quo that has existed for at least 50 years. Worryingly, the best predictor of the future is often the past and should "modern" Alzheimer's disease management remain confined to the ivory tower of academic clinics then there will be no or little progress to impact upon that tripling of numbers. If however, the modern way, as pointed to in this chapter, is allowed to expand globally then rather than no impact on the tripling of cases of dementia there may well be no dementia at all.

References

[1] Bondi MW, Edmonds EC, Salmon DP. Alzheimer's disease: past, present, and future. J Int Neuropsychol Soc: JINS 2017;23(9−10):818−31.

[2] Goudsmit J. The incubation period of Alzheimer's disease and the timing of tau versus amyloid misfolding and spreading within the brain. Eur J Epidemiol 2016;31(2):99−105.

[3] DeTure MA, Dickson DW. The neuropathological diagnosis of Alzheimer's disease. Mol Neurodegener 2019;14(1):32.

[4] Ritchie K, Ritchie CW, Yaffe K, Skoog I, Scarmeas N. Is late-onset Alzheimer's disease really a disease of midlife? Alzheimer's Dementia 2015;1(2):122−30.

[5] Rajan KB, Wilson RS, Weuve J, Barnes LL, Evans DA. Cognitive impairment 18 years before clinical diagnosis of Alzheimer disease dementia. Neurology 2015;85(10):898−904.

[6] Davis M, Thomas OC, Johnson S, Cline S, Merikle E, Martenyi F, et al. Estimating Alzheimer's disease progression rates from normal cognition through mild cognitive impairment and stages of dementia. Curr Alzheimer Res 2018;15(8):777−88.

[7] Chen X-Q, Mobley WC. Alzheimer disease pathogenesis: insights from molecular and cellular biology studies of oligomeric Aβ and tau species. Front Neurosci 2019;13:659.

[8] Bekris LM, Yu C-E, Bird TD, Tsuang DW. Genetics of Alzheimer disease. J Geriatr Psychiatr Neurol 2010;23(4):213−27.

[9] Livingston G, Huntley J, Sommerlad A, Ames D, Ballard C, Banerjee S, et al. Dementia prevention, intervention, and care: 2020 report of the lancet commission. Lancet 2020;396(10248):413−46.

[10] Ritchie CW, Muniz-Terrera G, Kivipelto M, Solomon A, Tom B, Molinuevo JL. The European prevention of Alzheimer's dementia (EPAD) longitudinal cohort study: baseline data release V500.0. J Prev Alzheimers Dis 2020;7(1):8−13.

[11] Ritchie CW, Ritchie K. The PREVENT study: a prospective cohort study to identify midlife biomarkers of late-onset Alzheimer's disease. BMJ Open 2012;2(6):e001893.

[12] Molinuevo JL, Gramunt N, Gispert JD, Fauria K, Esteller M, Minguillon C, et al. The ALFA project: a research platform to identify early pathophysiological features of Alzheimer's disease. Alzheimers Dement (N Y) 2016;2(2):82−92.

[13] Sperling RA, Aisen PS, Beckett LA, Bennett DA, Craft S, Fagan AM, et al. Toward defining the preclinical stages of Alzheimer's disease: recommendations from the National Institute on Aging-Alzheimer's Association workgroups on diagnostic guidelines for Alzheimer's disease. Alzheimers Dement 2011;7(3):280−92.

[14] Milne R, Bunnik E, Diaz A, Richard E, Badger S, Gove D, et al. Perspectives on communicating biomarker-based assessments of Alzheimer's disease to cognitively healthy individuals. J Alzheim Dis 2018;62:487−98.

[15] Milne R, Diaz A, Badger S, Bunnik E, Fauria K, Wells K. At, with and beyond risk: expectations of living with the possibility of future dementia. Sociol Health Illness 2018;40(6):969−87.

[16] Ritchie CW, Molinuevo JL, Truyen L, Satlin A, Van der Geyten S, Lovestone S. Development of interventions for the secondary prevention of Alzheimer's dementia: the European Prevention of Alzheimer's Dementia (EPAD) project. Lancet Psychiatr 2016;3(2):179−86.

[17] Cummings J, Aisen P, Barton R, Bork J, Doody R, Dwyer J, et al. Re-engineering Alzheimer clinical trials: global Alzheimer's platform network. J Prev Alzheimer's Dis 2016;3(2):114−20.

[18] Serrano-Pozo A, Frosch MP, Masliah E, Hyman BT. Neuropathological alterations in Alzheimer disease. Cold Spring Harb Perspect Med 2011;1(1):a006189−a.

[19] Iturria-Medina Y, Sotero RC, Toussaint PJ, Mateos-Pérez JM, Evans AC, Weiner MW, et al. Early role of vascular dysregulation on late-onset Alzheimer's disease based on multifactorial data-driven analysis. Nat Commun 2016;7(1):11934.

[20] Korte N, Nortley R, Attwell D. Cerebral blood flow decrease as an early pathological mechanism in Alzheimer's disease. Acta Neuropathol 2020;140(6):793−810.

[21] Márquez F, Yassa MA. Neuroimaging biomarkers for Alzheimer's disease. Mol Neurodegener 2019;14(1):21.

[22] Braak H, Braak E. Neuropathological staging of Alzheimer-related changes. Acta Neuropathol 1991;82(4):239−59.

[23] Heneka MT, Carson MJ, El Khoury J, Landreth GE, Brosseron F, Feinstein DL, et al. Neuroinflammation in Alzheimer's disease. Lancet Neurol 2015;14(4):388−405.

[24] Kinney JW, Bemiller SM, Murtishaw AS, Leisgang AM, Salazar AM, Lamb BT. Inflammation as a central mechanism in Alzheimer's disease. Alzheimer's Dementia 2018;4:575−90.

[25] Chen MK, Mecca AP, Naganawa M, Finnema SJ, Toyonaga T, Lin SF, et al. Assessing synaptic density in Alzheimer disease with synaptic vesicle glycoprotein 2A positron emission tomographic imaging. JAMA Neurol 2018;75(10):1215−24.

[26] Colom-Cadena M, Spires-Jones T, Zetterberg H, Blennow K, Caggiano A, DeKosky ST, et al. The clinical promise of biomarkers of synapse damage or loss in Alzheimer's disease. Alzheimer's Res Ther 2020;12(1):21.

[27] Calderon-Garcidueñas AL, Duyckaerts C. Alzheimer disease. Handb Clin Neurol 2017;145:325−37.

[28] Forner S, Baglietto-Vargas D, Martini AC, Trujillo-Estrada L, LaFerla FM. Synaptic impairment in Alzheimer's disease: a dysregulated symphony. Trends Neurosci 2017;40(6):347−57.

[29] Nation DA, Sweeney MD, Montagne A, Sagare AP, D'Orazio LM, Pachicano M, et al. Blood-brain barrier breakdown is an early biomarker of human cognitive dysfunction. Nat Med 2019;25(2):270−6.

[30] Sweeney MD, Sagare AP, Zlokovic BV. Blood−brain barrier breakdown in Alzheimer disease and other neurodegenerative disorders. Nat Rev Neurol 2018;14(3):133−50.

[31] Kisler K, Nelson AR, Montagne A, Zlokovic BV. Cerebral blood flow regulation and neurovascular dysfunction in Alzheimer disease. Nat Rev Neurosci 2017;18(7):419−34.

[32] Scheltens P, Blennow K, Breteler MM, de Strooper B, Frisoni GB, Salloway S, et al. Alzheimer's disease. Lancet 2016;388(10043):505−17.
[33] König A, Linz N, Baykara E, Tröger J, Ritchie C, Saunders S, et al. Screening over speech in unselected populations for clinical trials in AD (PROSPECT-AD): study design and protocol. J Prev Alzheimers Dis 2023;10(2):314−21.
[34] Liu C-C, Liu C-C, Kanekiyo T, Xu H, Bu G. Apolipoprotein E and Alzheimer disease: risk, mechanisms and therapy. Nat Rev Neurol 2013;9(2):106−18.
[35] Saddiki H, Fayosse A, Cognat E, Sabia S, Engelborghs S, Wallon D, et al. Age and the association between apolipoprotein E genotype and Alzheimer disease: a cerebrospinal fluid biomarker−based case−control study. PLoS Med 2020;17(8):e1003289.
[36] Qiu C, Kivipelto M, Agüero-Torres H, Winblad B, Fratiglioni L. Risk and protective effects of the APOE gene towards Alzheimer's disease in the Kungsholmen project: variation by age and sex. J Neurol Neurosurg Psychiatr 2004;75(6):828.
[37] Tang M-X, Stern Y, Marder K, Bell K, Gurland B, Lantigua R, et al. The APOE-ε4 allele and the risk of Alzheimer disease among African Americans, whites, and Hispanics. JAMA 1998;279(10):751−5.
[38] Sinha N, Berg CN, Tustison NJ, Shaw A, Hill D, Yassa MA, et al. APOE ε4 status in healthy older African Americans is associated with deficits in pattern separation and hippocampal hyperactivation. Neurobiol Aging 2018;69:221−9.
[39] Khan TK. Chapter 4 - genetic biomarkers in Alzheimer's disease. In: Khan TK, editor. Biomarkers in Alzheimer's disease. Academic Press; 2016. p. 103−35.
[40] Fernandez CG, Hamby ME, McReynolds ML, Ray WJ. The role of APOE4 in disrupting the homeostatic functions of astrocytes and microglia in aging and Alzheimer's disease. Front Aging Neurosci 2019;11(14).
[41] Blennow K, Zetterberg H. Biomarkers for Alzheimer's disease: current status and prospects for the future. J Intern Med 2018;284(6):643−63.
[42] Budelier MM, Bateman RJ. Biomarkers of Alzheimer disease. J Appl Lab Med 2020;5(1):194−208.
[43] Mayeux R. Biomarkers: potential uses and limitations. NeuroRx 2004;1(2):182−8.
[44] Nisticò R, Borg JJ. Aducanumab for Alzheimer's disease: a regulatory perspective. Pharmacol Res 2021;171:105754.
[45] Refusal of marketing authorisation for Aduhelm (aducanumab). 2021.
[46] Cummings J, Apostolova L, Rabinovici GD, Atri A, Aisen P, Greenberg S, et al. Lecanemab: appropriate use recommendations. J Prev Alzheimers Dis 2023;10(3):362−77.
[47] Sims JR, Zimmer JA, Evans CD, Lu M, Ardayfio P, Sparks J, et al. Donanemab in early symptomatic Alzheimer disease: the TRAILBLAZER-ALZ 2 randomized clinical trial. JAMA 2023;330(6):512−27.
[48] Hoffman M. Biogen announces 50% drop in aducanumab pricing amid feedback on costs. 2021.
[49] Anderson M, Sathe N, Polacek C, Vawter J, Fritz T, Mann M, et al. Site readiness framework to improve health system preparedness for a potential new Alzheimer's disease treatment paradigm. J Prev Alzheimers Dis 2022;9(3):542−9.
[50] Cummings JL, Morstorf T, Zhong K. Alzheimer's disease drug-development pipeline: few candidates, frequent failures. Alzheimer's Res Ther 2014;6(4):37.
[51] Cummings J, Morstorf T, Lee G. Alzheimer's drug-development pipeline: 2016. Alzheimer's Dementia: Transl Res Clin Inter 2016;2(4):222−32.

[52] Cummings J, Lee G, Ritter A, Sabbagh M, Zhong K. Alzheimer's disease drug development pipeline: 2020. Alzheimer's Dementia 2020;6(1):e12050.
[53] Cummings J, Lee G, Ritter A, Sabbagh M, Zhong K. Alzheimer's disease drug development pipeline: 2019. Alzheimers Dement (N Y) 2019;5:272−93.
[54] Rosen WG, Mohs RC, Davis KL. A new rating scale for Alzheimer's disease. Am J Psychiatr 1984;141(11):1356−64.
[55] Bucks RS, Ashworth DL, Wilcock GK, Siegfried K. Assessment of activities of daily living in dementia: development of the bristol activities of daily living scale. Age Ageing 1996;25(2):113−20.
[56] Sikkes SAM, Knol DL, Pijnenburg YAL, de Lange-de Klerk ESM, Uitdehaag BMJ, Scheltens P. Validation of the Amsterdam IADL Questionnaire©, a new tool to measure instrumental activities of daily living in dementia. Neuroepidemiology 2013;41(1):35−41.
[57] Francis PT. The interplay of neurotransmitters in Alzheimer's disease. CNS Spectr 2005;10(11 Suppl. 18):6−9.
[58] Ferreira-Vieira TH, Guimaraes IM, Silva FR, Ribeiro FM. Alzheimer's disease: targeting the cholinergic system. Curr Neuropharmacol 2016;14(1):101−15.
[59] Rogers SL, Friedhoff LT. The efficacy and safety of donepezil in patients with Alzheimer's disease: results of a US multicentre, randomized, double-blind, placebo-controlled trial. Dement Geriatr Cognit Disord 1996;7(6):293−303.
[60] Rogers SL, Doody RS, Mohs RC, Friedhoff LT, Group atDS. Donepezil improves cognition and global function in Alzheimer disease: a 15-week, double-blind, placebo-controlled study. Arch Intern Med 1998;158(9):1021−31.
[61] Rogers SL, Doody RS, Pratt RD, Ieni JR. Long-term efficacy and safety of donepezil in the treatment of Alzheimer's disease: final analysis of a US multicentre open-label study. Eur Neuropsychopharmacol 2000;10(3):195−203.
[62] Rogers SL, Farlow MR, Doody RS, Mohs R, Friedhoff LT. A 24-week, double-blind, placebo-controlled trial of donepezil in patients with Alzheimer's disease. Neurology 1998;50(1):136.
[63] Wilcock GK, Lilienfeld S, Gaens E. Efficacy and safety of galantamine in patients with mild to moderate Alzheimer's disease: multicentre randomised controlled trial. Galantamine International-1 Study Group. BMJ 2000;321(7274):1445−9.
[64] Rösler M, Anand R, Cicin-Sain A, Gauthier S, Agid Y, Dal-Bianco P, et al. Efficacy and safety of rivastigmine in patients with Alzheimer's disease: international randomised controlled trial. BMJ 1999;318(7184):633−8.
[65] Howard R, McShane R, Lindesay J, Ritchie C, Baldwin A, Barber R, et al. Donepezil and memantine for moderate-to-severe Alzheimer's disease. N Engl J Med 2012;366(10):893−903.
[66] Knapp M, King D, Romeo R, Adams J, Baldwin A, Ballard C, et al. Cost-effectiveness of donepezil and memantine in moderate to severe Alzheimer's disease (the DOMINO-AD trial). Int J Geriatr Psychiatr 2017;32(12):1205−16.
[67] Parsons C, Lim WY, Loy C, McGuinness B, Passmore P, Ward SA, et al. Withdrawal or continuation of cholinesterase inhibitors or memantine or both, in people with dementia. Cochrane Database Syst Rev 2021;(2).
[68] Birks J, Flicker L. Donepezil for mild cognitive impairment. Cochrane Database Syst Rev 2006;(3).
[69] Zhang X, Lian S, Zhang Y, Zhao Q. Efficacy and safety of donepezil for mild cognitive impairment: a systematic review and meta-analysis. Clin Neurol Neurosurg 2022;213:107134.

[70] Matsunaga S, Fujishiro H, Takechi H. Efficacy and safety of cholinesterase inhibitors for mild cognitive impairment:A systematic review and meta-analysis. J Alzheim Dis 2019;71:513−23.

[71] Hansen RA, Gartlehner G, Webb AP, Morgan LC, Moore CG, Jonas DE. Efficacy and safety of donepezil, galantamine, and rivastigmine for the treatment of Alzheimer's disease: a systematic review and meta-analysis. Clin Interv Aging 2008;3(2):211−25.

[72] Tricco AC, Ashoor HM, Soobiah C, Rios P, Veroniki AA, Hamid JS, et al. Comparative effectiveness and safety of cognitive enhancers for treating Alzheimer's disease: systematic review and network metaanalysis. J Am Geriatr Soc 2018;66(1):170−8.

[73] Meng YH, Wang PP, Song YX, Wang JH. Cholinesterase inhibitors and memantine for Parkinson's disease dementia and Lewy body dementia: a meta-analysis. Exp Ther Med 2019;17(3):1611−24.

[74] Guo J, Wang Z, Liu R, Huang Y, Zhang N, Zhang R. Memantine, donepezil, or combination therapy—what is the best therapy for Alzheimer's disease? A network meta-analysis. Brain Behavior 2020;10(11):e01831.

[75] Montero-Odasso M, Speechley M, Chertkow H, Sarquis-Adamson Y, Wells J, Borrie M, et al. Donepezil for gait and falls in mild cognitive impairment: a randomized controlled trial. Eur J Neurol 2019;26(4):651−9.

[76] Folch J, Busquets O, Ettcheto M, Sánchez-López E, Castro-Torres RD, Verdaguer E, et al. Memantine for the treatment of dementia: a review on its current and future applications. J Alzheim Dis 2018;62:1223−40.

[77] Areosa SA, Sherriff F. Memantine for dementia. Cochrane Database Syst Rev 2003;(3):Cd003154.

[78] Kishi T, Matsunaga S, Oya K, Nomura I, Ikuta T, Iwata N. Memantine for Alzheimer's disease: an updated systematic review and meta-analysis. J Alzheim Dis 2017;60:401−25.

[79] Alimoradzadeh R, Mirmiranpour H, Hashemi P, Pezeshki S, Salehi SS. Effect of memantine on oxidative and antioxidant indexes among elderly patients with prediabetes and mild cognitive impairment. J Neurol Neurophysiol 2019;10(1).

[80] Ma HM, Zafonte RD. Amantadine and memantine: a comprehensive review for acquired brain injury. Brain Inj 2020;34(3):299−315.

[81] Hernandez CM, Dineley KT. α7 nicotinic acetylcholine receptors in Alzheimer's disease: neuroprotective, neurotrophic or both? Curr Drug Targets 2012;13(5):613−22.

[82] Hoskin JL, Al-Hasan Y, Sabbagh MN. Nicotinic acetylcholine receptor agonists for the treatment of Alzheimer's dementia: an update. Nicotine Tob Res 2019;21(3):370−6.

[83] Hopkins TJ, Rupprecht LE, Hayes MR, Blendy JA, Schmidt HD. Galantamine, an acetylcholinesterase inhibitor and positive allosteric modulator of nicotinic acetylcholine receptors, attenuates nicotine taking and seeking in rats. Neuropsychopharmacology 2012;37(10):2310−21.

[84] Potasiewicz A, Krawczyk M, Gzielo K, Popik P, Nikiforuk A. Positive allosteric modulators of alpha 7 nicotinic acetylcholine receptors enhance procognitive effects of conventional anti-Alzheimer drugs in scopolamine-treated rats. Behav Brain Res 2020;385:112547.

[85] Coughlin JM, Rubin LH, Du Y, Rowe SP, Crawford JL, Rosenthal HB, et al. High availability of the α7-nicotinic acetylcholine receptor in brains of individuals with mild cognitive impairment: a pilot study using ^{18}F-ASEM PET. J Nucl Med 2020;61(3):423.

[86] Schedin-Weiss S, Inoue M, Hromadkova L, Teranishi Y, Yamamoto NG, Wiehager B, et al. Monoamine oxidase B is elevated in Alzheimer disease neurons, is associated with

γ-secretase and regulates neuronal amyloid β-peptide levels. Alzheimer's Res Ther 2017;9(1):57.
[87] Binde CD, Tvete IF, Gåsemyr J, Natvig B, Klemp M. A multiple treatment comparison meta-analysis of monoamine oxidase type B inhibitors for Parkinson's disease. Br J Clin Pharmacol 2018;84(9):1917−27.
[88] Birks J, Flicker L. Selegiline for Alzheimer's disease. Cochrane Database Syst Rev 2003;(1):Cd000442.
[89] Schneider LS, Geffen Y, Rabinowitz J, Thomas RG, Schmidt R, Ropele S, et al. Low-dose ladostigil for mild cognitive impairment: a phase 2 placebo-controlled clinical trial. Neurology 2019;93(15):e1474−84.
[90] Nave S, Doody RS, Boada M, Grimmer T, Savola JM, Delmar P, et al. Sembragiline in moderate Alzheimer's disease: results of a randomized, double-blind, placebo-controlled phase II trial (MAyflOwer RoAD). J Alzheimers Dis 2017;58(4):1217−28.
[91] Song IU, Im JJ, Jeong H, Na SH, Chung YA. Possible neuroprotective effects of rasagiline in Alzheimer's disease: a SPECT study. Acta Radiol 2020:284185120940264.
[92] ClinicalTrials.gov. Available from: https://clinicaltrials.gov/ct2/show/NCT02359552.
[93] Ismail Z, Nguyen M-Q, Fischer CE, Schweizer TA, Mulsant BH. Neuroimaging of delusions in Alzheimer's disease. Psychiatr Res Neuroimaging 2012;202(2):89−95.
[94] Sultzer DL, Leskin LP, Melrose RJ, Harwood DG, Narvaez TA, Ando TK, et al. Neurobiology of delusions, memory, and insight in Alzheimer disease. Am J Geriatr Psychiatr 2014;22(11):1346−55.
[95] Csernansky JG, Dong H, Fagan AM, Wang L, Xiong C, Holtzman DM, et al. Plasma cortisol and progression of dementia in subjects with Alzheimer-type dementia. Am J Psychiatr 2006;163(12):2164−9.
[96] Pietrzak RH, Laws SM, Lim YY, Bender SJ, Porter T, Doecke J, et al. Plasma cortisol, brain amyloid-β, and cognitive decline in preclinical Alzheimer's disease: a 6-year prospective cohort study. Biol Psychiatr: Cognitive Neuroscience and Neuroimaging 2017;2(1):45−52.
[97] Li G, Cherrier MM, Tsuang DW, Petrie EC, Colasurdo EA, Craft S, et al. Salivary cortisol and memory function in human aging. Neurobiol Aging 2006;27(11):1705−14.
[98] Karlamangla AS, Singer BH, Chodosh J, McEwen BS, Seeman TE. Urinary cortisol excretion as a predictor of incident cognitive impairment. Neurobiol Aging 2005;26(1, Suppl. ment):80−4.
[99] Udeh-Momoh CT, Su B, Evans S, Zheng B, Sindi S, Tzoulaki I, et al. Cortisol, amyloid-β, and reserve predicts Alzheimer's disease progression for cognitively normal older adults. J Alzheim Dis 2019;70(2):551−60.
[100] Wyrwoll CS, Holmes MC, Seckl JR. 11β-Hydroxysteroid dehydrogenases and the brain: from zero to hero, a decade of progress. Front Neuroendocrinol 2011;32(3):265−86.
[101] Seckl JR, Walker BR. Minireview: 11β-hydroxysteroid dehydrogenase type 1 - a tissue-specific amplifier of glucocorticoid action. Endocrinology 2001;142(4):1371−6.
[102] Tomlinson JW, Stewart PM. Mechanisms of Disease: selective inhibition of 11β-hydroxysteroid dehydrogenase type 1 as a novel treatment for the metabolic syndrome. Nat Clin Pract Endocrinol Metab 2005;1(2):92−9.
[103] Gregory S, Hill D, Grey B, Ketelbey W, Miller T, Muniz-Terrera G, et al. 11β-hydroxysteroid dehydrogenase type 1 inhibitor use in human disease-a systematic review and narrative synthesis. Metabolism 2020;108:154246.

[104] Marek GJ, Katz DA, Meier A, Nt G, Zhang W, Liu W, et al. Efficacy and safety evaluation of HSD-1 inhibitor ABT-384 in Alzheimer's disease. Alzheimers Dement 2014;10(5 Suppl. l):S364—73.

[105] de Quervain DJF, Poirier R, Wollmer MA, Grimaldi LME, Tsolaki M, Streffer JR, et al. Glucocorticoid-related genetic susceptibility for Alzheimer's disease. Hum Mol Genet 2004;13(1):47—52.

[106] Todd WD. Potential pathways for circadian dysfunction and sundowning-related behavioral aggression in Alzheimer's disease and related dementias. Front Neurosci 2020;14(910).

[107] Gehrman P, Marler M, Martin JL, Shochat T, Corey-Bloom J, Ancoli-Israel S. The relationship between dementia severity and rest/activity circadian rhythms. Neuropsychiatric Dis Treat 2005;1(2):155—63.

[108] Zee PC, Vitiello MV. Circadian rhythm sleep disorder: irregular sleep wake rhythm. Sleep Med Clin 2009;4(2):213—8.

[109] Moline M, Thein S, Bsharat M, Rabbee N, Kemethofer-Waliczky M, Filippov G, et al. Safety and efficacy of lemborexant in patients with irregular sleep-wake rhythm disorder and Alzheimer's disease dementia: results from a phase 2 randomized clinical trial. J Prev Alzheimers Dis 2021;8(1):7—18.

[110] Jean-Louis G, von Gizycki H, Zizi F. Melatonin effects on sleep, mood, and cognition in elderly with mild cognitive impairment. J Pineal Res 1998;25(3):177—83.

[111] Cruz-Aguilar MA, Ramírez-Salado I, Guevara MA, Hernández-González M, Benitez-King G. Melatonin effects on EEG activity during sleep onset in mild-to-moderate Alzheimer's disease: a pilot study. J Alzheimer's Dis Rep 2018;2:55—65.

[112] Ju Y-ES, Lucey BP, Holtzman DM. Sleep and Alzheimer disease pathology—a bidirectional relationship. Nat Rev Neurol 2014;10(2):115—9.

[113] Gehrman P, Gooneratne NS, Brewster GS, Richards KC, Karlawish J. Impact of Alzheimer disease patients' sleep disturbances on their caregivers. Geriatr Nurs 2018;39(1):60—5.

[114] van Dyck CH. Anti-Amyloid-β monoclonal antibodies for Alzheimer's disease: pitfalls and promise. Biol Psychiatr 2018;83(4):311—9.

[115] Vandenberghe R, Rinne JO, Boada M, Katayama S, Scheltens P, Vellas B, et al. Bapineuzumab for mild to moderate Alzheimer's disease in two global, randomized, phase 3 trials. Alzheimer's Res Ther 2016;8(1):18.

[116] Dominantly Inherited Alzheimer Network Trial. An Opportunity to Prevent Dementia. A Study of Potential Disease Modifying Treatments in Individuals at Risk for or With a Type of Early Onset Alzheimer's Disease Caused by a Genetic Mutation. Master Protocol DIAN-TU001 (DIAN-TU) 2021. Available from: https://clinicaltrials.gov/ct2/show/NCT01760005?term=solanezumab&draw=3&rank=12.

[117] Clinical Trial of Solanezumab for Older Individuals Who May be at Risk for Memory Loss (A4). 2021. Available from: https://clinicaltrials.gov/ct2/show/NCT02008357?term=solanezumab&draw=2&rank=5.

[118] AlzForum. Available from: https://www.alzforum.org/news/conference-coverage/dian-tu-gantenerumab-brings-down-tau-lot-open-extension-planned.

[119] Aisen PS, Bateman RJ, Carrillo M, Doody R, Johnson K, Sims JR, et al. Platform trials to expedite drug development in Alzheimer's disease: a report from the EU/US CTAD task force. J Prev Alzheimer's Dis 2021;8(3):306—12.

[120] Efficacy and Safety Study of Gantenerumab in Participants With Early Alzheimer's Disease (AD). Available from: https://clinicaltrials.gov/ct2/show/NCT03444870?term=gantenerumab&draw=4&rank=5.

[121] Sevigny J, Chiao P, Bussière T, Weinreb PH, Williams L, Maier M, et al. The antibody aducanumab reduces Aβ plaques in Alzheimer's disease. Nature 2016;537(7618):50−6.

[122] Cummings J, Aisen P, Lemere C, Atri A, Sabbagh M, Salloway S. Aducanumab produced a clinically meaningful benefit in association with amyloid lowering. Alzheimer's Res Ther 2021;13(1):98.

[123] Cummings J, Aisen P, Apostolova LG, Atri A, Salloway S, Weiner M. Aducanumab: appropriate use recommendations. J Prev Alzheimer's Dis 2021;8(4):398−410.

[124] Schneider L. A resurrection of aducanumab for Alzheimer's disease. Lancet Neurol 2020;19(2):111−2.

[125] Howard R, Liu KY. Questions EMERGE as Biogen claims aducanumab turnaround. Nat Rev Neurol 2020;16(2):63−4.

[126] Selkoe DJ. Alzheimer disease and aducanumab: adjusting our approach. Nat Rev Neurol 2019;15(7):365−6.

[127] Grill JD, Karlawish J. Implications of FDA approval of a first disease-modifying therapy for a neurodegenerative disease on the design of subsequent clinical trials. Neurology 2021. https://doi.org/10.1212/WNL.0000000000012329.

[128] Swanson CJ, Zhang Y, Dhadda S, Wang J, Kaplow J, Lai RYK, et al. A randomized, double-blind, phase 2b proof-of-concept clinical trial in early Alzheimer's disease with lecanemab, an anti-Aβ protofibril antibody. Alzheimer's Res Ther 2021;13(1):80.

[129] AHEAD 3-45 Study: A Study to Evaluate Efficacy and Safety of Treatment With Lecanemab in Participants With Preclinical Alzheimer's Disease and Elevated Amyloid and Also in Participants With Early Preclinical Alzheimer's Disease and Intermediate Amyloid. Available from: https://clinicaltrials.gov/ct2/show/NCT04468659?term=lecanemab&draw=2&rank=1.

[130] A Study to Confirm Safety and Efficacy of Lecanemab in Participants With Early Alzheimer's Disease (Clarity AD). Available from: https://clinicaltrials.gov/ct2/show/NCT03887455?term=lecanemab&draw=2&rank=2.

[131] van Dyck Christopher H, Swanson CJ, Aisen P, Bateman Randall J, Chen C, Gee M, et al. Lecanemab in early Alzheimer's disease. N Engl J Med 2023;388(1):9−21.

[132] Lowe SL, Willis BA, Hawdon A, Natanegara F, Chua L, Foster J, et al. Donanemab (LY3002813) dose-escalation study in Alzheimer's disease. Alzheimer's Dementia 2021;7(1):e12112.

[133] Mintun MA, Lo AC, Duggan Evans C, Wessels AM, Ardayfio PA, Andersen SW, et al. Donanemab in early Alzheimer's disease. N Engl J Med 2021;384:1691−704.

[134] Congdon EE, Sigurdsson EM. Tau-targeting therapies for Alzheimer disease. Nat Rev Neurol 2018;14(7):399−415.

[135] Novak P, Kontsekova E, Zilka N, Novak M. Ten years of tau-targeted immunotherapy: the path walked and the roads ahead. Front Neurosci 2018;12(798).

[136] Alam R, Driver D, Wu S, Lozano E, Key SL, Hole JT, et al. [O2−14−05]: preclinical characterization of an antibody [LY3303560] targeting aggregated tau. Alzheimer's Dementia 2017;13(7S_Part_11):P592−3.

[137] Vander Zanden CM, Chi EY. Passive immunotherapies targeting amyloid beta and tau oligomers in Alzheimer's disease. J Pharmaceut Sci 2020;109(1):68−73.

[138] Kerchner GA, Ayalon G, Brunstein F, Chandra P, Datwani A, Fuji RN, et al. [O2−17−03]: a phase I study to evaluate the safety and tolerability of RO7105705 in healthy volunteers and patients with mild-to-moderate AD. Alzheimer's Dementia 2017;13(7S_Part_12):P601−P.

[139] Gauthier S, Feldman HH, Schneider LS, Wilcock GK, Frisoni GB, Hardlund JH, et al. Efficacy and safety of tau-aggregation inhibitor therapy in patients with mild or moderate Alzheimer's disease: a randomised, controlled, double-blind, parallel-arm, phase 3 trial. Lancet 2016;388(10062):2873−84.

[140] Karikari TK, Pascoal TA, Ashton NJ, Janelidze S, Benedet AL, Rodriguez JL, et al. Blood phosphorylated tau 181 as a biomarker for Alzheimer's disease: a diagnostic performance and prediction modelling study using data from four prospective cohorts. Lancet Neurol 2020;19(5):422−33.

[141] Leuzy A, Cullen NC, Mattsson-Carlgren N, Hansson O. Current advances in plasma and cerebrospinal fluid biomarkers in Alzheimer's disease. Curr Opin Neurol 2021;34(2):266−74.

[142] Macfarlane S, Cecchi M, Moore D, Maruff P, Capiak KM, Missling CU. Safety and efficacy 31 Week data of Anavex 2-73 in a phase 2a study in mild-to-moderate Alzheimer's disease patients. Insight 2016;3:11.

[143] Alam J, Blackburn K, Neflamapimod PD. Clinical phase 2b-ready oral small molecule inhibitor of p38alpha to reverse synaptic dysfunction in early Alzheimer's disease. J Prev Alzheimers Dis 2017;4:273−8.

[144] Koenig AM, Mechanic-Hamilton D, Xie SX, Combs MF, Cappola AR, Xie L, et al. Effects of the insulin sensitizer metformin in Alzheimer disease: pilot data from a randomized placebo-controlled crossover study. Alzheimer Dis Assoc Disord 2017;31(2):107−13.

[145] Maes T, Molinero C, Antonijoan RM, Ferrero-Cafiero JM, Martínez-Colomer J, Mascaro C, et al. [P4−576]: first-in-human phase I results show safety, tolerability and brain penetrance of ORY-2001, an epigenetic drug targeting LSD1 and MAO-B. Alzheimer's Dementia 2017;13(7S_Part_32):P1573−4.

[146] Maccecchini ML, Chang MY, Pan C, John V, Zetterberg H, Greig NH. Posiphen as a candidate drug to lower CSF amyloid precursor protein, amyloid-β peptide and τ levels: target engagement, tolerability and pharmacokinetics in humans. J Neurol Neurosurg Psychiatr 2012;83(9):894−902.

[147] Turner RS, Hebron ML, Lawler A, Mundel EE, Yusuf N, Starr JN, et al. Nilotinib effects on safety, tolerability, and biomarkers in Alzheimer's disease. Ann Neurol 2020;88(1):183−94.

[148] Webster L, Groskreutz D, Grinbergs-Saull A, Howard R, O'Brien JT, Mountain G, et al. Core outcome measures for interventions to prevent or slow the progress of dementia for people living with mild to moderate dementia: systematic review and consensus recommendations. PLoS One 2017;12(6):e0179521−e.

[149] FDA UFaDA. Early Alzheimer's disease: developing drugs for treatment guidance for industry. 2018.

[150] Ema EMA. Guideline on the clinical investigation of medicines for the treatment of Alzheimer's disease. In: Committee for Medicinal Products for Human Use (CHMP); 2018. p. 1−36.

[151] Hughes CP, Berg L, Danziger WL, Coben LA, Martin RL. A new clinical scale for the staging of dementia. Br J Psychiatry 1982;140:566−72.

[152] Saunders S, Muniz-Terrera G, Watson J, Clarke CL, Luz S, Evans AR, et al. Participant outcomes and preferences in Alzheimer's disease clinical trials: the electronic Person-Specific Outcome Measure (ePSOM) development program. Alzheimers Dement (N Y) 2018;4:694−702.

[153] Saunders S, Muniz-Terrera G, Sheehan S, Ritchie CW, Luz S. A UK-wide study employing Natural Language Processing to determine what matters to people about brain health to improve drug development: the electronic person-specific outcome measure (ePSOM) programme. J Prev Alzheimer's Dis 2021;8:448−56.

[154] Chan D, Gallaher LM, Moodley K, Minati L, Burgess N, Hartley T. The 4 mountains test: a short test of spatial memory with high sensitivity for the diagnosis of pre-dementia Alzheimer's disease. J Vis Exp 2016;116.

[155] Tu S, Spiers HJ, Hodges JR, Piguet O, Hornberger M. Egocentric versus allocentric spatial memory in behavioral variant frontotemporal dementia and Alzheimer's disease. J Alzheimers Dis 2017;59(3):883−92.

[156] Ritchie K, Carrière I, Howett D, Su L, Hornberger M, O'Brien JT, et al. Allocentric and egocentric spatial processing in middle-aged adults at high risk of late-onset Alzheimer's disease: the PREVENT dementia study. J Alzheim Dis 2018;65:885−96.

[157] Bayat S, Babulal GM, Schindler SE, Fagan AM, Morris JC, Mihailidis A, et al. GPS driving: a digital biomarker for preclinical Alzheimer disease. Alzheimer's Res Ther 2021;13(1):115.

[158] Abdalla M, Rudzicz F, Hirst G. Rhetorical structure and Alzheimer's disease. Aphasiology 2018;32(1):41−60.

[159] de la Fuente Garcia S, Ritchie CW, Luz S. Artificial intelligence, speech, and language Processing approaches to monitoring Alzheimer's disease: a systematic review. J Alzheim Dis 2020;78:1547−74.

[160] Lim YY, Maruff P, Pietrzak RH, Ames D, Ellis KA, Harrington K, et al. Effect of amyloid on memory and non-memory decline from preclinical to clinical Alzheimer's disease. Brain 2014;137(Pt 1):221−31.

[161] Wilcockson TDW, Mardanbegi D, Xia B, Taylor S, Sawyer P, Gellersen HW, et al. Abnormalities of saccadic eye movements in dementia due to Alzheimer's disease and mild cognitive impairment. Aging 2019;11(15):5389−98.

[162] Davis R, Sikorskii A. Eye tracking analysis of visual cues during wayfinding in early stage Alzheimer's disease. Dement Geriatr Cognit Disord 2020;49(1):91−7.

[163] Opwonya J, Doan DNT, Kim SG, Kim JI, Ku B, Kim S, et al. Saccadic eye movement in mild cognitive impairment and Alzheimer's disease: a systematic review and meta-analysis. Neuropsychol Rev 2021;32(2):193−227.

[164] Öhman F, Hassenstab J, Berron D, Schöll M, Papp KV. Current advances in digital cognitive assessment for preclinical Alzheimer's disease. Alzheimers Dement (Amst). 2021;13(1):e12217−e.

[165] Souillard-Mandar W, Penney D, Schaible B, Pascual-Leone A, Au R, Davis R. DCTclock: clinically-interpretable and automated artificial intelligence analysis of drawing behavior for capturing cognition. Front Digit Health 2021;3:750661.

[166] Kawas CH, Kim RC, Sonnen JA, Bullain SS, Trieu T, Corrada MM. Multiple pathologies are common and related to dementia in the oldest-old: the 90+ Study. Neurology 2015;85(6):535−42.

[167] Kapasi A, DeCarli C, Schneider JA. Impact of multiple pathologies on the threshold for clinically overt dementia. Acta Neuropathol 2017;134(2):171−86.

[168] Zetterberg H, Blennow K. Moving fluid biomarkers for Alzheimer's disease from research tools to routine clinical diagnostics. Mol Neurodegener 2021;16(1):1—7.

[169] Zetterberg H. Blood-based biomarkers for Alzheimer's disease—an update. J Neurosci Methods 2019;319:2—6.

[170] Ritchie CW, Waymont JMJ, Pennington C, Draper K, Borthwick A, Fullerton N, et al. The Scottish brain health service model: rationale and scientific basis for a national care pathway of brain health services in Scotland. J Prev Alzheimer's Dis 2021;9(2):348—58.

[171] Lu C-H, Yang C-Y, Li C-Y, Hsieh C-Y, Ou H-T. Lower risk of dementia with pioglitazone, compared with other second-line treatments, in metformin-based dual therapy: a population-based longitudinal study. Diabetologia 2018;61(3):562—73.

[172] Burns DK, Chiang C, Welsh-Bohmer KA, Brannan SK, Culp M, O'Neil J, et al. The TOMMORROW study: design of an Alzheimer's disease delay-of-onset clinical trial. Alzheimer's Dementia 2019;5:661—70.

[173] ClinicalTrials.gov. Available from: https://clinicaltrials.gov/ct2/show/results/NCT01931566?term=TOMMORROW&draw=2&rank=1.

[174] Berry SM, Connor JT, Lewis RJ. The platform trial: an efficient strategy for evaluating multiple treatments. JAMA 2015;313(16):1619—20.

[175] Bateman RJ, Benzinger TL, Berry S, Clifford DB, Duggan C, Fagan AM, et al. The DIAN-TU Next Generation Alzheimer's prevention trial: adaptive design and disease progression model. Alzheimers Dement 2017;13(1):8—19.

[176] Rohrer JD, Nicholas JM, Cash DM, van Swieten J, Dopper E, Jiskoot L, et al. Presymptomatic cognitive and neuroanatomical changes in genetic frontotemporal dementia in the Genetic Frontotemporal dementia Initiative (GENFI) study: a cross-sectional analysis. Lancet Neurol 2015;14(3):253—62.

[177] Adams NE, Hughes LE, Phillips HN, Shaw AD, Murley AG, Nesbitt D, et al. GABA-Ergic dynamics in human frontotemporal networks confirmed by pharmaco-magnetoencephalography. J Neurosci 2020;40(8):1640—9.

[178] Murley AG, Rouse MA, Jones PS, Ye R, Hezemans FH, O'Callaghan C, et al. GABA and glutamate deficits from frontotemporal lobar degeneration are associated with disinhibition. Brain 2020;143(11):3449—62.

[179] Stanley K, Whitfield T, Kuchenbaecker K, Sanders O, Stevens T, Walker Z. Rate of cognitive decline in Alzheimer's disease stratified by age. J Alzheimers Dis 2019;69(4):1153—60.

[180] Inoue J, Hoshino R, Nojima H, Ishida W, Okamoto N. Investigation of responders and non-responders to long-term donepezil treatment. Psychogeriatrics 2010;10(2):53—61.

Chapter 8a(ii)

Non-pharmacological interventions

Jill G. Rasmussen[a], Shireen Sindi[b,c] and Miia Kivipelto[b,c,d,e]

[a]Psi-napse, Surrey, United Kingdom; [b]Division of Clinical Geriatrics, Center for Alzheimer Research, Department of Neurobiology, Care Sciences and Society (NVS), Karolinska Institutet, Stockholm, Sweden; [c]Aging Epidemiology (AGE) Research Unit, School of Public Health, Imperial College London, London, United Kingdom; [d]Institute of Public Health and Clinical Nutrition, University of Eastern Finland, Kuopio, Finland; [e]Theme Inflammation and Aging, Karolinska University Hospital, Stockholm, Sweden

Due to medical successes and advancements, there has been a decline in death rates at younger ages and consequently increase in longevity. Older adults are living healthier and more active lives and are contributing to society until an advanced age. However, aging populations pose medical and societal challenges due to degenerative conditions, comorbidities, and polypharmacy, frailty and dementia. Although the actual number of people with dementia is increasing (because of increased longevity), the age-specific incidence of dementia has fallen in several countries [1]. This is thought to be the result of improvements in education, nutrition, health care, and lifestyle.

Around 50 million people worldwide are living with dementia, and as the global population is aging, this number is estimated to increase to 152 million by 2050, accompanied by major societal and economic challenges [2]. The current global costs of dementia are more than US$ 1 trillion annually [2]. During the current ongoing COVID-19 pandemic, older adults have been particularly susceptible to the effects of COVID-19 with regards to the severity of symptoms, need for hospitalization and death [3,4]. Nevertheless, statistics from England (March 2021), where the COVID-19 vaccination rollout has been efficient, with high uptake levels among those aged 50+ years, showed that dementia and Alzheimer's disease (AD) were the leading cause of death (approximately 10.1% of registered deaths in March) [5]. This is in contrast to previous months when COVID-19 was the leading cause of death [5]. In the near future, these trends may be observed in other countries as vaccine programmes progress. Moreover, independent of age, dementia and Alzheimer's disease are risk factors for disease severity and death due to COVID-19 [6]. This further highlights the importance of dementia and AD in this age group during the pandemic circumstances.

Dementia and late-onset AD are multifactorial conditions, with various demographic, lifestyle, vascular, metabolic and environmental risk factors. Importantly, the literature has identified various risk factors that play a role throughout the life course. The most recent *Lancet Commission report on dementia prevention, intervention, and care* has suggested that the modification of 12 risk factors may prevent or delay 40% or more of dementias [7]. The identified risk factors were, early in life: low levels of education; in midlife: hypertension, traumatic brain injury, excessive alcohol consumption, hearing impairment, and obesity, while in late life: depression, smoking, social isolation, physical inactivity, diabetes and air pollution. This evidence highlights that different factors increase the risk for dementia during different life stages [7]. Other novel factors (e.g., sleep) were highlighted in the report, but more evidence is needed before they are added to the list of risk factors. Addressing these factors at a population level should be the objective of national public health strategies whilst the evaluation of these factors at an individual level should be the basis of a plan for non-pharmacological interventions for those at risk of dementia or with Mild Cognitive Impairment (MCI) or early/prodromal AD.

The World Health Organization (WHO) published the first guidelines for "*Risk reduction of cognitive decline and dementia*" in 2019 [8]. The Guidelines evaluate the quality of the evidence for individual interventions to decrease cognitive decline and the risk for dementia, and makes recommendations based on their value. The strongest recommendations are for physical exercise, smoking cessation, a balanced diet and management of diabetes mellitus. Conditional recommendations are made for cognitive interventions, Mediterranean diet and decreasing alcohol. Although there was a lack of evidence for the value of social activity in risk reduction, the guidelines recognized the contribution of social participation and support to overall health and well-being. Based on their review of the evidence, diet supplementation with vitamins and polyunsaturated fatty acids is not recommended. The WHO guidelines focused on individual risk and protective factors, while it did not consider multi-domain interventions.

In the absence of disease modifying drugs, it is important to simultaneously address modifiable lifestyle, vascular and psychosocial risk factors for early dementia, and to integrate them into preventive lifestyle interventions. The main objectives/aims of non-pharmacological interventions for people with

MCI and early AD are to reduce the rate of progression of disease and maintain quality of life by:

► Mitigating risk factors and enhancing protective factors
► Maintaining cognitive abilities and cognitive reserve
► Optimizing physical and psychological health
► Complementing pharmacological interventions

Overview of risk and protective factors

An overview of risk and protective factors for cognitive decline and dementia and the optimal time for intervention is shown in Fig. 1 below.

Identification of "at risk" individuals

Individuals who are at increased risk for dementia, and who can benefit from non-pharmacological interventions can be identified from their medical history, investigations in primary and/or secondary care and calculation of an individualized "dementia risk score" (Table 1).

A validated instrument to estimate an individual risk score is the Cardiovascular Risk Factors, Aging and Dementia (CAIDE) risk score [9]. The

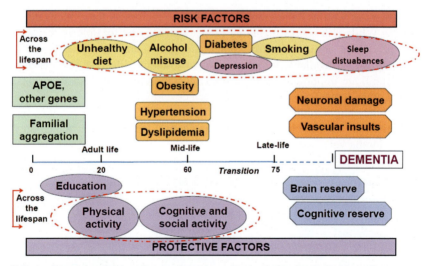

FIG. 1 Overview of risk and protective factors for dementia. *Credit: Evidence from recent epidemiological studies have linked several modifiable risk and protective factors to AD and dementia. Thus, besides genetic factors, exposure to various environmental and lifestyle risk factors during the whole life course influences the risk for disease and its timing. For multifactorial conditions, there is a need to target several risk factors simultaneously for optimal preventive effects.*

TABLE 1 Factors predictive of dementia.

Medical history	Investigations
Co-morbid conditions: diabetes, hypertension, history of depression, delirium, excess alcohol, frailty, smoking family history of dementia	ApoE4 status MCI subtype: Amnestic or multidomain 0.5 SD from expected ability for function or neuropsychological testing Neuroimaging/CSF: White matter hyperintensities, hippocampal and/or global cerebral atrophy and ventricular enlargement; abnormal brain $A\beta_{1-42}$, Tau on PET or CSF

CAIDE score identifies demographic and modifiable vascular risk factors in midlife that are highly predictive of the likelihood of developing dementia 20 years later. The tool uses parameters that are readily available in primary care, and it has been externally validated in a multi-ethnic population [10]. The addition of other parameters associated with dementia risk (obesity, smoking, pulmonary function, and depression) did not improve predictability; however, the inclusion of diabetes improved the predictability score for Asian participants, but not for black or white participants [10] (Fig. 2).

In the future, assessments of "risk" are likely to include Polygenic Risk, biomarkers as well as demographic and lifestyle factors as those included in CAIDE. While risk scores including biomarkers may be feasible in research and specialized clinical settings, pragmatic risk scores (such as CAIDE) including widely available information (demographics, lifestyle and vascular risk factors) are important for wider implementation in primary care settings. Several other dementia risk scores have been developed [11], and they differ in whether they target midlife or late-life risk factors, the prediction duration, the proportions of modifiable or non-modifiable risk factors and whether they include biomarkers.

Non-pharmacological management strategies

Lifestyle interventions

Diet

It is generally acknowledged that nutrition can influence the risk for dementia and AD and the rate of progression [12]. Adoption of a Mediterranean diet (low meat and dairy; high fruit, vegetables, fish) reduces vascular risk factors, the risk for MCI and the risk for MCI conversion to AD dementia [7,13]. The mechanism is thought to be related to decreases in plasma glucose, serum

Factors		Points
Age years	< 47	0
	47-53	3
	>53	4
Education years	≥10	0
	7-9	2
	0-6	3
Sex	Women	0
	Men	1
Systolic BP mmHg	≤ 140	0
	> 140	2
BMI kg/m^2	≤ 30	0
	> 30	2
Total cholesterol mmol/l	≤ 6.5	0
	> 6.5	2
Physical activity	Active	0
	Inactive	1

FIG. 2 Risk factors included in the CAIDE dementia risk score.

insulin concentrations, insulin resistance, and markers of oxidative stress and inflammation [7].

The association between nutrition, diet and healthy brain aging has been the basis for the investigation of supplements to improve cognitive function in patients with MCI or AD dementia. Clinical trial evidence to date does not support the use of single-agent nutrients (Vitamins B, C, D or E, flavonoids, carotenoids, omega-3 fatty acids) to modify the course of cognitive decline in patients with MCI or early AD [14]. However, a broader nutritional approach combining multiple nutrition factors (such as the product Fortasyn Connect) has shown promise in early AD [15]. The constituents of Fortasyn Connect are a mixture of precursors and cofactors (long-chain omega-3 fatty acids, uridine, choline, B vitamins, vitamin C, vitamin E, and selenium) that were selected because of their profile of biological and neuroprotective properties relevant to early AD and the function of neuronal membranes and synapses [15]. Evidence from randomized controlled trials showed improvements in memory in mild AD, decrease in hippocampal atrophy and improvement in Clinical Dementia Rating-Sum of Boxes (CDR-SB) and AD Composite Score (ADCOMS) in prodromal AD [16,17]. This profile may support the use of Souvenaid in the management of patients with early AD, including MCI.

Exercise

A meta-analysis of 19 controlled studies with 23 interventions that included an exercise-only intervention compared with a non-diet, non-exercise control group and assessed cognitive function suggests that exercise, particularly aerobic exercise, delays the decline in cognitive function in people at risk for AD or have AD [18]. Exercise should be of moderate intensity, performed approximately 3 days per week for about 45 min per session.

Multi-domain interventions

Important evidence for the benefits of multidomain non-pharmacological interventions in "at risk" populations comes from the *Finnish Geriatric Intervention Study to Prevent Cognitive Impairment and Disability* (FINGER) study. This demonstrated the benefits of a 2-year, intensive multidomain intervention on global cognition, executive function and processing speed in older adults aged 60—77 years who were at increased risk of dementia [19]. The intervention comprised of dietary counseling, individualized physical exercise program, cognitive training, and management of metabolic and vascular risk factors.

Another two large multidomain lifestyle-based intervention studies have been conducted, the Multidomain Alzheimer Preventive Trial (MAPT) [20], and Prevention of Dementia by Intensive Vascular Care (PreDIVA) trial [21] (See Ref. [22]) (Table 2).

There have been important lessons learned from these trials:

- Interventions need to be targeted to "at risk" individuals i.e., right population at the right time and individualized to their risk profiles.
- Importance of starting interventions early enough.
- The content of the interventions is critical; exercise must be of sufficient intensity, duration and frequency, and all intervention components need to be accompanied by motivational support and coaching.
- Especially among older adults, social interactions are important for maintaining motivation levels and providing additional cognitive stimulation.

Psychological symptoms

Depression, particularly late-onset, is associated with increased risk of dementia, and depressive symptoms are often found in the prodrome and early stages of dementia [23]. Other psychological symptoms also have a negative impact among individuals with MCI or early AD dementia. Apathy, which occurs frequently in association with depression in AD, has a negative impact on caregiver burden, activities of daily living and morbidity [24]. In people with MCI it is associated with increased risk of conversion to AD dementia

TABLE 2 Key features and results of multi-domain intervention studies.

Study (country)	Multidomain intervention	Participants	Outcome	Primary outcome results
FINGER (Finland)	**Intervention components:** Diet, physical activity, cognitive training management, vascular risk **Control group:** Regular health advice. **Duration:** 2 years. **Additional follow-up:** 7 and 10 years	N = 1260 participants **Age:** 60–77 years **Randomization:** Individual **Characteristics:** Elevated dementia risk based on CAIDE score ≥6 points, and average cognitive function or slightly below average. Participants were recruited from previous population-based national surveys.	Cognition measured using NTB (a composite measure of 14 standard cognitive tasks)	FINGER Intervention had beneficial effect on NTB: Difference in NTB change was 0.022 ($P = .030$) per year between the intervention and control groups
MAPT (France)	**Intervention components:** Multidomain Intervention (including diet, physical activity, cognitive training, preventive consultation) + omega-3 PUFAs versus multidomain versus omega-3 PUFAs **Control group:** Omega-3 PUFAs placebo capsule **Duration:** 3 years	N = 1680 participants **Age:** ≥70 years **Randomization:** Individual **Characteristics:** With memory complaints, IADL limitation or slow gait speed. Recruited using multiple strategies including advertisements and patient databases.	Cognition measured using a composite Z score that combines four cognitive tests	No significant difference between the three intervention groups and placebo. Between-group difference = 0.093 ($P = .142$) for multidomain + PUFA, 0.079 ($P = .179$) for multidomain and 0.011 ($P = .812$) for PUFA compared with placebo
PreDIVA (The Netherlands)	**Intervention components:** Advice on multidomain cardiovascular factors **Control group:** Usual care **Duration:** 6 years	N = 3526 participants. **Age:** 70–78 years. **Randomization:** Cluster randomization of 116 general practice clinics **Characteristics:** Recruited from general practice clinics.	Incidence of dementia	No effect intervention effect on dementia: HR 0.92 ($P = .54$)

[24]. Such evidence highlights the importance of evidence-based interventions for depression, such as cognitive behavioral therapy (CBT) [25].

Depression and anxiety are common in people with MCI and early AD dementia, and psychological therapies (CBT, interpersonal therapy, counseling, or multimodal interventions including a specific psychological therapy) were effective in reducing depressive and anxiety symptoms [23,26]. Mild/moderate depression may improve without specific treatment. However, there is benefit in identifying and addressing predisposing situational factors such as:

- Loneliness.
- Under-stimulation from lack of activity.
- Being cared for by a depressed carer.
- Improve communication - treat sensory impairment (auditory, visual).

For those people who do not respond to these interventions, behavioral activation can be considered [7,27].

Sleep

Sleep has an important role in the maintenance, disease prevention, repair, and restoration of physical and cognitive functions. More recently it has been recognized that sleep disturbances (circadian rhythm disturbance, altered duration, insomnia, poor quality, sleep apnea) are associated with an increased risk for all-cause dementia [28]. Both sleep disturbance and dementia are common in the elderly. Evidence from a review and meta-analysis found that participants who reported sleep disturbances had a higher risk of incident all-cause dementia, AD, and vascular dementia [28]. Insomnia was associated with incident AD, while sleep disordered breathing was also a risk factor of all-cause dementia, AD, and vascular dementia [28]. Sleep disturbances may be improved through lifestyle changes (sleepy hygiene), exercise and CBT for insomnia (CBTi) [29].

It is important to identify conditions that are associated with sleep disturbances and sleep disorders, as they are significant risk factors for cognitive decline and dementia [7,30], and may benefit from various targeted interventions.

Factors associated with sleep disturbances are:

- Circadian rhythm disturbance.
- Obstructive Sleep Apnea.
- Pain, physical health conditions, anxiety, lack of activity.
- Lifestyle factors contributing to poor sleep hygiene such as excessive alcohol. consumption, late heavy meals, and excessive/late caffeine consumption.

Interventions that may benefit sleep are:

- Exercise.
- Low-intensity physical/mental activities can improve self-reported sleep quality.
- Bright light—helpful in circadian rhythm disorders although more data are required about timing, mode of delivery, wavelength of light.
- CBT—helpful in reducing anxiety, and CBTi which is specifically tailored for insomnia.
- Educating carers and patients about the value of improving and maintaining good sleeping patterns.

Hearing impairment

The recent Lancet Commission on Dementia highlighted hearing impairment is an established risk factor for dementia [7]; midlife hearing impairment is associated with greater volume loss of the temporal lobe and associated structures (the hippocampus and entorhinal cortex) [7,31]. The use of hearing aids appears to mitigate the excess risk of this impairment.

People at risk of impairment from excessive noise exposure should use protection, and those with a known impairment should be encouraged to use hearing aids. All older adults should have regular checks for early identification of hearing loss, including easily remedied causes such as ear wax.

Cognitive training/stimulation

There is mixed evidence for the value of cognitive interventions or cognitive treatments in early AD/MCI due to issues of quality, heterogeneity and lack of long-term data [7]. The only randomized control trial of behavioral activation compared with supportive therapy in people with amnestic MCI found a benefit on memory decline after 2 years [32]. A review on interventions aiming to prevent cognitive decline reported that for cognitive training the strength of the evidence was moderate; cognitive training improves performance in the cognitive domains that were targeted by interventions (e.g., memory, processing speed, reasoning), but these benefits did not transfer to other non-targeted cognitive domains [33]. Moreover, the benefits did not extend beyond 2 years, nor was it beneficial among individuals who already have Alzheimer's type dementia [33].

The evidence for the benefit of non-pharmacological management strategies for people with MCI and early AD is strongest for:

- Lifestyle—includes diet, alcohol use, nutrition, physical exercise, smoking.
- Depression.
- Sleep.

Role of health and care professionals

Health and care professionals have an important role in educating people about the importance of good control of long-term conditions, and appreciation of the importance of compliance with management recommendations and medications.

The following guidelines/suggestions promote healthy lifestyle:

Lifestyle factors (non-pharmacological) [34]:

- **Alcohol:** intake should be limited to 14 units per week spread over two to 3 days. For people who misuse alcohol, interventions should be offered to achieve abstinence or moderate drinking as appropriate and prevent relapse, in community-based settings. All interventions should be the subject of routine outcome monitoring and should inform decisions about ongoing psychological and pharmacological treatments [35].
- **Smoking:** The most effective way to quit smoking is with expert behavioral support from local stop smoking services combined with stop smoking aids [36].
- **Healthy diet and weight management**: Diet should include five portions of different fruit and vegetables per day, oily fish twice per week; sugar and salt should be kept to a minimum. Weight management strategies should combine healthy diet, physical activity and motivational support.
- **Physical activity:** Activity should consist of aerobic, resistance and flexibility elements, including 75 min of vigorous exercise [37].
- **Social Engagement:** Keeping in contact with friends/family, joining a community group or doing some volunteering can promote social engagement.
- **Mental Stimulation:** This includes doing puzzles, learning a new language, reading, playing card or board games, using electronic games available from the App store, or other activities that offer mental stimulation.

Conclusions

Dementia and late-onset AD are heterogenous and multifactorial conditions, with various metabolic, lifestyle, mental health and environmental risk factors. Importantly, most of the risk factors, as those listed in the Lancet Commission, are modifiable factors. With no disease modifying treatments for AD, early modification and targeting of these risk factors is crucial, both at the individual and societal level, in order to prevent or delay the onset of these conditions.

Evidence from the former generation of unimodal short-term and small clinical trials, which focused on individual lifestyle factors (e.g., diet, exercise, cognitive training) generally had moderate quality of evidence [8], although more recent trials are addressing limitations that were found in previous trials (inadequate sample size/trial duration). The WHO Guidelines focused on such individual factors, and provided strong recommendations for physical exercise, smoking cessation, a healthy balanced diet and diabetes management [8]. The assessment of individuals at increased risk for dementia can be carried out through the use of dementia risk scores, which quantify the risk and can be used as an educational and motivational tool. While some risk factors focus on midlife risk factors for long-term prediction of dementia (e.g., 20 years when using the CAIDE dementia risk score), others are developed for late-life risk factors, and for short-term prediction [3–10] years, such as the Australian National University Alzheimer's Disease Risk Index (ANU-ADRI) (predicts dementia up to 6 years). Risk scores also differ in the extent to which they target modifiable factors. For example, the Lifestyle for Brain Health (LIBRA) risk score only focuses on modifiable factors, whereas others include genetics, age and other non-modifiable factors [38]. More recently, dementia risk scores have been proposed as surrogate outcomes that may reflect the intervention response [39].

Recent multidomain lifestyle interventions (FINGER, PreVDIVA, MAPT) have demonstrated the importance of selecting individuals who are at increased dementia risk for participation in the trial. For example, the FINGER trial used the CAIDE dementia risk score selected participants at higher risk for dementia, while PreDIVA showed in subgroup analyses that participants with untreated baseline hypertension showed most benefit. Similarly, in the MAPT trial in France, the intervention was effective among those with an increased risk for dementia at baseline (CAIDE score ≥ 6) or have elevated beta-amyloid in the brain [20]. In the PreDIVA trial in the Netherlands, lower dementia incidence was found among participants with untreated hypertension at baseline, and who were then prescribed treatments for their hypertension [21]. Collectively, these trials highlight the importance of selecting individuals at increased risk for dementia.

These trials have also demonstrated the value in simultaneously targeting multiple risk factors in order to prevent or postpone multifactorial conditions such as dementia and AD. By targeting lifestyle factors such as diet and exercise, and managing vascular and metabolic risk factors, such interventions also have the potential to reduce other health outcomes. For example, in the FINGER trial, those in the intervention group developed fewer chronic diseases during the 2-year trial, compared to the control group [40]. The trial also showed benefits for health-related quality of life [41] and daily functioning [42]. The Healthy aging through internet counseling in the elderly (HATICE) trial, which also simultaneously targeted multiple lifestyle factors improved cardiovascular risk factors including systolic blood pressure, body mass index and low density lipoprotein (LDL) cholesterol.

Most if not all of the risk factors summarized in this chapter tend to be chronic, and interventions are most optimal when they commence early, prior to negative physiological, neurological and health outcomes occur. For effective results, interventions also need to be of sufficient duration and intensity. As shown with the aforementioned multidomain lifestyle interventions, whereas lifestyle advice may be insufficient, coaching and more frequent visits with health professionals and research nurses may increase the level of engagement and adherence to the interventions. Internet counseling - as that used in the HATICE trial - is also providing a promising tool that allows older adults to receive lifestyle interventions at the population level, while reducing costs and burdens on participants. Various mobile applications have been developed for specific lifestyle factors, and some are based on RCT evidence, such as the digital CBT App for insomnia *Sleepio* (also used by the National Health Service (NHS), England) [43,44], and the *Sleep Healthy Using the Internet (SHUTi)* [45]. It will be important for future evidence-based eHealth tools to integrate multiple lifestyle factors, similar to the multidomain trials.

While the multidomain trials to date have been conducted in high-income countries in Europe, it is important to test these interventions in middle- and low-income countries, while adapting the interventions to local settings. This is the goal of the World-Wide FINGERS network [46], which was launched in 2017. The aim of this network is to adapt the FINGER intervention model to numerous countries and world regions, and offer a platform that allows for data sharing, joint analyses, and collaborative opportunities.

In conclusion, evidence has been accumulating regarding several risk and protective factors for dementia, and targeting these factors can contribute to substantial dementia risk reduction. Evidence from lifestyle interventions suggest that they may offer a promising approach to reducing the risk for cognitive impairment and dementia, and important lessons have been learned on how to optimize the trial methodologies. Finally, ongoing and future work will support with wider-scale implementation of dementia risk reduction initiatives for various populations and settings.

Funding sources

This work was supported by Stiftelsen Stockholms sjukhem, the Joint Program - Neurodegenerative Disease Research (JPND) EURO-FINGERS grant; Alzheimerfonden, Swedish Research Council; Region Stockholm (ALF); Center for Innovative Medicine (CIMED). Additionally, Shireen Sindi was supported by Swedish Research Council (Dnr: 2020-02325), Alzheimerfonden, The Rut and Arvid Wolff Memorial Foundation, The Center for Medical Innovation (CIMED) Network Grant (Karolinska Institutet), The Foundation for Geriatric Diseases at Karolinska Institutet, Erik Rönnbergs Stipend - Riksbankens Jubileumsfond, and Loo and Hans Osterman Foundation for Medical Research.

References

[1] Gao S, Burney HN, Callahan CM, Purnell CE, Hendrie HC. Incidence of dementia and Alzheimer disease over time: a meta-analysis. J Am Geriatr Soc 2019;67(7):1361−9.

[2] Alzheimers Disease International. World Alzheimer report 2015: the global impact of dementia. 2015.

[3] Ho FK, Petermann-Rocha F, Gray SR, Jani BD, Katikireddi SV, Niedzwiedz CL, et al. Is older age associated with COVID-19 mortality in the absence of other risk factors? General population cohort study of 470,034 participants. PLoS One 2020;15(11):e0241824.

[4] Wang C, Wang Z, Wang G, Lau JY, Zhang K, Li W. COVID-19 in early 2021: current status and looking forward. Signal Transduct Targeted Ther 2021;6(1):114.

[5] Electronic Publication Office of National Statistics, Data and Analysis from Census 2021. https://www.ons.gov.uk/peoplepopulationandcommunity/birthsdeathsandmarriages/deaths/bulletins/monthlymortalityanalysisenglandandwales/march2021.

[6] Tahira AC, Verjovski-Almeida S, Ferreira ST. Dementia is an age-independent risk factor for severity and death in COVID-19 inpatients. Alzheimers Dement 2021;17(11):1818−31. https://doi.org/10.1002/alz.12352.

[7] Livingston G, Huntley J, Sommerlad A, Ames D, Ballard C, Banerjee S, et al. Dementia prevention, intervention, and care: 2020 report of the Lancet Commission. Lancet 2020;396(10248):413−46.

[8] World Health Organization. Risk reduction of cognitive decline and dementia. WHO Guidelines 2019.

[9] Kivipelto M, Ngandu T, Laatikainen T, Winblad B, Soininen H, Tuomilehto J. Risk score for the prediction of dementia risk in 20 years among middle aged people: a longitudinal, population-based study. Lancet Neurol 2006;5(9):735−41.

[10] Exalto LG, Quesenberry CP, Barnes D, Kivipelto M, Biessels GJ, Whitmer RA. Midlife risk score for the prediction of dementia four decades later. Alzheimers Dement 2014;10(5):562−70.

[11] Goerdten J, Cukic I, Danso SO, Carriere I, Muniz-Terrera G. Statistical methods for dementia risk prediction and recommendations for future work: a systematic review. Alzheimers Dement (N Y) 2019;5:563−9.

[12] Vandewoude MB-GP, Cederholm T, Mecocci P, Salvà A, Sergi G, Topinkova E, Van Asselt D. Healthy brain ageing and cognition: nutritional factors. Eur Geriatr Med 2016;7(1):77−85.

[13] Scarmeas N, Anastasiou CA, Yannakoulia M. Nutrition and prevention of cognitive impairment. Lancet Neurol 2018;17(11):1006−15.

[14] Munoz Fernandez SS, Ivanauskas T, Lima Ribeiro SM. Nutritional strategies in the management of Alzheimer disease: systematic review with network meta-analysis. J Am Med Dir Assoc 2017;18(10):897 e13−e30.

[15] Cummings J, Passmore P, McGuinness B, Mok V, Chen C, Engelborghs S, et al. Souvenaid in the management of mild cognitive impairment: an expert consensus opinion. Alzheimer's Res Ther 2019;11(1):73.

[16] Soininen H, Solomon A, Visser PJ, Hendrix SB, Blennow K, Kivipelto M, et al. 24-month intervention with a specific multinutrient in people with prodromal Alzheimer's disease (LipiDiDiet): a randomised, double-blind, controlled trial. Lancet Neurol 2017;16(12):965−75.

[17] Soininen H, Solomon A, Visser PJ, Hendrix SB, Blennow K, Kivipelto M, et al. 36-month LipiDiDiet multinutrient clinical trial in prodromal Alzheimer's disease. Alzheimers Dement 2021;17(1):29—40.
[18] Panza GA, Taylor BA, MacDonald HV, Johnson BT, Zaleski AL, Livingston J, et al. Can exercise improve cognitive symptoms of Alzheimer's disease? J Am Geriatr Soc 2018;66(3):487—95.
[19] Ngandu T, Lehtisalo J, Solomon A, Levalahti E, Ahtiluoto S, Antikainen R, et al. A 2 year multidomain intervention of diet, exercise, cognitive training, and vascular risk monitoring versus control to prevent cognitive decline in at-risk elderly people (FINGER): a randomised controlled trial. Lancet 2015;385(9984):2255—63.
[20] Andrieu S, Guyonnet S, Coley N, Cantet C, Bonnefoy M, Bordes S, et al. Effect of long-term omega 3 polyunsaturated fatty acid supplementation with or without multidomain intervention on cognitive function in elderly adults with memory complaints (MAPT): a randomised, placebo-controlled trial. Lancet Neurol 2017;16(5):377—89.
[21] Moll van Charante EP, Richard E, Eurelings LS, van Dalen JW, Ligthart SA, van Bussel EF, et al. Effectiveness of a 6-year multidomain vascular care intervention to prevent dementia (preDIVA): a cluster-randomised controlled trial. Lancet 2016;388(10046):797—805.
[22] MSmfdrparffwAsr K, Mangialasche F, Ngandu T. Lifestyle interventions to prevent cognitive impairment, dementia and Alzheimer disease. Nat Rev Neurol 2018;14(11):653—66.
[23] Ismail Z, Elbayoumi H, Fischer CE, Hogan DB, Millikin CP, Schweizer T, et al. Prevalence of depression in patients with mild cognitive impairment: a systematic review and meta-analysis. JAMA Psychiatr 2017;74(1):58—67.
[24] Breitve MH, Bronnick K, Chwiszczuk LJ, Hynninen MJ, Aarsland D, Rongve A. Apathy is associated with faster global cognitive decline and early nursing home admission in dementia with Lewy bodies. Alzheimer's Res Ther 2018;10(1):83.
[25] Jayasekara R, Procter N, Harrison J, Skelton K, Hampel S, Draper R, et al. Cognitive behavioural therapy for older adults with depression: a review. J Ment Health 2015;24(3):168—71.
[26] Orgeta V, Qazi A, Spector A, Orrell M. Psychological treatments for depression and anxiety in dementia and mild cognitive impairment: systematic review and meta-analysis. Br J Psychiatry 2015;207(4):293—8.
[27] Orgeta V, Qazi A, Spector AE, Orrell M. Psychological treatments for depression and anxiety in dementia and mild cognitive impairment. Cochrane Database Syst Rev 2014;(1):CD009125.
[28] Shi L, Chen SJ, Ma MY, Bao YP, Han Y, Wang YM, et al. Sleep disturbances increase the risk of dementia: a systematic review and meta-analysis. Sleep Med Rev 2018;40:4—16.
[29] Davidson JR, Dickson C, Han H. Cognitive behavioural treatment for insomnia in primary care: a systematic review of sleep outcomes. Br J Gen Pract 2019;69(686):e657—64.
[30] Miller MA. The role of sleep and sleep disorders in the development, diagnosis, and management of neurocognitive disorders. Front Neurol 2015;6:224.
[31] Golub JS, Brickman AM, Ciarleglio AJ, Schupf N, Luchsinger JA. Association of subclinical hearing loss with cognitive performance. JAMA Otolaryngol Head Neck Surg 2020;146(1):57—67.
[32] Rovner BW, Casten RJ, Hegel MT, Leiby B. Preventing cognitive decline in black individuals with mild cognitive impairment: a randomized clinical trial. JAMA Neurol 2018;75(12):1487—93.

[33] Kane RL, Butler M, Fink HA, Brasure M, Davila H, Desai P, et al. Interventions to prevent age-related cognitive decline, mild cognitive impairment, and clinical Alzheimer's-type dementia. In: AHRQ Comparative Effectiveness Reviews. Rockville (MD); 2017.

[34] Alzheimer's Society (UK). How to reduce your risk of dementia. https://www.alzheimers.org.uk/about-dementia/risk-factors-and-prevention/how-reduce-your-risk-dementia#:~:text=Doing%20regular%20physical%20activity%20is,and%20build%20it%20up%20gradually.

[35] National Institute for Health and Care Excellence (NICE). Alcohol-use disorders: diagnosis, assessment and management of harmful drinking (high-risk drinking) and alcohol dependence Clinical guideline (CG115) Published: 23 February 2011. https://www.nice.org.uk/guidance/cg115.

[36] Public Health England. Health matters: stopping smoking - what works?. 2018. https://publichealthmatters.blog.gov.uk/2018/09/25/health-matters-stopping-smoking-what-works/.

[37] National Health Service (NHS). https://www.nhs.uk/live-well/exercise/exercise-as-you-get-older/.

[38] Schiepers OJG, Kohler S, Deckers K, Irving K, O'Donnell CA, van den Akker M, et al. Lifestyle for Brain Health (LIBRA): a new model for dementia prevention. Int J Geriatr Psychiatry 2018;33(1):167–75.

[39] Coley N, Hoevenaar-Blom MP, van Dalen JW, Moll van Charante EP, Kivipelto M, Soininen H, et al. Dementia risk scores as surrogate outcomes for lifestyle-based multi-domain prevention trials-rationale, preliminary evidence and challenges. Alzheimers Dement 2020;16(12):1674–85.

[40] Marengoni A, Rizzuto D, Fratiglioni L, Antikainen R, Laatikainen T, Lehtisalo J, et al. The effect of a 2-year intervention consisting of diet, physical exercise, cognitive training, and monitoring of vascular risk on chronic morbidity-the FINGER randomized controlled trial. J Am Med Dir Assoc 2018;19(4):355–360 e1.

[41] Strandberg TELE, Ngandu T, Solomon A, Kivipelto M, Lehtisalo J, Laatikainen T, Soininen H, Strandberg T, Antikainen R, Jula A. Health-related quality of life in a multi-domain intervention trial to prevent cognitive decline (FINGER). Eur. Geriatr. Med. 2017;8(2):164–7.

[42] Kulmala J, Ngandu T, Havulinna S, Levalahti E, Lehtisalo J, Solomon A, et al. The effect of multidomain lifestyle intervention on daily functioning in older people. J Am Geriatr Soc 2019;67(6):1138–44.

[43] Espie CA, Kyle SD, Williams C, Ong JC, Douglas NJ, Hames P, et al. A randomized, placebo-controlled trial of online cognitive behavioral therapy for chronic insomnia disorder delivered via an automated media-rich web application. Sleep 2012;35(6):769–81.

[44] NHS, England. Wellbeing Apps. Sleepio. https://www.england.nhs.uk/supporting-our-nhs-people/support-now/wellbeing-apps/sleepio/.

[45] Ritterband LM, Thorndike FP, Ingersoll KS, Lord HR, Gonder-Frederick L, Frederick C, et al. Effect of a web-based cognitive behavior therapy for insomnia intervention with 1-year follow-up: a randomized clinical trial. JAMA Psychiatr 2017;74(1):68–75.

[46] Kivipelto M, Mangialasche F, Snyder HM, Allegri R, Andrieu S, Arai H, Baker L, Belleville S, Brodaty H, Brucki SM, Calandri I. World-Wide FINGERS Network: a global approach to risk reduction and prevention of dementia. Alzheimer's Dement July 2020;16(7):1078–94.

Chapter 8b

Next generation technologies for diagnosis

Arlene J. Astell[a,b,c], Tamlyn Watermeyer[a,d] and James Semple[e]
[a]Psychology Department, Northumbria University, Newcastle upon Tyne, United Kingdom; [b]Department of Occupational Sciences & Occupational Therapy & Department of Psychiatry, University of Toronto, Toronto, ON, Canada; [c]KITE Research, University Health Network, Toronto, ON, Canada; [d]Edinburgh Dementia Prevention, Centre for Clinical Brain Sciences, College of Medicine & Veterinary Sciences, University of Edinburgh, Edinburgh, United Kingdom; [e]Formerly GSK Clinical Pharmacology Unit, Addenbrookes Hospital, Cambridge, United Kingdom

Introduction

Looking into the future of dementia diagnosis we can envisage more sophisticated, rapid or scalable extensions of existing approaches. We can also look outside our traditional domains for emerging developments that have promise and potential for advancing our capacity to detect dementia at earlier stages. With that in mind this section briefly considers likely advances in current approaches plus the potential of Artificial Intelligence (AI) and Neurotechnologies. While AI has already started to make an impact in dementia diagnosis, neurotechnology for dementia is very much in its infancy. This allows for consideration of these emerging sectors in the political, ethical and social context that informs where and how we want these to develop.

Advancing current approaches

The cumulative weight of the scientific evidence collected to date leaves no doubt that early indicators of accumulating Alzheimer-type pathology can be reliably detected in vivo in the "silent" preclinical stage of the condition [1]. Refinements in current methods and the development of new methods stemming from our ever-increasing knowledge of the molecular processes underpinning the pathological cascade, will push this detection earlier and earlier. A conservative interpretation of existing data, points to being able to predict with some accuracy the development of clinically significant signs and symptoms at

The Management of Early Alzheimer's Disease. https://doi.org/10.1016/B978-0-12-822240-9.00018-1
Copyright © 2025 Elsevier Inc. All rights reserved, including those for text and data mining, AI training, and similar technologies.

least 9 years before their appearance (e.g., Swaddiwudhipong et al. [2]). This would open a window of opportunity to modify disease progression at a time when individuals are to all intents and purposes functionally unimpaired. If we accept this, the next question becomes the economics and logistics of setting up a mass screening program to identify individuals at risk [3]. The need to identify such individuals has been galvanized by regulatory approval of the first drugs, lecanemab (Leqembi, Biogen), aducanemab (Aduhelm, Eisai & Biogen) and donanemab (Kisunla, Lily), shown to have a significant effect on brain amyloid load (Alzheimer's Society 2023). These drugs are expensive, complex to administer and may have, as yet undiscovered side effects. Blanket treatment of all individuals over a certain age is not an option so the identification of high risk individuals in the silent preclinical phase becomes imperative (e.g., Choi et al. [4]).

As the previous chapters of this book attest, the early diagnosis of Alzheimer's disease is a rich and vigorous field of enquiry. This chapter will try to gaze into the future by considering how and to what extent the various approaches described could contribute to the identification of asymptomatic individuals who are nevertheless in the initial stage of disease progression. With the rapid global rise in dementia and countries different availability of healthcare infrastructure, future approaches need to be scalable and affordable to address health inequalities [5,6].

Imaging

Despite their pivotal role in advancing our knowledge of the nature of Alzheimer-type pathology and its corrosive effects on the brain—for example the potential of the default-mode network to predict future dementia [7]—neuroimaging procedures such as PET and MRI are unlikely in their current format to be in the vanguard in any widespread screening program. This is because both technologies make heavy infrastructural demands in terms of equipment and high levels of professional expertise to run. No doubt there will be technical developments that reduce costs, simplify and streamline data capture, automate data processing and make possible the use of non-local expertise via telemedicine, and in the case of PET, reduce radiation exposure. The development of portable MRI that can come to the bed rather than requiring the individual to come to the scanner, mean mobile services analogous to X-ray scanning for breast cancer are not inconceivable. In February 2020, the FDA granted its first clearance for clinical use for low field portable imaging system that can be deployed at point of care to Hyperfine for their Swoop system. A study of patients with head injury imaged using this equipment found that while image resolution is somewhat poorer than on a standard clinical scanner this is only the beginning of the development of such machines and image processing software [8]. At a projected cost of 20 times lower than a standard clinical scanner, using 35 times less power supplied from a single wall plug and weighing a 10th of 1.5T scanner [9] it seems highly likely that this equipment will be extensively used in community settings, and it can be adapted for use with contrast agents for Alzheimer pathology.

Another promising approach is near infrared scanning [10]. Equipment demands are modest, it can be used to study brain functioning in real time and in the real world, and we have seen the development of near-infrared fluorescent molecular markers of Alzheimer pathology [11]. However, this is a branch of science that, despite the enthusiasm of its advocates, has a long way to go to move from the periphery to become mainstream clinical practice. The requirement for screening from the earth's magnetic field is a major barrier to it holding a major place in screening. Centers capable of carrying it out are few and far between.

EEG, however, has greater potential in that EEG expertise is more widely available because of its central role in the diagnosis and management of epilepsy. Once exclusively in the clinical domain, EEG is emerging as a potential man-machine interface. Commercial systems are already available as a control gateway for computer games (e.g., MindWave Mobile 2). Whether these devices could be adapted to provide clinically relevant data remains to be seen. Nonetheless, mobile EEG devices may emerge as a valuable tool for diagnosing and monitoring dementia, offering a non-invasive, portable, and efficient means of assessing brain activity. The collection of such data in real-world settings arguably provides a more accurate representation of a patient's cognitive and functional capacities than sporadic clinic-based assessments. Among the latest innovations in this field are mobile EEG devices designed to be placed in the inner ear canal offering a discreet and comfortable option for continuous monitoring. This technology holds particular promise for dementia due to its unobtrusive nature, where it can capture electrical activity from the brain with minimal discomfort and interference. One such device is the cEEGrid, which employs electrode arrays positioned around the ear to detect brain signals [12]. A recent study by Bleichner and Debener [13] demonstrated the feasibility and effectiveness of using ear-EEG for monitoring brain activity in various cognitive tasks. Changes in alpha and theta wave patterns have been linked to cognitive decline [14], highlighting mobile EEG's potential for early screenings of potential cognitive issues and prompting of timely medical evaluations and interventions. For the latter, these devices can be used to evaluate the effectiveness of cognitive training programs and other therapeutic interventions, allowing for personalised treatment plans [15]. The incorporation of ongoing data collection may be particularly useful for managing the long-term care of dementia patients [16] as it can allow for the detection of transient or sporadic changes in brain activity that might be missed during scheduled clinical visits. As research continues to validate and refine these technologies, inner ear EEG devices are poised to become a standard tool in dementia care, enhancing the ability to monitor and manage this challenging condition.

Optical markers

While the importance of fluid biomarkers is covered elsewhere in this book, the potential of optical markers in the detection of early indications of Alzheimer-type pathology is emerging. Whilst currently on the periphery of diagnostic efforts, two approaches - optic coherence tomography (OCT; e.g., Chan et al. [17]; Singh and Verma [18]) and frequency doubling flicker fusion contrast sensitivity (e.g., Wu et al. [19]; Risacher [20])—in particular, are showing promise. If ophthalmology were to become better substantiated and more widely accepted it could well be an important vehicle for detecting and monitoring individuals at risk given that eyesight testing is so widespread and moreover carried out at regular intervals.

The use of retinal imaging by ophthalmologists holds significant potential for the early detection and monitoring of Alzheimer's disease. Techniques such as OCT and retinal fundus photography offer non-invasive, cost-effective, and accessible means for identifying retinal biomarkers associated with neurodegeneration. Individuals with preclinical Alzheimer's disease exhibit significant retinal layer thinning, suggesting that retinal imaging could identify those at risk before clinical symptoms emerge [21], while amyloid-beta plaques, a hallmark of Alzheimer's, can be detected in the retina using non-invasive imaging techniques [22]. Retinal imaging may therefore provide a proxy for cerebral amyloid pathology in lieu of traditional and costly imaging and lumbar puncture techniques. Moreover, by tracking changes in retinal structure and vasculature over time, clinicians can assess the efficacy of therapeutic interventions and make informed decisions about treatment adjustments [23]. Ophthalmologists are well-positioned to play a crucial role in the early detection of AD, offering a first line of prevention by identifying individuals at risk and prompting timely referrals to neurologists for further evaluation [24]. The widespread availability of retinal imaging devices in eye clinics and the relatively low cost of the procedure make it a practical option for large-scale screening programs. Given that many individuals undergo regular eye examinations, incorporating retinal imaging into routine screenings could significantly increase the detection rate of early-stage AD.

Related to ophthalmology, eye tracking technology has emerged as a powerful tool for assessing cognitive functions and providing valuable insights into their neural mechanisms. Eye movements are closely linked to cognitive processes such as attention, memory, and visual perception [25]. Individuals with dementia often show altered gaze patterns, characterized by shorter fixation durations and more frequent shifts in gaze [26,27]. These changes are thought to reflect impairments in sustained attention and difficulties in processing visual information. For example, when asked to observe a complex scene, people living with dementia may exhibit scattered fixations and fail to focus on relevant details, indicating a decline in cognitive control and visual processing. Saccades are rapid eye movements that shift the focus from one point to another. Abnormalities in saccadic movements, such as increased latency (delay in initiating movement), decreased accuracy, and reduced velocity, have been observed in individuals with Alzheimer's disease and other forms of dementia [28]. These abnormalities can be attributed to dysfunctions in the neural circuits that control eye movements, which are also implicated in cognitive processes. Smooth pursuit movements allow the eyes to follow moving objects smoothly. In people living with dementia, these movements are often disrupted, resulting in a jerky pursuit or the inability to maintain a steady gaze on moving targets [29]. This impairment reflects deficits in the coordination between visual perception and motor control, which are essential for performing daily activities.

Advancements in eye tracking technology have led to the development of portable and user-friendly devices that can be integrated into everyday digital tools, such as computers, tablets, and smartphones. These devices use infrared cameras and sophisticated algorithms to precisely measure eye movements in real-time. Eye tracking can be seamlessly incorporated into routine assessments and daily activities, making it suitable for continuous monitoring of cognitive health in individuals at risk of dementia. By analyzing cognitive performance metrics along with eye movement data, clinicians can identify early signs of cognitive decline [28]. By example, Neurotrack combines digital cognitive assessments with machine learning to predict the risk of developing dementia. Its assessments are based on validated neuropsychological tests, such as a visual paired comparison task and eye-tracking technology, providing a non-invasive method to detect early cognitive decline. Research has supported its accuracy and predictive power in identifying individuals at risk for Alzheimer's disease [30]. Thus, eye movement analysis offers a promising avenue for early detection and monitoring of dementia. By leveraging advanced eye tracking technology, researchers and clinicians can gain valuable insights into the cognitive function of individuals at risk of or diagnosed with dementia.

Behavioral markers

Self-awareness of cognitive failures is often the trigger for referral to medical practitioners, but objective evidence of impaired cognition at this stage often eludes the most commonly employed neuropsychological tools. The ability to collect reliable cognitive data through unsupervised testing [31], as in the Novel Assessment of Nutrition and Aging (NANA) [32], increases accessibility and convenience of capturing cognitive data on a daily basis. To address this, apps for capturing micro-longitudinal data from older adults with and without dementia are becoming more common (e.g., Astell et al. [33]; Fox et al. [34]). The DataDay app, developed by Astell et al. [33]; is designed to capture detailed daily cognitive and behavioral data from older adults. Users complete short cognitive tasks and answer questions about their mood, physical activity, and social interactions multiple times a day. The app's frequent data collection helps identify early signs of cognitive decline and provides insights into the daily life of individuals with and without dementia. The Digital Memory Assessment (DIMA) app, as discussed by Fox et al. [34]; focuses on assessing memory and cognitive function through daily tasks and exercises. The app adjusts the difficulty of tasks based on user performance, ensuring a personalised and engaging experience. It collects data on user responses, task completion times, and error rates, which are analyzed to detect changes in function. These examples of mobile and personalised monitoring can lead to more tailored interventions, improving the overall management of dementia while real-time data entry ensures more accurate and reliable data by reducing reliance on user recall—a particular challenge for individuals with cognitive impairment. Mobile applications equipped with speech analysis tools can also continuously monitor verbal interactions.

Speech biomarkers have emerged as promising proxies for early detection and monitoring of dementia. These biomarkers involve analyzing various aspects of speech, including linguistic content [35], acoustic features [36], and conversational dynamics, to identify patterns indicative of cognitive decline [37]. Language samples, for example describing a picture of a scene are easy to collect and more readily incorporated into routine examinations [38]. This is already leading to commercially available products such as the tablet-based assessment from Winterlight Labs [39] and the emerging UK assessment CognoSpeak [40,41]. This tool uses advanced natural language processing (NLP) algorithms to evaluate linguistic features such as speech rate, coherence, and vocabulary usage. Studies have shown that this tool can effectively identify early signs of cognitive decline, making it a valuable addition to routine clinical examinations [39]. Additionally, novel research combing EEG with language testing has been shown to distinguish individuals with MCI from age-matched controls without cognitive impairment. In a recent study, the integration of EEG and language testing provided a more robust and reliable method for identifying early cognitive impairments compared to either modality alone [42]. The use of speech biomarker technologies in the detection and monitoring of dementia offers significant advantages. Automated analysis provides quick and objective evaluation of speech samples, ensuring consistent and accurate assessments while reducing the time and potential errors associated with manual evaluations. These technologies feature user-friendly interfaces designed for ease of use by both patients and clinicians, facilitating regular and effective data collection (e.g., O'Malley et al. [43]). Additionally, they generate detailed reports that highlight key linguistic markers of cognitive health, offering comprehensive insights into an individual's cognitive status and aiding in early diagnosis and monitoring of dementia progression.

Related to mobile applications are Digital Twin Technologies. Digital twins, a concept originating from the industrial and manufacturing sectors, is increasingly being applied to healthcare. A digital twin is a virtual replica of a physical entity that is continuously updated with real-time data and simulations. In healthcare, this can mean a comprehensive digital model of a patient, integrating data from various sources to monitor health, predict disease progression, and personalise treatments [44]. Explored the application of digital twin technology in healthcare settings, specifically focusing on its potential to predict cognitive decline in people living with dementia. The study demonstrated that digital twins, which are virtual replicas of individual patients, can integrate data from electronic health records, wearable devices, genomic information, and environmental sensors to create comprehensive and dynamic models. These models accurately simulate disease progression and forecast future cognitive states, enabling early detection of cognitive decline and proactive healthcare management. By evaluating the potential impact of various interventions, such as medications and lifestyle changes, digital twins might help personalise treatment plans, optimize care delivery, and improve patient outcomes. The study highlighted the transformative potential of digital twins in providing a holistic, data-driven approach to managing dementia, paving the way for more effective and personalised healthcare solutions. Additionally, Walsh et al. [45] proposed that digital twins can accelerate RCTs, by reducing the number of participants required, also reducing costs, and timelines in dementia drug studies.

Virtual and augmented reality markers

As covered in previous chapters (see Chapter 7a), there is also growing exploration of new and emerging technologies for early assessment of cognitive impairment including Mixed Reality Technologies, i.e., Virtual and Augmented Reality Technologies. While the potential of VR for identifying cognitive impairment was recognized in the early days of the technology (e.g., Rizzo et al. [46]; Rose et al. [47]), strides have only been made in the last few years. For example, immersive virtual reality (iVR) involving movement in a range of environments and completing a variety of activities (e.g., ATM withdrawal) can distinguish people with AD or prodromal AD from older adults without cognitive impairment [48]. Benefits of VR cognitive assessment include testing in different environment and the possibility to test visuospatial ability [49]. These are exploited in a recently launched VR system that assesses functional cognition across a range of tasks, such as ordering and paying for a cup of coffee, which aims to provide alternate tools to current pen-and paper tests [50].

Studies have shown that VR environments can effectively identify deficits in spatial navigation and memory recall, which are early indicators of cognitive impairments associated with Alzheimer's disease. VR tasks can successfully detect spatial disorientation in participants with mild cognitive impairment (MCI), correlating with traditional neuropsychological tests [51] using VR found subtle deficits in spatial behavior, associated with impaired path integration in the entorhinal cortex, is associated with midlife risk of developing Alzheimer's disease. Cogné et al. [52] highlighted the diagnostic accuracy of VR cognitive tasks for early detection of Alzheimer's disease, emphasizing their effectiveness in identifying spatial memory and executive function impairments [52]. Laczó et al. [53] reported that performance in VR navigation tasks could predict cognitive decline, differentiating between healthy elderly participants and those with MCI, thus providing a sensitive tool for early diagnosis. Finally, Howett et al. [54] demonstrated that VR-based spatial navigation tasks were highly effective in distinguishing individuals with MCI from healthy controls, offering a more engaging and ecologically valid measure of cognitive function compared to traditional tests [54].

Apart from diagnostic utility, VR technology is increasingly being explored as a promising tool for the intervention and management of dementia. By creating immersive and interactive environments, VR can simulate everyday situations and tasks, providing a safe and controlled setting for memory training exercises. Recent studies have shown that VR-based cognitive training can improve memory function in individuals with MCI and early-stage dementia [55]. Patients able to navigate through a virtual neighborhood or building, practicing spatial orientation and memory recall. Routine practice may improve their ability to remember routes and locations, potentially translating to better real-world navigation skills [56]. Similarly, VR-based tasks might enhance executive function and attention in people living with dementia [57], potentially supporting the preservation of these cognitive domains. For example, a VR scenario where patients follow a recipe, selecting ingredients and performing cooking steps, may help improve planning, sequencing, and task execution abilities [58].

Apart from direct cognitive benefits, VR can create calming and engaging experiences that help reduce agitation and anxiety, common problems for people living with dementia. Immersive environments, such as a serene beach or a familiar hometown, can provide a sense of comfort and relaxation [59]. Individuals can explore virtual natural environments, which have been shown to reduce stress and improve mood [60]. Furthermore, VR can facilitate reminiscing by enabling people to revisit familiar places and experiences from their past. This can evoke positive memories and emotions, enhancing their sense of identity and well-being [61]. Appel et al. [62] used virtual reality (VR) to provide recreations of their childhood homes or significant places from their past to people living with dementia. This VR experience successfully sparked memories and facilitated conversations about their life history, demonstrating the potential of VR to enhance reminiscing and improve the emotional well-being of people living with dementia.

Finally, VR can provide opportunities for social interaction, both with virtual characters and through shared VR experiences with other people living with dementia or caregivers. This can help combat loneliness and social isolation, which are prevalent among people living with dementia across care environments [63]. Lindner et al. [64] explored the use of VR to facilitate group activities such as virtual bingo and painting classes for people living with dementia. They found that the VR-based activities promoted social engagement and interaction, helping to combat loneliness and improve the overall social well-being of participants. Moreover, the training enhanced caregivers' skills in responding appropriately to difficult situations, ultimately reducing their burden and improving the quality of care they provided [64]. Virtual reality may therefore offer a multifaceted approach that enhances not only the quality of life of people living with dementia but also supports their caregivers in delivering effective care at home.

Digital biomarkers

Digital biomarkers represent a form of objective, quantifiable physiological and behavioral data collected and measured using digital devices. The widespread availability of smartphones and digital wearables, such as smartwatches and fitness trackers, equips researchers and clinicians with an extensive array of sensors capable of monitoring body physiology, sleep patterns, daily activities, and social interactions. The capacity to collect data during everyday activities, both inside and outside the home, offers significant potential for detecting early signs of cognitive decline.

One notable example of this approach is the Oregon Center for Aging and Technology (ORCATECH) at Oregon Health & Science University (OHSU), which has been installing arrays of sensors in people's homes since 2004. This initiative led to the development of the Collaborative Aging Research using Technology (CART) platform, currently deployed in over 230 homes. The CART platform exemplifies the use of technology to assess daily activity and detect meaningful changes over time, often spanning several years [65]. These sensors continuously gather data on movement, sleep, and other behaviors, providing valuable insights into the health and cognitive function of older adults in real world settings [66]. More recent initiatives have expanded on this foundation. For example, the RADAR-AD project in Europe uses a combination of home-based sensors and wearable devices to monitor daily activities and detect early signs of cognitive decline in patients with Alzheimer's disease [67]. This project integrates data from multiple sources, including smartphones, wearables, and environmental sensors, to create a comprehensive picture of a patient's cognitive health. The digital biomarkers collected via remote measurement technologies (RMTs) provide more sensitive indices of functional decline than traditional methods [68]. For example, the potential of measuring navigation and gait in everyday environments using wearables, shows great promise for earlier detection of various dementias [69].

However, the authors note that there is a need to address validation, regulatory, and integration challenges to fully realize the potential of these technologies in clinical research as obtaining regulatory acceptance for digital endpoints remains difficult due to the lack of standardized guidelines and frameworks.

Early research at OHSU also explored digital biomarkers through computer interactions [70]. developed algorithms to infer cognitive performance from monitoring data related to computer games, keyboard entry, and mouse movement. This research aimed to classify changes in performance and provide personalised support by adapting device interfaces accordingly. Although this work preceded the advent of smart devices, it laid the groundwork for utilizing everyday computer interactions as indicators of cognitive health. More recently, studies have continued to explore this approach. For instance Stringer et al. [71], used digital biomarkers derived from computer interactions to differentiate between individuals with subjective cognitive decline and mild cognitive impairment (MCI). They found that tasks such as basic desktop navigation, word processing, email tasks, and browsing could reveal significant differences in cognitive function. Additionally, another study by Stringer et al. [72] analyzed keystroke dynamics and mouse movements to detect cognitive decline. Their research demonstrated that specific patterns in typing speed, accuracy, and mouse movement could serve as reliable indicators of early-stage dementia.

When combined with Internet footprints and analysis of communication networks and content, digital biomarkers can potentially cover many aspects of daily life. This data can be automatically collected in the background, minimizing the need for active participation. The challenge remains to effectively sample, store, and process the vast quantities of data available. Big data approaches and machine learning (ML) techniques are already being applied to existing clinical datasets, and it is reasonable to expect that these methods will extend to current and emerging digital biomarkers, informing new diagnostic approaches. For example, perhaps machine learning algorithms may be able to analyze social media interactions and digital communication patterns, identifying changes that correlate with cognitive decline. These methods could detect subtle shifts in language use, response times, and interaction frequency, providing a non-intrusive way to monitor cognitive health.

Artificial Intelligence

AI was originally conceived as an approach to understanding "intelligent beings", but today is mostly concerned with interpreting huge amounts of data to support decision-making [73]. The development of AI is described in terms of two contrasting approaches, one top-down and one bottom-up. The traditional, top-down, symbolic approach developed from the premise that humans use symbols to define and describe actions, items and events in the world. Using symbols as the basis of creating AI models, this approach has been responsible for various breakthroughs including natural language processing (NLP) which allows machines to understand human language and is used in search engines, predictive text and voice-activated systems such as Google Home Hub, Siri and Alexa. However, limitations in symbolic AI relating to reliance on rule-based learning and requiring continuous human input of both modified rules and new data, led to the emergence of a bottom-up approach creating artificial neural networks based on the structure of the human brain, initially known as connectionist models.

Artificial neural networks are composed of units which are essentially artificial neurons arranged in a series of layers, traveling from an input layer through one or more hidden layers leading to an output layer. The links between units are weighted to represent the strength of the connection between units. The hidden layer extracts some of the most relevant patterns from the inputs and sends them on to the next layer for further analysis. Artificial neural networks are able to learn and model non-linear relationships, can generalize beyond their original inputs and do not constrain the input variables [74].

The applications of this massive computational ability are growing daily from traffic management [75], changing the retail experience [76], and of course healthcare, with a dedicated journal: Artificial Intelligence in Medicine. The ability to assign multidimensional inputs to specific categories is both a significant achievement and a continuously progressing aspect of which the field of medical diagnostics—essentially the process whereby empirical input interacts with clinical knowledge to produce a classificatory output—is but one example. Both medical experts and AI systems, derive their competence from training and experience. For humans training takes years, the experience of a career. AI training is much faster, and typically employs more training exemplars than any clinician would encounter in a lifetime, though the richness of the data is heavily constrained. Human brains are limited in their capacity to process large amounts of data and extract marginal signals. AI, in contrast, thrives on large data sets and is capable of discerning patterns amongst nuggets of information that are too dispersed and small to be picked out of the background noise by humans.

Existing applications of AI in dementia are focusing on training diagnostic models (see Pellegrini et al. [77] for a review), with some predictive modeling of incident dementia (e.g., Nori et al. [78,79]) and the course of dementia using imaging and cognitive score datasets (e.g., Rudovic et al. [80]). Combining multidimensional data from traditionally disparate fields is emerging as a powerful direction of effort. For example El-Sappagh et al. [81], employed 28 features measuring cognitive and functional assessment, MRI, PET, genetic markers and medical history. Another paper attracted attention for predicting who would develop dementia within 2-years with 92% accuracy [82]. The data were collected from 15,300 people who attended 30 National Alzheimer's Coordinating Center memory clinics in the US between 2005—15. Interestingly, the AI modeling also identified that one in eight dementia diagnoses were erroneous, signaling another potential benefit of machine learning algorithms [82].

Data collected from people attending memory clinics can be expected to have a high proportion with prodromal or confirmed dementia. However, identifying those at risk and not yet the subject of medical concern, is more challenging as the data sets will be much sparser and most likely devoid of high tech-high value data such as MRI, PET etc. What will be available are basic demographics, electronic medical records and histories of insurance claims and hopefully some of the easily collectable molecular biomarkers such as blood and urine. For example, machine learning models trained on a large data set of administrative claims to private health care insurers and the Korean National Health Insurance Service [78,83,84] enabled prediction of the onset of dementia, including that due to Alzheimer's disease several years in advance of diagnosis. Similarly, natural language processing has been used to identify key specific terms and sentiments in clinical notes incorporated into electronic health records [85]. This additional information augmented predictive accuracies obtained on the basis of medical claims and structured clinical data [78] and extended the predictive period to up to 8 years before diagnosis.

Looking beyond analysis of big data, evidence from other populations is starting to demonstrate the success of deep neural networks (DNN), a type of model trained on large datasets to reproduce human-like behavior, in delivering some aspects of cognition, such as visual object recognition. DNNs are a subset of neural networks characterized by multiple hidden layers between the input and output layers. This depth allows DNNs to model complex, hierarchical representations of data, making them particularly effective for tasks such as image and speech recognition. The potential of DNNs to elucidate more complex cognitive functions such as Theory of Mind is also being explored [86]. Additionally, AI is being used to model cognitive function in children who struggle to learn [87], as well as people who have experienced stroke [88], or are living with MS [89]. A team from Cambridge University have recently reported a predictive prognostic model with highly accurate identification of whether individuals with MCI would go on to develop dementia or not validated against clinical data [90].

While AI is very much a vanguard element in the current zeitgeist, its achievements impressive and its future bright, it is not without its limitations. These models draw largely upon multimodal data sets derived from state-of-the-art medical technologies such as imaging, genomics, molecular biomarkers etc. In the real world such data would not be available for the vast majority of individuals. Moreover, the operations performed by AI to arrive at their output are generally opaque and obscure and not interrogable by human users [91]. Black boxes need to be opened, so-called Explainable Artificial Intelligence or XAI [81], so that algorithms provide accounts of how they arrived at their decision in a manner that can be understood by clinicians not trained in machine learning, as demonstrated in the 2024 paper by Lee and colleagues. Alternately we can expect to see the establishment of more transdisciplinary centers such as the Temerty Center for AI Research and Education in Medicine (T-CAIREM; University of Toronto), Cambridge Center for AI in Medicine (University of Cambridge) and Center for Artificial Intelligence in Medicine and Imaging (Stanford University) bringing together health professionals and AI researchers to maximize the benefits and mitigate the challenges of AI in dementia.

Direct brain technologies

If we look further ahead, we can anticipate developments based around direct technological intervention in the brain—neurotechnologies and nanotechnologies, chemogenetics and optogenetics - that could transcend the gap between diagnosis and treatment to provide early detection and personalised, adaptable responses. Neurotechnologies fall into the recently established field of Neuroengineering, Assistive and Rehabilitation Technologies (NARTs [92]): which encompasses five classes of technology: prosthetics, brain-computer interfaces (BCIs), functional electrical stimulation (FES) techniques, devices for robot-assisted training and powered mobility aids. Currently developed as interventions to repair or remediate neural damage or loss, NARTs incorporating AI could provide early detection of changes in the brain with the potential to deliver corrective action, adapting as neurodegeneration advances. The possibility to "utilize miniaturized implantable devices that automatically detect and selectively control neuronal activity as an engineering-based treatment without side effects encountered with traditional medicines" (page 115 [93]), is emerging as a potential treatment option for diseases causing dementia as well as other neurological and neuropsychiatric conditions.

As an approach to correcting cognitive impairment, investigation of implantable cognitive prostheses such as a micro-chip for the hippocampus to address memory problems have been explored [94]. This neural prosthesis approach replicates the signal processing functions of the hippocampus, a critical brain region for converting short-term memory into long-term memory, by employing electrodes implanted in the hippocampus to record neural activity between the CA3 and CA1 subregions. These recordings are subsequently used to mimic the memory encoding process through an electronic system [94].

In rodent models, the prosthesis demonstrated the ability to restore memory functions in rats whose hippocampal activity was pharmacologically inhibited. Activation of the device enabled the rats to recall learned tasks, despite disruptions in their natural hippocampal activity, thereby highlighting the prosthesis's efficacy in real-time memory manipulation [94]. Additionally, the prosthesis showed potential in enhancing memory in normal rats, suggesting its utility for cognitive enhancement [94].

In humans, Neuralink is advancing BCI technology through its N1 Implant, designed to record and stimulate neural activity. The N1 Implant, currently under investigation in the Precision Robotic Implantation Brain-Computer Interface (PRIME) study, aims to restore motor and sensory functions in individuals with severe neurological impairments [95]. The device comprises 1024 electrodes distributed across 64 threads and utilizes advanced materials and microfabrication techniques to ensure biocompatibility and durability. Its implantation involves a custom-built surgical robot for precision and minimal brain tissue damage [95]. While primarily designed for paraplegia, Neuralink's technology holds promise for cognitive enhancements too, as similar devices could theoretically be developed to support memory and cognitive functions. The ability to record and stimulate specific neural circuits could enable the restoration of cognitive processes lost due to neurodegenerative diseases [95]. Of course, the use of these technologies raises significant ethical and practical concerns, such as issues such as privacy, long-term safety, and implications of enhancing human cognition. Regardless, these technologies might inform future applications in neurodegenerative diseases.

Neuromodulation, which was originally developed in the area of pain relief and management, describes technology that acts directly on nerves and includes both external and implantable devices. Examples of neurological applications include deep brain stimulation for Parkinson's disease and other movement disorders and vagus nerve stimulation for epilepsy [96]. Memory is another popular neuromodulation target with various efforts focusing on entorhinal cortex, basal ganglia, and fornix [97] although more work is required to demonstrate clear benefits for people with dementia. For example, deep brain stimulation of the fornix of people with Alzheimer's disease was found to be safe and well tolerated but delivered no measurable impact on cognition [98]. However, these patients already had significant cognitive impairment and presumed widespread neurodegeneration More recently [99], proposed that non-invasive neuromodulation could be an effective approach for cognitive rehabilitation. It remains to be seen if interventions of this kind may have greater benefit in early disease stages, before dementia onset and possibly even before the occurrence of symptoms, when there is less in the way of irreversible brain damage and thus potentially greater capacity for treatment response.

Nanotechnology offers a very different option for direct delivery of interventions to the brain. Nanomedicine for example, is concerned with diagnosis and precise delivery of drugs [100]. For AD and other brain disorders, nanotechnology provides a solution to the blood-brain barrier which restricts the entry to the brain of exogenous agents for protective purposes [101]. In their article, Teleanu and colleagues explored the potential of nanotechnology to deliver nanoparticles, liposomes, dendrimers, micelles, and carbon nanotubes to treat brain disorders. One example is the use of berberine, an isoquinoline alkaloid, absorbed onto the surface of multi-walled carbon nanotubes. Drug absorption was significantly improved by this method of administration, demonstrated the potential of this approach to reducing β-amyloid induced AD [102]. Integration of a drug into or onto a polymeric and/or lipidic nanoparticle (NP) increases the bioavailability of a drug [103]. Nanoparticles have been used to deliver AD therapeutic protein or peptide functions [104]. Other researchers are exploring mitochondrion-specific nanotechnology (e.g., Kwon et al. [105]), and anti-amyloid therapies (e.g., Wang and Wang [106]), which deliver antioxidants and other therapeutic agents specifically to mitochondria or disrupt Aβ aggregation and enhance the clearance of Aβ peptides, respectively, offering novel and targeted therapeutic approaches.

Optogenetics and chemogenetics both permit direct control of neurons. Optogenetics, a technique that applies light to control neurons or other cells, is starting to be explored in relation to dementia. For example, Etter et al. [107]; "rescued" spatial memory of J20-APP AD mice using optogenetic stimulation of medial septal parvalbumin neurons. A comprehensive review of the potential of optogenetics for Alzheimer's disease by Mirzayi et al. [108] identified revealing pathogenesis and disease treatment as the two core areas of future contribution.

Chemogenetics, delivers "remote control" of cells via systemic injection or microinfusion of ligands [109]. This includes designer receptors exclusively activated by designer drugs (so-called DREADDs). Whilst chemogenetics has lower temporal resolution than optogenetics [109], argue that it can be useful for functional mapping, and multiplexed control of neurons. In their review of both opto- and chemogenetics [110], found both approaches had been applied to determine Alzheimer's pathogenesis. However, only a few studies explored the transcriptomic/proteomic profile [111]. In their review, of chemogenetics as a neuromodulation approach [112], concluded that "the potential of current chemogenetic receptors as a method for precision neuromodulation ... remains largely conceptual" (page 996 [112]).

Advances in neurotechnology, nanotechnology, optogenetics and chemogenetics, plus AI with preclinical, healthy and impaired populations, are increasingly advancing our understanding of the brain. These developments are also opening up the possibilities for developing next generation neurotechnology for people living with dementia. Applying Deep Learning to AD and similar disorders has the potential to open up neurotechnology to repair, restore or enhance their capabilities. This could transform the experience of people who develop dementia but also throws up unprecedented technical, social, and ethical questions. To address these, we need to explore the potential of AI-empowered neurotechnology for dementia from both a practical (i.e., what is possible) and a socio-ethical (what is acceptable) standpoint.

What do we want?

The extension of data collection into all aspects of our lives raises questions about whether we want to live our lives in the focus of such intensive scrutiny, how access to such sensitive material will be governed and our privacy is protected. We are all waiting for the development of effective treatments, but however frustrating this situation is it does offer an opportunity to reflect on what exactly is hoped for. Until there is a breakthrough in our understanding of the mechanisms underpinning AD and the causal relationship between AD molecular pathology and the clinical manifestations, discovery of interventions capable of preventing or curing AD will be challenging. In their absence, a slowing of disease progression represents a pragmatic next best outcome. A report from the Personal Social Services Research Unit [113] commissioned by Alzheimer's Research UK, employing the NICE standard of justifiable expenditure of £20,000 per QALY (quality-adjusted life-year), gained has examined the impact of a number of treatment scenarios including provision of disease modifying agents to pre-symptomatic individuals in their 50s. Their predictions based on treating all individuals in this cohort (some 93,7000 people) indicate that delays in disease onset by 1, 3 and 5 years would avert 16,900, 47,000 and 76,000 cases and reduce care costs by 10.8%, 30.1%, and 46% respectively. These figures are based on blanket treatment of this cohort and do not take account of screening to target only individuals at risk. Their analysis clearly shows that treatment with a disease modifying agent is much less effective if individuals already have established Alzheimer's disease dementia, even if mild, or indeed if they are classified has having MCI. Indeed, treating people with already established mild Alzheimer's disease increases the prevalence of the condition and associated healthcare costs by prolonging their lifespan.

Looking into the near and middle future there are a range of new interventional strategies and analytic approaches, utilizing advances in technology and data science, that may have transformative benefits for AD diagnosis and treatment. However, work in these fields is in its infancy and extensive future work will be needed to determine the effectiveness of the approaches outlined in this chapter. As this work is conducted, it is imperative that consideration is given in parallel to associations including ethical and health economic implications plus the need to ensure inclusivity and the avoidance of two-tier healthcare provision.

References

[1] Zolochevska O, Taglialatela G. Non-demented individuals with Alzheimer's disease pathology: resistance to cognitive decline may reveal new treatment strategies. Curr Pharmaceut Des 2016;22(26):4063—8.

[2] Swaddiwudhipong N, Whiteside DJ, Hezemans FH, Street D, Rowe JB, Rittman T. Pre-diagnostic cognitive and functional impairment in multiple sporadic neurodegenerative diseases. Alzheimer's Dementia 2022. https://doi.org/10.1002/alz.12802.

[3] Lafortune L, Khan A, Martin S, Fox C, Callum S, Dening T, Rait G, Katona C, Brayne C. A systematic review of costs and benefits of population screening for dementia. Lancet 2013;382:S56. https://doi.org/10.1016/S0140-6736(13)62481-2.

[4] Choi H, Kim H-J, Kim KH, Oh S-I, Kim SH. The consideration about usefulness of mass screening for dementia. Dementia Neurocogn Dis 2014;13:117−20. https://doi.org/10.12779/dnd.2014.13.4.117.

[5] Wiese LAK, Gibson A, Guest MA, Nelson AR, Weaver R, Gupta A, Carmichael O, Lewis JP, Lindauer A, Lo S, Peterson R, et al. Global rural health disparities in Alzheimer's disease and related disorders: state of the science. Alzheimer's Dementia 2023;19(9):4204−25.

[6] Aranda MP, Kremer IN, Hinton L, Zissimopoulos J, Whitmer RA, Hummel CH, et al. Impact of dementia: health disparities, population trends, care interventions, and economic costs. J Am Geriatr Soc 2021;69(7):1774−83. https://doi.org/10.1111/jgs.17345.

[7] Ereira S, Waters S, Razi A, et al. Early detection of dementia with default-mode network effective connectivity. Nat Mental Health 2024;2:787−800. https://doi.org/10.1038/s44220-024-00259-5.

[8] Sneth KN, Mazurek MH, et al. Assessment of brain injury using portable low-field magnetic resonance imaging at the bedside of critically ill patients. JAMA Neurol Jamaneurol 2020;78:41−7. https://doi.org/10.1001/jamaneurol.2020.3263. In this issue.

[9] Cho A. MRI for all: Portable low-field scanners could revolutionize medical imagingin nations rich and poor—if doctors embrace them. Science 2023;379(6634):748−51.

[10] Hoshi Y. Near-Infrared spectroscopy for studying higher cognition. In: Neural correlates of thinking. Berlin: Springer; 2009. p. 83−93.

[11] Gyasi I Y, Pang Y-P, Li X-R, Gu J-X, Cheng X-J, Liu J, et al. Biological applications of near infrared fluorescence dye probes in monitoring Alzheimer's disease. Eur J Med Chem 2020;187:1−16. 111982.

[12] Debener S, Emkes R, De Vos M, Bleichner M. Unobtrusive ambulatory EEG using a smartphone and flexible printed electrodes around the ear. Sci Rep 2015;5:1−11. 16743.

[13] Bleichner MG, Debener S. Concealed, unobtrusive ear-centered EEG acquisition: cEE-Grids for transparent EEG. Front Hum Neurosci 2017;11:1−14. https://doi.org/10.3389/fnhum.2017.00163.

[14] Cassani R, Estarellas M, San-Martin R, Fraga FJ, Falk TH. Systematic review on resting-state EEG for Alzheimer's disease diagnosis and progression assessment. Dis Markers 2018;2018:5174815.

[15] Looney D, Kidmose P, Park C, Ungstrup M, Rank ML, Mandic DP. The in-the-ear recording concept: user-centered and wearable brain monitoring. IEEE Pulse 2012;3(6):32−42.

[16] Byrom B, McCarthy M, Schueler P, Muehlhausen W. Brain monitoring devices in neuroscience clinical research: the potential of remote monitoring using sensors, wearables, and mobile devices. Clin Pharmacol Therapeut 2018;104(1):59−71.

[17] Chan VTT, et al. Spectral domain OCT measurements in Alzheimer's disease: a systematic review and meta-analysis. Ophthalmology 2019;126(4):497−510.

[18] Singh AK, Verma S. Use of ocular biomarkers as a potential tool for the early diagnosis of Alzheimer's disease. Indian J Opthalmol 2020;68:555−61.

[19] Wu SZ, Masurkar AV, Balcer LJ. Afferent and efferent visual markers of Alzheimer's disease: a review and update in early stage disease. Front Aging Neurosci 2020;12:572337.

[20] Risacher SL, et al. Visual contrast sensitivity is associated with the presence of cerebral amyloid and tau deposition. Brain Commun 2020;2(1):fcaa019. https://doi.org/10.1093/braincomms/fcaa019.

[21] Ferrari L, Huang SC, Magnani G, Ambrosi A, Comi G, Leocani L. Optical coherence tomography reveals retinal neuroaxonal thinning in frontotemporal dementia as in Alzheimer's disease. J Alzheim Dis 2017;56(3):1101−7.

[22] Koronyo Y, Biggs D, Barron E, Boyer DS, Pearlman JA, Au WJ, Kile SJ, Blanco A, Fuchs DT, Ashfaq A, Frautschy S, Cole GM, Miller CA, Hinton DR, Verdooner SR, Black KL, Koronyo-Hamaoui M. Retinal amyloid pathology and proof-of-concept imaging trial in Alzheimer's disease. JCI Insight August 17, 2017;2(16):e93621. https://doi.org/10.1172/jci.insight.93621.

[23] Santos CY, Johnson LN, Sinoff SE, Festa EK, Heindel WC, Snyder PJ. Change in retinal structural anatomy during the preclinical stage of Alzheimer's disease. Alzheimer's Dement 2018;10:196−209.

[24] Frost S, Martins RN, Yogesan K. Retinal screening for early detection of Alzheimer's disease. In: Digital teleretinal screening: teleophthalmology in practice. Springer Berlin Heidelberg; 2012. p. 91−100.

[25] Rayner K. Eye movements and cognitive processes in reading, visual search, and scene perception. In: Studies in visual information processing, vol. 6. North-Holland; 1995. p. 3−22.

[26] Anderson TJ, MacAskill MR. Eye movements in patients with neurodegenerative disorders. Nat Rev Neurol 2013;9(2):74−85.

[27] Hutton JT, Nagel JA, Loewenson RB. Eye tracking dysfunction in Alzheimer-type dementia. Neurology 1984;34(1). 99-99.

[28] Molitor RJ, Ko PC, Ally BA. Eye movements in Alzheimer's disease. J Alzheim Dis 2015;44(1):1−12.

[29] Deravet N, Orban de Xivry JJ, Ivanoiu A, Bier JC, Segers K, Yüksel D, Lefèvre P. Frontotemporal dementia patients exhibit deficits in predictive saccades. J Comput Neurosci 2021;49:357−69.

[30] Glenn JM, Bryk K, Myers JR, Anderson J, Onguchi K, McFarlane J, Ozaki S. The efficacy and practicality of the Neurotrack Cognitive Battery assessment for utilization in clinical settings for the identification of cognitive decline in an older Japanese population. Front Aging Neurosci 2023;15:1206481.

[31] Brown LJ, Adlam T, Hwang F, Khadra H, Maclean L, Rudd B, Smith T, Timon C, Williams EA, Astell AJ. Computer-based tools for assessing micro-longitudinal patterns of cognitive function in older adults. Age 2016;38(4):335−50.

[32] Astell AJ, Hwang F, Brown LJE, Timon C, Maclean LM, Smith T, et al. Validation of the NANA (Novel Assessment of Nutrition and Ageing) touch screen system for use at home by older adults. Exp Gerontol 2014;(60),:100−7.

[33] Astell AJ, Dove E, Morland C, Donovan S. Using the TUNGSTEN approach to co-create DataDay: a self-management app for dementia. In: Kenning G, Brankaert R, editors. HCI and Design in the context of dementia. New York: Springer; 2020.

[34] Fox S, Brown L, Antrobus S, Brough D, Drake RJ, Jury F, Leroi I, Parry-Jones AR, Machin M. Co-Design of a smartphone app for people living with dementia by applying agile, iterative Co-design principles: development and usability study. JMIR mHealth and uHealth 2022;10(1):e24483. https://doi.org/10.2196/24483.

[35] Thomas JA, Burkhardt HA, Chaudhry S, Ngo AD, Sharma S, Zhang L, Hosseini Ghomi R. Assessing the utility of language and voice biomarkers to predict cognitive impairment in the framingham heart study cognitive aging cohort data. J Alzheim Dis 2020;76(3):905−22.

[36] Luz S, Haider F, de la Fuente Garcia S, Fromm D, MacWhinney B. Alzheimer's dementia recognition through spontaneous speech. Front Comput Sci 2021;3:780169.

[37] Wang N, Luo F, Peddagangireddy V, Subbalskshmi KP, Chandramouli R. Personalized early-stage Alzheimer's disease detection: a case study of President Reagan's speeches. 2020. https://www.aclweb.org/anthology/2020.bionlp- 1.14.pdf.

[38] Beltrami D, Gagliardi G, Roussini Favretti R, Ghidoni E, Tamburini F, Calza L. Speech analysis by Natural Language Processing techniques: a possible tool for very early detection of cognitive decline? Front Aging Neurosci 2018;10:369.

[39] Winterlight labs. https://winterlightlabs.com/.

[40] O'Malley RPD, Mirheidari B, Harkness K, Reuber M, Venneri A, Walker T, et al. Fully automated cognitive screening tool based on assessment of speech and language. J Neurol Neurosurg Psychiat 2021;92(1):12−5.

[41] Blackburn DJ, Sproson L, Egbuta C, Bell SM, O'Malley R, Mirheidari B, et al. Developing an automated Cognitive assessment based on language; CognoSpeak-working with an ethnic minority group. Alzheimer's Dementia 2023;19. e076244.

[42] Segaert K, Poulisse C, Markiewicz R, Wheeldon L, Marchment D, Adler Z, Howett D, Chan D, Mazaheri A. Detecting impaired language processing in patients with mild cognitive impairment using around-the-ear cEEgrid electrodes. Psychophysiology 2021;00:e13964. https://doi.org/10.1111/psyp.13964.

[43] O'Malley RPD, Mirheidari B, Harkness K, Reuber M, Venneri A, Walker T, Christensen H, Blackburn D. Fully automated cognitive screening tool based on assessment of speech and language. J Neurol Neurosurg Psychiatry November 20, 2020. https://doi.org/10.1136/jnnp-2019-322517.

[44] Fuller A, Fan Z, Day C, Barlow C, Lu Y. Digital twin: enabling technologies, challenges and open research. IEEE Access 2020;8:108952−71. https://doi.org/10.1109/ACCESS.2020.2998358.

[45] Walsh JR, Roumpanis S, Bertolini D, Delmar P. Evaluating digital twins for Alzheimer's disease using data from a completed phase 2 clinical trial. Alzheimer's Dement 2022;18:e065386. https://doi.org/10.1002/alz.065386.

[46] Rizzo AA, Buckwalter JG, Neumann U. Virtual reality and cognitive rehabilitation: a brief review of the future. J Head Trauma Rehabil 1997;12(6):1−15.

[47] Rose FD, Attree EA, Brooks BM. Virtual environments in neuropsychological assessment and rehabilitation. Stud Health Technol Inf 1997;44:147−55.

[48] Clay F, Howett D, FitzGerald J, Fletcher P, Chan D, Price A. Use of immersive virtual reality in the assessment and treatment of Alzheimer's disease: a systematic review. J Alzheimers Dis 2020;75(1):23−43. https://doi.org/10.3233/JAD-191218.

[49] Jin R, Pilozzi A, Huang X. Current cognition tests, potential virtual reality applications, and serious games in cognitive assessment and NonPharmacological therapy for neurocognitive disorders. J Clin Med October 2020;9(10):3287. https://doi.org/10.3390/jcm9103287.

[50] Porffy LA, Mehta MA, Patchitt J, Boussebaa C, Brett J, D'Oliveira T, Mouchlianitis E, Shergill SS. A novel virtual reality assessment of functional cognition: validation study. J Med Internet Res 2022;24(1):e27641.

[51] Newton C, Pope M, Rua C, et al. For the PREVENT Dementia Research Programme. Entorhinal-based path integration selectively predicts midlife risk of Alzheimer's disease. Alzheimer's Dement 2024:1−15. https://doi.org/10.1002/alz.13733.

[52] Cogné M, Taillade M, N'Kaoua B, Tarruella A, Klinger E, Larrue F, et al. The contribution of virtual reality to the diagnosis of spatial navigation disorders and to the study of the role of navigational aids: a systematic literature review. Ann Phys Rehabil Med 2017;60(3):164−76. https://doi.org/10.1016/j.rehab.2015.12.004. Epub 2016 Mar 24. PMID: 27017533.

[53] Laczó M, Martinkovic L, Lerch O, Wiener JM, Kalinova J, Matuskova V, et al. Different profiles of spatial navigation deficits In Alzheimer's disease biomarker-positive versus biomarker-negative older adults with amnestic mild cognitive impairment. Front Aging Neurosci 2022;14:886778. https://doi.org/10.3389/fnagi.2022.886778. PMID: 35721017; PMCID: PMC9201637.

[54] Howett D, Castegnaro A, Krzywicka K, Hagman J, Marchment D, Henson R, et al. Differentiation of mild cognitive impairment using an entorhinal cortex-based test of virtual reality navigation. Brain 2019;142(6):1751−66. https://doi.org/10.1093/brain/awz116. PMID: 31121601; PMCID: PMC6536917.

[55] Papaioannou T, Voinescu A, Petrini K, Stanton Fraser D. Efficacy and moderators of virtual reality for cognitive training in people with dementia and mild cognitive impairment: a systematic review and meta-analysis. J Alzheimer's Dis 2022;88(4):1341−70. https://doi.org/10.3233/JAD-210672.

[56] Optale G, Urgesi C, Busato V, Marin S, Piron L, Priftis K, Gamberini L, Capodieci S, Bordin A. Controlling memory impairment in elderly adults using virtual reality memory training: a randomized controlled pilot study. Neurorehabilitation Neural Repair 2010;24(4):348−57. https://doi.org/10.1177/1545968309353328.

[57] Makmee P, Wongupparaj P. Virtual reality-based cognitive intervention for enhancing executive functions in community-dwelling older adults. Psychosoc Interv 2022;31(3):133.

[58] Manera V, et al. Kitchen and cooking,' a serious game for mild cognitive impairment and Alzheimer's disease: a pilot study. Front Aging Neurosci 2016;8:1−9. https://doi.org/10.3389/fnagi.2016.00058.

[59] Moyle W, Jones C, Dwan T, Petrovich T. Effectiveness of a virtual reality forest on people with dementia: a mixed methods pilot study. Gerontol 2018;58(3):478−87.

[60] Gaggioli A, Gianotti E, Chirico A. Psycho-physiological effects of a virtual reality relaxation experience after acute stressor exposure. Ann Rev Cybertherapy Telemed 2020;2020:123.

[61] Tominari M, Uozumi R, Becker C, Kinoshita A. Reminiscence therapy using virtual reality technology affects cognitive function and subjective well-being in older adults with dementia. Cogent Psychol 2021;8(1):1968991.

[62] Appel L, Appel E, Bogler O, Wiseman M, Cohen L, Ein N, Campos JL. Older adults with cognitive and/or physical impairments can benefit from immersive virtual reality experiences: a feasibility study. Front Med 2020;6:329. https://doi.org/10.3389/fmed.2020.00329.

[63] Wiederhold BK, Riva G. Virtual reality therapy: emerging topics and future challenges. Cyberpsychol, Behav Soc Netw 2019;22(1):3−6.

[64] Lindner P, Miloff A, Fagernäs S, Andersen J, Sigeman M, Enander J, Andersson G. Therapist-led and self-led one-session virtual reality exposure therapy for public speaking anxiety with consumer hardware and software: a randomized controlled trial. J Anxiety Disord 2017;61:45−54. https://doi.org/10.1016/j.janxdis.2018.07.003.

[65] Beattie Z, Miller LM, Almirola C, Au-Yeung W-TM, Bernard H, Cosgrove KE, et al. The collaborative aging research using technology initiative: an open, sharable, technology-agnostic platform for the research community. Digit Biomark 2020;4(Suppl. 1):100−18. https://doi.org/10.1159/00051220.

[66] Thomas N, Beattie Z, Riley T, Hofer S, Kaye J. Home-based assessment of cognition and health measures: the collaborative aging research using technology (CART) initiative and international collaborations. IEEE Instrumentation Measurement Magazine September 2021;vol 24(6):68−78. https://doi.org/10.1109/MIM.2021.9513638.

[67] Muurling M, de Boer C, Kozak R, Religa D, Koychev I, Verheij H, Nies VJM, et al. Remote monitoring technologies in Alzheimer's disease: design of the RADAR-AD study. Alzheimer's Res Ther 2021;13(1):89.

[68] Grammatikopoulou M, Lazarou I, Alepopoulos V, et al. Assessing the cognitive decline of people in the spectrum of AD by monitoring their activities of daily living in an IoT-enabled smart home environment: a cross-sectional pilot study. Front Neurol 2024;16. https://doi.org/10.3389/fnagi.2024.1375131.

[69] Čepukaitytė G, Newton C, Chan D. Early detection of diseases causing dementia using digital navigation and gait measures: a systematic review of evidence. Alzheimers Dement 2024;20(4):3054−73. https://doi.org/10.1002/alz.13716.

[70] Jimison HB, Pavel M, McKanna J, Pavel J. Unobtrusive monitoring of computer interactions to detect cognitive status in elders. IEEE Trans Inf Technol Biomed 2004;8(3):248−52. https://doi.org/10.1109/TITB.2004.835539.

[71] Stringer G, Couth S, Brown LJE, Montaldi D, Gledson A, Mellor J, Sutcliffe A, et al. Can you detect early dementia from an email? A proof of principle study of daily computer use to detect cognitive and functional decline. Int J Geriatr Psychiatr 2018;33(7):867−74.

[72] Stringer G, Couth S, Heuvelman H, Bull C, Gledson A, Keane J, Leroi I. Assessment of non-directed computer-use behaviours in the home can indicate early cognitive impairment: a proof of principle longitudinal study. Aging Ment Health 2023;27(1):193−202.

[73] Dignum V. Responsible Artificial Intelligence: how to develop and use AI in a responsible way. Springer; 2019.

[74] Singh G. Introduction to artificial neural networks. Data Sci Blogathon 2024;(Nov). https://www.analyticsvidhya.com/blog/2021/09/introduction-to-artificial-neural-networks/.

[75] Susteková D, Knutelská M. How is the artificial intelligence used in applications for traffic management?. In: Scientific Proceedings XXIII International Scientific-Technical Conference "Trans & Motauto '15"; 2015. https://transmotauto.com/sbornik/2015/3/26.HOW%20IS%20THE%20ARTIFICIAL%20INTELLIGENCE%20USED%20IN%20APPLICATIONS%20FOR%20TRAFFIC%20MANAGEMENT.pdf.

[76] Weber F, Schütte R. State-of-the-art and adoption of artificial intelligence in retail. Digit Policy Regul Gov 2019;21(2). https://doi.org/10.1108/DPRG-09-2018-0050.

[77] Pellegrini E, Ballerini L, Hernandez MDCV, Chappell FM, González-Castro V, Anblagan D, et al. Machine learning of neuroimaging for assisted diagnosis of cognitive impairment and dementia: a systematic review. Alzheimers Dement (Amst) 2018;10:519−35. https://doi.org/10.1016/j.dadm.2018.07.004. PMID: 30364671; PMCID: PMC6197752.

[78] Nori VS, Hane CA, Martin DC, Kravetz AD, Sanghavi DM. Identifying incident dementia by applying machine learning to a very large administrative claims dataset. PLoS One 2019. https://doi.org/10.1371/journal.pone.0203246.

[79] Nori VS, Hane CA, Crown WH, Au R, Burke WJ, Sanghavi DM, et al. Machine learning models to predict onset of dementia: a label learning approach. Alzheimer's Dement 2019;5:918−25.

[80] Rudovic O, Utsumi Y, Guerrero R, Peterson K, Rueckert D, Picard RW. Meta-weighted Gaussian process experts for personalized forecasting of AD cognitive changes. In: Machine Learning for Healthcare Conference (ML4HC2019), arXiv:1904.09370v1 [cs.LG]. Arxiv; 2019. p. https://www.analyticsvidhya.com/blog/2021/09/introduction-to-artificial-neural-networks/. In this issue.

[81] El-Sappagh S, Alonso JM, Riazul-Islam SM, Sultan AM, Kwak KS. A multilayer multimodal detection and prediction model based on explainable artificial intelligence for Alzheimer's disease. Sci Rep 2021;11:2660.

[82] James C, Ranson JM, Everson R, Llewellyn DJ. Performance of machine learning algorithms for predicting progression to dementia in memory clinic patients. JAMA Netw Open 2021;4(12):e2136553. https://doi.org/10.1001/jamanetworkopen.2021.36553.

[83] Albrecht JS, Hanna M, Kim D, Perfetto EM. Predicting diagnosis of Alzheimer's disease and related dementias using administrative claims. J Manag Care Spec Pharm 2018;24:1138−45.

[84] Park KV, Oh KH, Jeong YJ, Rhee J, Han MS, Han SW, Choi J. Machine learning models for predicting hearing prognosis in unilateral idiopathic sudden sensorineural hearing loss. Clin Exp Otorhinolaryngol 2020;13(2):148−56.

[85] Hane CA, Nori VS, Crown WH, Sanghavi DM, Bleicher P. Predicting onset of dementia using clinical notes and machine learning: case-control study. JMIR Med Inform 2020;8(6):e17819. https://doi.org/10.2196/17819.

[86] Aru J, Vicente R. What deep learning can tell us about higher cognitive functions like mindreading? Arxiv 2019. https://arxiv.org/pdf/1803.10470.pdf.

[87] Astle DE, Bathelt J, CALM Team, Holmes J. Remapping the cognitive and neural profiles of children who struggle at school. Dev Sci 2019;22(1):e1274. https://doi.org/10.1111/desc.12747.

[88] Chauhan S, Vig L, Dr Grazia M de F, Corbetta M, Ahmad S, Zorzi M. A comparison of shallow and deep learning methods for predicting cognitive performance of stroke patients from lesion images. Front Neuroinf 2019;13:53. https://doi.org/10.3389/fninf.2019.00053.

[89] Kiiski, et al. Machine learning EEG to predict cognitive functioning and processing speed over a 2-year period in Multiple Sclerosis patients and controls. Brain Topogr 2018;31(3):346−63.

[90] Lee LY, et al. Robust and interpretable AI-guided marker for early dementia prediction in real-world clinical settings. eClinicalMedicine 2024:1−13. https://doi.org/10.1016/j.eclinm.2024.102725.

[91] Ekmecki PE, Arda B. History of artificial intelligence. Artificial intelligence and bioethics (SpringerBriefs in Ethics). New York: Springer; 2020. p. 1−15.

[92] Ienca M, Kressig RW, Jotterand F, Elger B. Proactive ethical design for neuroengineering, assistive and rehabilitation technologies: the Cybathlon lesson. J NeuroEng Rehabil 2017;14:115. https://doi.org/10.1186/s12984-017-0325-z.

[93] Won SM, Wong E, Reeder JT, Rogers JA. Emerging modalities and implantable technologies for neuromodulation. Cell 2020;181:115−35.

[94] Berger TW, Hampson RE, Song D, Goonawardena A, Marmarelis VZ, Deadwyler SA. A cortical neural prosthesis for restoring and enhancing memory. J Neural Eng 2011;8(4):046017.

[95] Neuralink. PRIME study progress update. Retrieved from Neuralink website. 2024.

[96] International Neuromodulation Society. Conditions that may be treated with Neuromodulation. 2024. https://www.neuromodulation.com/conditions.
[97] Bick SKB, Eskander EN. Neuromodulation for restoring memory. Neurosurg Focus 2016;40(5):1−12. E5.
[98] Lozano A.M., Fosdick L., Chakravarty M.M., Leoutsakos J.M., Munro C., Oh E., et al. A phase II study of fornix deep brain stimulation in mild Alzheimer's disease. J Alzheimers Dis 2016;54(2):777-787. https://doi.org/10.3233/JAD-160017. PMID: 27567810; PMCID: PMC5026133.
[99] Hampstead BM, Lengu K, Goldenkoff ER, Vesia M. Noninvasive brain stimulation as a rehabilitation tool for cognitive impairment. In: Brown GG, Crosson B, Haaland KY, King TZ, editors. APA handbook of neuropsychology: neuroscience and neuromethods. American Psychological Association; 2023. p. 449−72. https://doi.org/10.1037/0000308-022.
[100] Patra JK, Das G, Fraceto LF, et al. Nano based drug delivery systems: recent developments and future prospects. J Nanobiotechnol 2018;16:71. https://doi.org/10.1186/s12951-018-0392-8.
[101] Teleanu DM, Chircov C, Grumezescu AM, Volceanov A, Teleanu RI. Blood-brain delivery methods using nanotechnology. Pharmaceutics 2018;10(4):269. https://doi.org/10.3390/pharmaceutics10040269. PMID: 30544966; PMCID: PMC6321434.
[102] Lohan S, Raza K, Mehta SK, Bhatti GK, Saini S, Singh B. Anti-Alzheimer's potential of berberine using surface decorated multi-walled carbon nanotubes: a preclinical evidence. Int J Pharm 2017;530:263−78.
[103] Liu Y, Liang Y, Yuhong J, Xin P, Han JL, Du Y, et al. Advances in nanotechnology for enhancing the solubility and bioavailability of poorly soluble drugs. Drug Des Devel Ther 2024:1469−95.
[104] Radhakrishnan K, Senthil Kumar P, Rangasamy G, Ankitha K, Niyathi V, Manivasagan V, et al. Recent advances in nanotechnology and its application for neuro-disease: a review. Appl Nanosci 2023;13(9):6631−65.
[105] Kwon HJ, Cha MY, Kim D, et al. Mitochondria-targeting ceria nanoparticles as antioxidants for Alzheimer's disease. ACS Nano 2016;10(2):2860−70.
[106] Wang EC, Wang AZ. Nanoparticles and their applications in cell and molecular biology. Integr Biol 2014;6(1):9−26.
[107] Etter G, van der Veldt S, Manseau F, et al. Optogenetic gamma stimulation rescues memory impairments in an Alzheimer's disease mouse model. Nat Commun 2019;10:5322. https://doi.org/10.1038/s41467-019-13260-9.
[108] Mirzayi P, Shobeiri P, Kalantari A, et al. Optogenetics: implications for Alzheimer's disease research and therapy. Mol Brain 2022;15:20. https://doi.org/10.1186/s13041-022-00905-y.
[109] Campbell EJ, Marchant NJ. The use of chemogenetics in behavioural neuroscience: receptor variants, targeting approaches and caveats. Br J Pharmacol April 2018;175(7):994−1003. https://doi.org/10.1111/bph.14146.
[110] Ying Y, Wang J-Z. Illuminating neural circuits in Alzheimer's disease. Neurosci Bull 2021;37:1203−17. https://doi.org/10.1007/s12264-021-00716-6.
[111] Claes M, De Groef L, Moons L. The DREADDful hurdles and opportunities of the chronic chemogenetic toolbox. Cells March 25, 2022;11(7):1110. https://doi.org/10.3390/cells11071110.

[112] Song J, Patel RV, Sharif M, Ashokan A, Michaelides M. Chemogenetics as a neuromodulatory approach to treating neuropsychiatric diseases and disorders. Mol Ther 2022;30(3):990−1005. https://doi.org/10.1016/j.ymthe.2021.11.019.

[113] Anderson R, Knapp M, Wittenberg R, Handels R, Schott JM. Economic modelling of disease-modifying therapies in Alzheimer's disease. In: Personal Social Services Research Unit. London School of Economics. LSE; 2018. https://www.lse.ac.uk/cpec/assets/documents/EconomicmodellingAD.pdf.

Index

'*Note:* Page numbers followed by "f" indicate figures, "t" indicate tables and "b" indicate boxes.'

A

Acetylcholinesterase inhibitors (AChEIs), 223–224
Aducunamab, 222
Adversaries, 209–210
Allocentric spatial working memory, 75
α7 nicotinic receptor agonists and modulators, 225
Alzheimer's clinical syndrome, 21
Alzheimer's continuum, 21
Alzheimer's disease (AD). *See also* Dementia
 amyloid-beta (Aβ) plaques, 35
 amyloid hypothesis, 15–16
 anterior-temporal memory system, 162
 artificial intelligence (AI), 274–277
 atherosclerosis, 37
 behavioral markers, 269–271
 biomarkers, 17–20
 blood-based testing, 1–2
 cerebral amyloid angiopathy (CAA), 37
 cerebral infarction, 10–11
 clinical-pathological correlation, 11–14
 clinical stages, 2–3
 clinic-based cognitive testing, 4
 co-pathologies, 37
 cortical tau pathology, 162
 diagnosis, 4–5
 digital biomarkers, 273–274
 direct brain technologies, 277–279
 disease modifying treatments (DMTs), 227–230
 early detection. *See* Digital technology
 early pathology, 35
 genetics, 221
 genetics and molecular biology, 15–16
 historical phases, 10f
 late-stage clinical presentation, 218–219
 microhaemorrhages, 37
 mild cognitive impairment (MCI). *See* Mild cognitive impairment (MCI)
 monoclonal antibodies, 1
 neuropathology, 10, 219–220
 non-pharmacological management. *See* Non-pharmacological management
 object memory, 162–163
 optical markers, 268–269
 parahippocampal cortex, 162
 preclinical stage, 219
 prodromal stage, 219
 progression, 162
 research framework, 22–24
  Aβ and tau accumulation, 23
  biological definition, 20–22
  biomarkers, 22–24
  clinical adoption, 23–24
 spatio-temporal progression pattern, 162
 structural brain imaging, 12–13
 symptomatic treatments, 223–227
  acetylcholinesterase inhibitors, 223–224
  α7 nicotinic receptor agonists and modulators, 225
  11β-HSD1 inhibitors, 226–227
  cognition and everyday function, 223
  monoamine oxidase B (MOA-B) inhibitors, 225–226
  NMDA receptor antagonist, 224–225
 tau protein, 36
 tech-based approach, 5
 virtual and augmented reality markers, 271–273
Alzheimer's Disease Composite Score (ADCOMS), 228–229
Alzheimer's Disease Research Centers (ADRCs), 11–12
Ambulatory Research in Cognition (ARC) smartphone app, 169
Amyloid-beta precursor protein (APP), 35
Amyloid hypothesis, 15–16
Amyloid-related imaging abnormalities (ARIA), 154–155
Anterior-temporal memory system, 162
Anti-inflammatories, 44
Anxiety, 77

289

Index

Apathy, 77
Apolipoprotein E (APOE)-ε4 genotype, 54–55
Arterial spin labeling (ASL), 130
Artificial intelligence (AI), 274–277
Aspirin, 44
Atherosclerosis, 37
Awareness of cognitive decline index (ACDI), 70–71

B

Behavioral markers, 269–271
Behavioral variant of FTD (bvFTD), 78
Big data, 105–108
Biologics, 222–223
Biomarkers, 17–20
 early disease identification, 218–221
 predictive value, 218–221
Boston Remote Assessment for Neurocognitive Health (BRANCH), 169
Brain Health Assessment (BHA), 166
Brain reserve, 125

C

Cambridge Neuropsychological Test Automated Battery (CANTAB), 166
Cardiovascular Risk Factors, Aging and Dementia (CAIDE) risk score, 251–252, 253f
Cerebral amyloid angiopathy (CAA), 37
Cerebral infarction, 10–11
Cerebrospinal fluid (CSF), 79–80
Cerebrovascular imaging, 128
Chemogenetics, 279
Cholinesterase inhibitors, 43–44
Clinical trials
 database discoveries, 234
 outcome measures
  clinical meaningfulness, 231
  computerized assessment methods, 232–233
  electronic Person Specific Outcome Measure (ePSOM) development program, 231–232
  treatment success measurement, 230–231
 platform trials, 234–236
 randomized controlled trial (RCT), 233–234
 stratified medicine, 236

Clock Drawing Test, 75–76
Cognitive domains test
 anxiety and depression, 77
 delayed recall tests, 74–75
 dysexecutive syndrome, 76–77
 episodic memory impairment, 74
 executive function test, 76–77
 fluency task, 75
 language abilities assessment, 75
 language presentation, 77
 semantic memory tasks, 75
 visuo-spatial and visuo-constructional function, 75–76
Cognitive impairment, 223
Cognitive reserve, 125–126
Cognitive tests
 associative memory, 163–164
 clinical standards, 165
 digital cognitive assessments, 165–170
 memory binding, 163–164
 object memory tasks, 162–163
 passive sensors, 171
 spatial memory and navigation, 164–165
 speech-based markers, 170–171
Cognitive training/stimulation, 45, 257
Cogstate Brief Battery (CBB), 166
Collaborative Aging Research using Technology (CART) platform, 273
Computerized assessment methods, 232–233
Convolutional neural networks (CNNs), 108–109
COVID-19 pandemic, 249–250

D

Database discoveries, 234
DataDay app, 269–270
Deep learning algorithms, 108–109
Delayed recall tests, 74–75
Dementia. *See also* Alzheimer's disease
 Cardiovascular Risk Factors, Aging and Dementia (CAIDE) risk score, 251–252
 cognitive decline, 251
 costs, 57–60, 57f
 dementia with Lewy Bodies (DLB), 51–52
 diagnosis access, 60–61, 61t–62t
 early diagnosis, 59t
 epidemiology, 51–52
 global time trends and inequalities, 52–54
 incidence, 51
 predictive factors, 252t
 prevalence, 3–4, 51, 53f

protective factors, 251
risk factors, 54−57, 250
subtypes, 51−52
Dementia with Lewy Bodies (DLB), 51−52
Depression, 77
Diagnostic strategies
  cerebrospinal fluid (CSF), 79−80
  clinical assessment, 70−72
  cognitive domains testing, 74−78
  functional assessment instruments, 78
  magnetic resonance imaging (MRI), 80−81
  neurological signs/symptoms, 72t
  neuropsychological testing, 73−78
  positron emission tomography (PET) imaging, 82−83
  preclinical AD, 69−70
  prodromal AD, 69−70
  screening tools, 73−74
Differential privacy, 211−212
Digital biomarkers, 197
Digital cognitive assessments
  in-clinic supervised, 165−166
  remote unsupervised assessments
    face-to-face cognitive assessments, 167
    high-frequency assessments, 167
    long-term memory and forgetting, 168−169
    practice effects, 168
    repeated assessments, 168
  validation, 169−170
Digital Memory Assessment (DIMA) app, 269−270
Digital technology
  active testing, 101−102
  big data, 105−108
  digital biomarkers, 99
  digital devices, 100
  machine learning, 108−110
  passive testing, 102−105
    accelerometers, 102−103
    automated signal extraction tools, 105
    Behapp smartphone app, 103
    data-related complexities, 104−105
    global positioning system (GPS), 102−103
    real-world behaviors, 103
    user burden, 104−105
  personalized diagnostics, 110−112
  wireless communications, 99
Digital Twin technologies, 270−271
Direct brain technologies, 277−279

Disease modifying treatments (DMTs), 124
  amyloid targeting treatments, 227−229
  biomarkers targeting treatments, 230
  tau targeting treatments, 229−230
Dominantly Inherited Alzheimer Network Trials Unit (DIAN-TU), 227−228, 235
Donanemab, 1, 229
Dysexecutive syndrome, 76−77

E

Early detection, ethical challenges
  clinically vulnerable people, 197
  digital biomarkers, 197
  vs. early diagnosis, 188−189
  and genetic risk, 195−197
  justice and equity considerations, 194−195
  predictive information harms, 190−192
  premature adoption risk, 198
  screening programmes, 197−198
  social consequences, 192−195
  stigma and discrimination, 192−193
  value of knowing, 189−190
Electronic Person Specific Outcome Measure (ePSOM) development program, 231−232
Encryption-based privacy protection mechanism, 213
European Prevention of Alzheimer's Dementia (EPAD) project, 235
Everyday Cognition scale, 71
Executive function test, 76−77
Exercise, 254

F

Face-Name Associative Memory Exam (FNAME), 164
Face-Name Associative Recognition Task (FNART), 164
Flortaucipir (FTP), 18
Fluid-based biomarkers
  Aβ pathology, 150−151
  limitations, 154
  neurodegeneration, 153
  tau pathology, 151−153
Four Mountains task, 164
Free and Cued Selective Reminding Test (FCSRT), 74−75
Frontotemporal dementia (FTD), 51−52
Frontotemporal lobar degenerations (FTLD), 16

G

Genetic FTD Initiative (GENFI), 236
Genome-wide association studies (GWAS), 54—55
Glymphatic imaging, 131

H

Healthy aging through internet counseling in the elderly (HATICE) trial, 259
Hearing impairment, 257
Homomorphic encryption (HE), 213

I

Immersive virtual reality (iVR), 271
In-clinic supervised digital cognitive assessments, 165—166
Inferior frontal sulcal hyperintensities (IFSH), 133—135
Irregular sleep-wake rhythm disorder (ISWRD), 226—227

L

Lecanemab, 1
Lifestyle for Brain Health (LIBRA) risk score, 258—259
Lifestyle interventions, 45, 252—253
Logopenic/phonological variant of primary progressive aphasia (lv-PPA), 77

M

Machine learning (ML)
  deep learning algorithms, 108—109
  ethical issues, 109—110
  models implementation, 109
  supervised, 108
  unsupervised learning, 108
Magnetic resonance imaging (MRI), 80—81
  arterial spin labeling (ASL), 130
  atrophy subtypes, 123—124
  brain reserve, 125
  cerebral perfusion and permeability, 129—130
  cerebrovascular imaging, 128
  cognitive performance, 124
  cognitive reserve, 125—126
  glymphatic imaging, 131
  inferior frontal sulcal hyperintensities (IFSH), 133—135
  neurodegeneration, 123—124
  perivascular spaces (PVS), 131—133
  structural traits, 126
  synaptic dysfunction, 127
  time of flight (TOF) imaging, 128—129
  white matter hyperintensities, 135—136
McKhann Criteria, 11—12, 19—20
Memantine, 224—225
Mental stimulation, 258
Microhaemorrhages, 37
Mild behavioral impairment (MBI), 39
Mild cognitive impairment (MCI), 14
  affective disturbance, 44
  clinical assessment, 41—43, 41b
  degenerative group, 40
  dementia conversion, 40
  diagnostic criteria, 38, 38b
  imaging modalities, 42
  incidence and prevalence, 39
  mild behavioral impairment (MBI), 39
  neuropsychological assessment, 41—42
  non-degenerative causes, 40
  non-pharmacological treatment, 44—46
  pharmacological treatment, 43—44
  structural brain imaging, 42
  sub-classification, 39
  subjective cognitive decline (SCD), 39
  syndrome, 38—39
Monoamine oxidase B (MOA-B) inhibitors, 225—226
Multidomain lifestyle interventions, 259
Multi-infarct dementia, 10—11

N

Nanomedicine, 278—279
Nanotechnology, 278—279
National Institute of Health Toolbox-Cognition Battery (NIH-TB), 166
National Institute on Aging (NIA), 19—20
National Institute on Aging-Alzheimer's Association (NIA-AA) criteria, 38
Neuralink, 278
Neurodegeneration, 20, 153
Neurofilament light chain (NfL), 80
Neuroimaging procedures, 266—267
Neuromodulation, 278
NMDA receptor antagonist, 224—225
Non-pharmacological management, 1—2
  cognitive training/stimulation, 257
  diet, 252—253
  exercise, 254
  health and care professionals role, 258
  hearing impairment, 257
  lifestyle interventions, 252—253

Index **293**

multi-domain interventions, 254, 255t
psychological symptoms, 254–256
sleep, 256–257
Non-steroidal anti-inflammatory drugs (NSAIDs), 44
Novel Assessment of Nutrition and Aging (NANA), 269–270

O

Object memory tasks, 162–163
Optical markers, 268–269
Optogenetics, 279

P

Paired Associates Learning (PAL) task, 166
Perivascular spaces (PVS), 131–133
Personalized diagnostics, 110–112
Physical exercise, mild cognitive impairment (MCI), 45
Pittsburgh compound B (PiB), 18
Plasma biomarkers, 19
Polygenic risk score (PRS), 54–55
Positron emission tomography (PET) imaging, 82–83
Precision Robotic Implantation Brain-Computer Interface (PRIME) study, 278
Privacy concerns
  adversaries, 209–210
  attacks and mitigation techniques, 210–213
  location tracking study, 208
  mobile health setting, 208–209, 209f
  mobile health system, 208
  privacy leaks in apps, 212
  re-identification attack, 210–211
  statistical inference attack, 211–212
Privacy protection framework, 211

Q

Quasi-identifiers (QID), 210

R

Random forest algorithms, 108
Randomized controlled trial (RCT), 233–234

S

Sea Hero Quest, 101, 164–165
Semantic memory tasks, 75
Social engagement, 258
Spatial behavioral tests, 101
Spatial memory and navigation, 164–165
Speech biomarkers, 170–171, 270
Statistical inference attack, 211–212
Stratified medicine, 236
Structural brain imaging, 12–13
Subjective cognitive decline (SCD), 39
Supervised machine learning, 108
Synaptic dysfunction, 127

T

Tau pathology, 151–153
  P-tau181, 152
  P-tau217, 152–153
  total tau (T-tau), 151

U

Unsupervised learning, 108

V

Vascular dementia (VaD), 51–52
Vessel distance mapping at 7T (VDM), 129
Virtual and augmented reality markers, 271–273

W

Washington University Clinical Dementia Rating Scale, 71
White matter hyperintensities (WMH), 135–136, 136f

Printed in the United States
by Baker & Taylor Publisher Services